WIND ENERGY CONVERSION 1983

Wind Energy Conversion
1983

PROCEEDINGS OF THE FIFTH BWEA WIND ENERGY CONFERENCE
READING 23-25 MARCH 1983

Edited by
PETER MUSGROVE
Department of Engineering, University of Reading

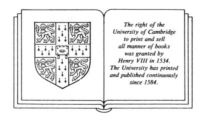

The right of the
University of Cambridge
to print and sell
all manner of books
was granted by
Henry VIII in 1534.
The University has printed
and published continuously
since 1584.

CAMBRIDGE UNIVERSITY PRESS
Cambridge
London   New York   New Rochelle
Melbourne   Sydney

Published by the Press Syndicate of the University of Cambridge
The Pitt Building, Trumpington Street, Cambridge CB2 1RP
32 East 57th Street, New York, NY 10022, USA
296 Beaconsfield Parade, Middle Park, Melbourne 3206, Australia

First published 1984

Printed in Great Britain at the University Press, Cambridge

Library of Congress catalogue card number: 83-20878

British Library Cataloguing in Publication Data
BWEA Wind Energy Conference (5th: 1983:Reading)
Wind energy conversion 1983
1. Wind power—Congresses
I. Title II. Musgrove, Peter
621.4'5 TK1541
ISBN 0 521 26250 X

The Fifth British Wind Energy Association Wind Energy Conference was organized by the BWEA.

Readers wishing to contact members of the Organising Committee or officers of the British Wind Energy Association will find their addresses at the back of this volume.

CONTENTS

Page

The UK Programme on Wind Energy
R. G. S. Skipper and L. A. W. Bedford 1

Recent Developments on Wind Energy Utilization in
the Netherlands   H. J. M. Beurskens and G. G. Piepers 10

The Wind Energy Programmes of the French Agency for
the Management of Energy (AFME)   M. Bremont and L. Drouot . . . 19

Wind Energy Projects in Austria   A. Szeless 23

Italian Wind Energy Activities   G. Gaudiosi 30

The Economics of Existing Wind Turbines in the Size
Range 10 to 100 metres diameter   P. J. Musgrove 34

The Economic Viability and Competitiveness of
Small Scale Wind Systems   P. L. Fraenkel and J. P. Kenna 46

The Cost of Electricity from Wind Turbines:
the Through Life Cost Concept   J. C. Riddell 54

Recent Developments and Results of the Reading/RAL
Grid Simulation Model   E. A. Bossanyi and J. A. Halliday 62

Operating Reserve, Utility Size and Power Variance
J. C. Dixon and R. J. Lowe 75

Simplified Dynamic Behaviour of a Straight Bladed
Vertical Axis Wind Turbine   M. S. Courtney 83

Some Effects of Speed on the Rotor Dynamics of the
Rutherford 6 metre VGVAWT   J. H. Webb, A. J. Pretlove and
L. W. Lack 91

Aspects of the Dynamics of a Vertical Axis Wind Turbine
I. P. Ficenec and A. Saia 100

Turbulence Induced Loads in a Wind Turbine Rotor
A. D. Garrad and U. Hassan 108

A Numerical Simulation of the Response of a Large
Horizontal Axis Wind Turbine to Real Wind Data
S. J. R. Powles, E. J. Fordham and M. B. Anderson 119

The Influence of Turbulence on the Dynamic Behaviour of
Large WTGs   R. H. Swansborough 129

Three Years Performance Record of a Small (10 kW) WECS
in the East of Ireland   T. O'Flaherty 137

Dynamics and Control of Autonomous Wind-Diesel Generating
Systems   A. Tsitsovits and L. L. Freris 143

Small Scale Wind/Diesel Systems for Electricity Generation
in Isolated Communities   *D. G. Infield* 151

A Transportable Wind–Battery System for Mongolia
G. R. Watson 163

The Fair Isle Wind Power System   *W. G. Stevenson and
W. M. Somerville* 171

Wind Power on Lundy Island   *W. M. Somerville and J. Puddy* 185

Measurement of Atmospheric Turbulence: An Assessment of Laser
Doppler Anemometry   *R. J. Delnon, R. Johnson,
C. W. A. Maskell and J. M. Vaughan* 198

The Distortion of Large–Scale Turbulence by a Wire Gauze Disc
E. J. Fordham 206

An Analysis of the Aerodynamic Forces on a Variable Geometry
Vertical Axis Wind Turbine   *M. B. Anderson* 224

Survey of Materials Suitable for Use in MW sized Aerogenerators
L. M. Wyatt, U. R. Lenel and M. A. Moore 235

An Appraisal of Straight Bladed Vertical and Horizontal Axis
Windmills   *K. McAnulty* 245

The 3MW Orkney Wind Turbine   *P. B. Simpson, D. Lindley,
A. S. Lee and A. Caesari* 253

Laboratory Performance Measurements of a Model Vertical Axis
Wind Turbine   *S. Gair, W. S. Bannister and M. Millar* 274

Interpretation and Analysis of Field Test Data   *M. B. Anderson* . . 283

Construction, Commissioning and Operation of the 300 kW Wind
Turbine at Carmarthen Bay   *T. Young, D. McLeish and D. Rees* . . 296

Disturbances in Electricity Supply Networks caused by a Wind
Energy System Equipped with a Static Converter
C. J. Looijesteijn 303

Wind Power for Ship Propulsion   *C. T. Nance* 318

Aerodynamics and Structural Compliance   *J. C. Dixon and
R. H. Swift* 326

Offshore Wind–Turbine Sub-Structure Design   *R. H. Swift and
J. C. Dixon* 334

Experimental and Theoretical Performance Analysis of the
Tornado Concentrator   *F. Haers and E. Dick* 342

Optimising the Designs of Flexible–Sail Wind–Turbines
P. D. Fleming and S. D. Probert 350

Wind Turbine Runaway Speeds   *D. J. Milborrow* 358

List of Delegates 367

THE U.K. PROGRAMME ON WIND ENERGY

R.G.S. SKIPPER

Head of Renewables and International Branch,
Energy Technology Division
Department of Energy
London, England.

and

L.A.W. BEDFORD

Programme Manager, Wind Energy
Energy Technology Support Unit,
A.E.R.E., U.K.A.E.A.
Harwell, England.

Abstract

The paper explains the UK Government's role in wind energy R&D, the reasons for Government support and the aims and objectives of the Department of Energy programme. It reviews briefly how the programme is administered. The paper concludes with a summary of the programmes and progress in 1982/83 and the forward programme.

The Government's Role In Wind Energy R&D

Within the UK programme of Research and Development on Renewable Energy which was started in 1974, the relative position of wind energy has steadily improved.

The March 1982 review of the Programme by the Advisory Council on Research and Development for Fuel and Power (ACORD) has resulted in wind energy taking an increased proportion of the Department of Energy's planned expenditure (about one-third in 1983/84) and there is also increasing interest from the Department of Industry, The Science and Engineering Research Council, the Electricity Boards, and from Industry. Wind energy R&D also is playing an increasing role in the activities of the European Commission and the International Energy Agency.

The Strategic Review of the Renewable Energy Technologies (1) was an important input to the deliberations of ACORD. Wind Energy, like the other technologies examined in the Review, was evaluated against three basic criteria, which were considered to be necessary, but not sufficient, conditions for Government funding. These three criteria are —

- that the resource itself is big enough to be of national significance

- that the prospects for the technology are economically attractive from the national point of view, within a reasonable timescale, or have the potential to become so if reasonable R&D targets are met.

- that market forces alone are not likely to be sufficient to ensure that the economic prize to the nation is won soon after the onset of cost effectiveness.

The Strategic Review considers these points both for Onshore and Offshore Wind Power. It concludes that Onshore Wind Power could, in theory, produce 50TWh/year or about 20% of the current UK electricity requirement. It notes that the constraints are more to do with environmental intrusions and economics than with the resource size.

Turning to Offshore Wind Energy, a comprehensive study by Taywood Engineering and CEGB concludes that for sites more than 5km offshore and with water depths 10–50m, the total resource could be in excess of 250 TWh/year, excluding any restrictions needed for the fishing industry. The Strategic Review suggests a tentative figure of 140 TWh/year after making allowance for fishing interests. Even this lower potential would represent more than half the present UK electricity requirement and would be well above the proportion of supply with wind power's characteristics that the electricity system could accept.

The resource is large for both onshore and offshore wind energy, so the first criterion is met.

Turning to the second, economic attractiveness: Onshore Wind Power shows itself, according to the Strategic Review, to be economically attractive under most scenarios and to be one of the most attractive of the electricity producing renewable energy technologies. The economic case for wind power is covered in greater detail in the Contribution of Renewable Energy Technologies (2) which attempts to identify as realistically as possible the contribution which renewables can be expected to make to energy requirements up to 2025. It was developed against a range of scenarios of future possible levels of fuel prices and economic activity developed by the Department for its Energy Projection. The contribution of onshore wind energy was found to be economic in five of the eight scenarios considered. The study discusses the limits to the rate of introduction in these cases and concludes that, with a maximum size of wind turbine between 2.5 and 5MW, an installation rate of 0.5 GW/year or up to four units a week was not unreasonable. The maximum installed capacity by 2025 was estimated to be 7.5 GW. This would represent some 1,500 to 3,000 machines. As the paper points out, it is difficult to predict the public acceptance of wind turbines on this scale, although it appears, in comparison with other visual intrusions (eg the approximately 22,000 towers of the supergrid) that it may well be environmentally acceptable with careful choice of sites.

Onshore wind energy therefore meets the second criterion. Offshore wind energy has higher estimated costs, but the prospects for eventual cost effectiveness look promising. Many of the developments needed to secure this effectiveness are common with onshore wind energy. In particular the Strategy Review points out that the economics of offshore generation might be crucially dependent on the costs achieved being towards the lower end of its estimates. It is therefore important to explore the possibly more cost effective vertical axis machines and compliant designs for horizontal axis machines.

With the third criterion the more difficult judgment arises with on-shore wind energy which is seen as approaching the development stage although further R&D is needed in some aspects. The Generating Boards are interested potential customers and manufacturing industry is developing its input. These interests, however, are not yet sufficiently firmly established to give either the Boards or the industry adequate incentive to carry out all the necessary R&D themselves. However the question of the Government's role is bound to arise once again when the present round of projects, which involve a considerable future commitment, nears completion. With offshore wind energy, the risks and uncertainties are greater, and the third criterion is met without any doubt.

The Aims and Objectives of the Department of Energy Programme

These can be summarised as follows:-

General Aim

To establish the technical feasibility and economic potential of wind energy, both on land and offshore, and assess the size and timing of the contribution which it might make to the UK energy supply in the foreseeable future.

Objectives

To produce, by about the mid-1980's, estimates of the amount of wind energy resources which might be utilised, taking account of factors limiting its incorporation into the UK energy supply:-

Direct Support

- evaluate existing meteorological data; collect and analyse new data.

- assess effects of variation in local topography on wind speeds, including a review of mathematical prediction models.

Collaborative Work

- determine the optimum wind turbine array configurations for different locations by studies of spacing, layout and, where necessary, physical testing and wake effects both in wind tunnels and with operating machines.

- evaluate environmental factors which may restrict wind turbine development.

- estimate the realisable wind energy resource for each of the various possible sites, including offshore.

To obtain experience of the problems of designing, constructing and operating large wind turbines while connected to an electricity grid network:-

Direct Support

- design, build and test a 3MW ratee horizontal wind turbine (HAWT) on Burgar Hill, Orkney.

- monitor the performance of a 250kW rated HA wind turbine on the same site as above.

- evaluate environmental factors such as noise, EM interference, visual impact, etc.

Collaborative Work

- take advantage of progress made abroad, by taking account of published information and by encouraging UK participation in international agreements.

To understand the comparative advantages and disadvantages under UK conditions of HAWTs and vertical axis wind turbines, (VAWTs) estimating the likely cost of energy from each:-

Direct Support

- assess potential of VAWTs by designing, and if justified by results, building and testing, an intermediate size prototype, followed by a MW version, based on the Musgrove concept.

- obtain capital cost estimates from UK industry for production versions of wind turbine designs evolving within the UK.

- evaluate, taking account of UK costs, published estimates from overseas programmes.

- maintain close liaison with the electricity utilities, particularly NSHEB with their main demonstration units, and with CEGB in their progress towards building and testing large wind turbines on low annual wind speed sites.

Collaborative Work

- examine variants of MW sized HAWTs.

- take advantage of progress made abroad on other types of VAWTs especially Darrieus machines in order to compare these with the Musgrove type.

To ensure that the major wind turbine development projects are supported by an adequate level of underpinning scientific research:-

Direct Support

- identify key problem areas and fundamental gaps in knowledge by discussion with industry, universities, and research establishments.

- set up groups of experts to advise on specific research needs.

- carry out the required research on a timescale determined by the needs of development projects.

Collaborative Work

- encourage industry and utilities to provide as much support as possible, financial as well as in kind.

- establish strong links with SERC to encourage fundamental and relevant research to be carried out in universities.

- disseminate results to UK industry.

To examine the technical and economic potential of offshore siting of wind turbine generators in order to determine whether there is sufficient justification for a large offshore demonstration unit: —

Direct Support

- carry out the broadly based study of offshore siting to assess main cost centres.

- prepare an outline design of a large offshore HAWT, and determine the most promising and economic foundation and tower configuration.

- determine the key parameters affecting the cost of electricity from offshore sited machines.

- determine the extent to which energy costs might be reduced.

- examine the effect of offshore siting on VAWTs and identify any type specific problems or advantages.

Collaborative Work

- explore possibilities and, where justified, set up collaborative programmes with other countries.

In conjunction with Department of Industry and with industry, to encourage the build up of a wind energy plant industry in the UK if it is shown to be commercially viable.

Administration of the Programme

This list of Aims and Objectives is being kept under close review by ACORD. In common with the other renewable energy sources the R&D programme is operated on the Customer/Contractor principle. The Department provides a Programme Supervisor within its Energy Technology Division who carries out the Customer role for R&D and the Energy Technology Support Unit at Harwell provides a Programme Manager and Project Officers who carry out the Contractor Role. The UKAEA as Contractor is advised by the Wind Energy Steering Committee whose terms of reference are listed in Appendix I. This is chaired by the Harwell Non-Nuclear Research Director and includes the Programme Supervisor and Manager and members from the Electricity Boards, The Departments of Industry and the Environment, manufacturing industry, and the Department's Consultants. The Department's Chief Scientist, in his customer role, is aided by the Energy R&D Committee which gives consideration to major project proposals and receives overall guidance from ACORD. This arrangement allows detailed evaluation of the Wind Energy Programme and enables the Department to seek the views regularly of the other organisations funding Wind Energy R&D and also provides for more general evaluation of the wider issues of the programme and especially its comparison with the other demands on the renewables R&D Budget.

The Programme and Progress in 1982/3

Several features of the programme are discussed in detail in other papers at this conference and this paper attempts only a summary updating of the main features.

Development of Horizontal Axis Wind Turbine

This project area covers building and testing a 3MW, 60m diameter wind turbine, preceded by the testing of a 250Kw 20m diameter version. The site for both these machines is provided by the North of Scotland Hydro-Electric Board at Burgar Hill, Orkney where the average wind speed at hub height is greater than 10m/sec. Both machines are designed and constructed by the Wind Energy Group. The smaller machine has now reached the stage where foundations and tower base are installed, manufacture of all major components is under way and first rotation is expected this year. The 60m machine is at an advanced stage of design. The data aquisition and analysis package is common to both turbines and is expected to be commissioned by the middle of this year. Wind conditions over the test site are being monitored using two masts of 28m and 80m height and the results are being compared with wind tunnel tests from a scale model of Burgar Hill at Oxford University.

Other work on horizontal axis machines includes the commissioning of a 200kW HA wind turbine, supplied and erected for the CEGB by James Howden Ltd at Carmarthen Bay. The North of Scotland Hydro-Electric Board have also provided engineering support to the installation of a 55kW 14m diameter HA machine.

Development of Vertical Axis Wind Turbine

The building and testing of a 25m machine based on the Musgrove Variable geometry concept is the subject of proposals from the Consortium led by Sir Robert McAlpine and Sons Ltd. The CEGB have indicated their willingness to collaborate in this project at their Carmarthen Bay site. The Department of Energy funded work has been completed on the 5m diameter Darrieus machine at Newcastle University.

Generic Work

The Department's Generic programme is increasingly being concentrated on support of the main lines of development described above. A considerable amount of work has been carried out on the choice of materials for large wind turbines including a survey by the Fulmer Research Institute. Other studies include work on wind data and resource size. These include the completion of an evaluation, made jointly with other countries under an IEA Implementing Agreement, of mathematical models for predicting wind flow over complex terrain. Five different models were tested by comparing predictions with observations from various sites. Several other studies are in hand including a new IEA study of flow over hills based on a site in the Hebrides where the first of two major field trials has been successfully completed. The Meteorological Office, ERA Technology Ltd, and the National Maritime Institute are collaborating on studies aimed at providing a better understanding of offshore wind conditions. The CEGB has continued its work on wake studies and has reached agreement with Denmark, Holland and Ireland for a joint measurement trial using the 40m HA axis machines at Nibe, Denmark.

Generic work on offshore potential includes the completion of a comprehensive study by Taywood Engineering and CEGB on the size of the resource. The CEGB is also collaborating with Sweden, Holland and Denmark in studies covering meteorology, outline design of an offshore wind farm, specification of a prototype offshore machine and generic studies.

The Forward Programme

The programme of the Department of Energy in 1982/83 involved an expenditure of about £2.25m. Expenditure on wind in 1983/84 is likely to be more than twice this. The major effort will continue to be directed towards obtaining experience with large machines connected to the grid. For horizontal axis machines this experience will start to build up this year while with the vertical axis machine it should start to be available from 1984. The supporting generic programme on large machines will continue to be concentrated on under-pinning these two main lines of approach. The Department also attaches importance to the need to maintain and reinforce contacts with international collaborative projects.

It is considered important to prove large wind turbines on land before building offshore turbines but offshore studies will continue especially in collaboration with international partners.

The views expressed in this paper are those of the authors and do not necessarily represent those of either the Department of Energy or the Energy Technology Support Unit.

WIND ENERGY STEERING COMMITTEE (WISC)

TERMS OF REFERENCE

In General

To advise the UKAEA, through the Harwell Non-Nuclear Energy Research Director or his deputy, on the technical and commercial aspects of the programme of research, development and demonstration on wind energy which the Energy Technology Support Unit at Harwell is implementing for the Department of Energy.

To provide a forum for technical liaison between Government Departments and other organisations involved with wind energy.

To advise on how the UK might obtain maximum benefit from international collaboration through the EEC, the IEA or bilaterally.

Specifically

To advise the Harwell Non-Nuclear Energy Research Director and the Wind Energy Programme Manager in developing and formulating, against programme aims and objectives set by the Department of Energy, programme plans for the Department's Wind Energy programme, taking due note of international programmes and any other relevant work.

To consider the technical and commercial merits of proposals for projects of wind energy closely related topics in the light of approved programme and to make recommendations to the UKAEA for support of these proposals.

To monitor progress in the Department's Wind Energy programme and of projects within the programme.

To advise the Programme Manager in producing reports for submission to the Department of Energy and ACORD.

References

(1)   Strategic Review of the Renewable Energy Tenologies, ETSU R13,
Two Volumes, £5.00 and £15.00, HMSO.

(2)   Contributions of Renewable Energy Technologies to Future Energy
Requirements ETSU R14, £5.00, HMSO.

RECENT DEVELOPMENTS OF WIND ENERGY UTILIZATION IN THE NETHERLANDS

H.J.M. Beurskens and G.G. Piepers
Project Office for Energy Research (BEOP)
Netherlands Energy Research Foundation (ECN)

Abstract

Most of the Research, Development and Demonstration activities on wind
energy in the Netherlands take place within the framework of the
National Development Programme for Wind Energy (NOW-2).
This programme can be divided into 3 parts:

1) Centralized applications;
2) Decentralized applications;
3) General applied research.

The most important elements of part 1 are the studies on the integration
of wind energy and conventional power plants, studies on control and conversion systems, the construction of an experimental 10 MW wind farm and
the design of a large wind turbine (1-3 MW). Part 2 consists of the operation of a test station for commercial wind turbines with an installed
power upto 100 kW, some 12 technical demonstration projects for selected,
most promising, applications, the development of special systems such as
hybrid diesel-wind plants, heat generating wind turbines for water treatment and cooling. Part 3 comprises among others meteorological research,
the development of the tipvane concept, the measurements on the experimental 25m HAWT and the 15m Darrieus turbine and activities to remove
the barriers that act as serious constraints in the application of wind
energy systems on a large scale.

The Dutch wind energy programmes

The current National Development Programme for Wind Energy (NOW-2) is a
continuation and an expansion of the Netherlands Research Programme on
Wind Energy that was carried out during the period March '76 - March '81.
The results of this programme justified a follow up 9-years programme.
In December 1981 the Minister of Ecnomic Affairs approved the first
phase (1982 - 1984) of this programme and allocated Dfl. 46 million as
a government contribution to the total estimated cost of Dfl. 58.6
million. The remaining Dfl 12.6 million have to be funded by industries,
the users of the developed WECS and the utilities.
The aim of NOW-2 is to provide for the knowledge and technology to meet
the long term goals of the government:

- 2000 MW of installed wind power by the year 2000, realised by large
  wind turbines, located in wind farms;
- 450 MW of installed wind power by the year 2000, realised by e.g.
  15,000 wind turbines of an average installed power of 30 kW each.

Centralized applications of WECS

R, D and D

In January 1983 the Minister of Economic Affairs has contracted FDO
Engineering Consultants, cooperating with Fokker Aircraft Comp. and
Holec, to design a 1 to 3 MW wind turbine based on the most advanced
WECS technology. Research institutes and technical universities will be
involved in the project.

In parallel with this project a 10 MW demonstration wind farm, consisting of 250 kW to 500 kW machines, will be built. In July 1982 the
Ministry of Economic Affairs and the Cooperating Electricity Producers
(SEP) allocated Dfl. 35 million each for the design and construction.
SEP intends to order the equipment by the end of 1983.
It is planned to start the experiments with the completed farm by mid
1985. The aim of the experiment is to investigate aerodynamical and
electrical interference phonemena between the individual wind turbines,
to determine the farm-efficiency, to gather operating experience, to
evaluate cost aspects, to monitor environmental effects and the social
acceptance of a wind farm and to stimulate the Dutch industry in developing wind turbines. The results from the 25m HAT experimental wind
turbine in Petten are a very important input for the design team. See
figures 13 and 14.

It has generally been accepted that 1000 MW of installed power can be
integrated in the national electrical power supply system, which has an
installed power of 15.000 MW, without storage facilities. The wind farm
experiments will produce data on the basis of which more accurate calculations on the penetration degree of WECS can be made. Not only storage
systems (overground-, underground pumped hydro storage, compressed air
in underground caves), but also regulating strategies of WECS and the
predictability of the wind speed will be considered to determine the
penetration degree of WECS.

Commercial developments

Presently two medium size Dutch made wind turbines are commercially
available, produced by Polenko and FDO Engineering Consultants.
The total investment cost, including installation, foundation and
connection to the grid, excluding tax and subsidies, vary from Dfl
1700,-- to Dfl 1900,-- per m^2 swept rotor area. This is about twice as
much as the commercially available 10m and 16m diameter machines.
In terms of kWh-cost the price difference of course is less than a factor 2, because rotors of large WECS are mounted on higher towers where
the wind speed is higher. Moreover maintenance and operating costs per
kWh of large units will be lower that those of small units. It is expected that investment costs per m^2 rotor area of large units will approach
those of small machines as soon as the large units are produced in small
series (e.g. for the 10 MW wind farm).
End 1982 the Utility of the Province of Zeeland (PZEM) at Middelburg
ordered the first commercial FDO machine. PZEM intends to purchase more
machines from different manufacturers in order to gain experience in
operating wind turbines and to compare the quality of the different
makes.

Industries with a relatively large electricity consumption (typical 300.000 kWh/a) show great interest in 25m diameter machines. The purchase of wind turbines is being stimulated by subventions from the government, varying from about 25% to 40%, depending on the amount of the investment. (Small scale applications are favoured by higher subventions). The subventions only apply to enterprises and not to private persons.

Decentralized applications of WECS

R, D and D

The decentralized application of wind energy is characterized by the individual energy consumer or by small groups of energy consumers using their own wind turbine. The generating capacity of the wind turbine mostly being used for this kind of application has a diameter of 7 to 16m. Compared to large WECS the application of these small and medium size WECS has made most progress.
The most promising applications of decentralized wind energy conversion systems in the Netherlands are:

- electricity supply to groups of family houses, small industries, farms and utility buildings;
- energy supply for cold stores;
- generation of heat (e.g. for green houses);
- driving of drainage and irrigation pumps.

Presently there are about 10 Dutch manufacturers. The products of a number of them are shown in the figures 1 to 8.
Five manufacturers have left the prototype stage and are producing wind turbines in small series for both the domestic and the export market. The domestic market for these types of wind turbines is growing gradually but slowly. This is mainly due to the relatively high investment cost of wind turbines.
Besides the economic threshold, there are several other constraints that prevent the domestic market to grow rapidly.
The most important constraint is the difficulty of obtaining licenses from the municipality to install wind turbines within a reasonable period.

Wind turbines form a new, rapidly growing, appearance in society not yet covered by existing laws and regulations. Officials responsible for licensing do not know how to act. Safety is one of the major problems in the licensing procedures. This was the main reason to found a test station for small wind turbines at the ECN research institute at Petten. This station came into operation at the end of 1980. (See figure 12).
The testing includes:

1. the operation of the wind turbine under normal conditions (starting, stopping, automatic operation);
2. a test on the proper functioning of the safety systems (external failures such as a voltage drop, internal failures such as a defect of one safety system);
3. performance measurements according to the IEA recommendations;

4. quality of electrical power;
5. acoustic measurements;
6. dynamic loading of the blades and tower.

If in the testing period failures are detected the manufacturer is invited to modify the wind turbine. For this reason the ECN test station has a relatively limited history of accidents or failures.
The failures that occured until January 1983 can be summarized as follows:

Wind turbine component	Serious	Inconvenient
1. Blades and hub	5	3
2. Yawing mechanism	5	2
3. Nacelle	0	4
4. Electrical components	3	11
5. Mechanical failures of generator, transmission and rotor shaft	3	0
6. Brake	2	3
7. Activation of safety systems	3	6

This record relates to 7 wind turbines

In January 1983 the tests of 2 wind turbines were completed (the wind turbines from fig. 1 and 3) and 4 others were under test (the wind turbines from fig. 2, 4, 6 and a Stork wind turbine).
A comprehensive test report is offered to the manufacturer, who is free to publish (integrally!) or not. The fact that a test report is completed and a summary of the test results are published by ECN.

In parallel with the test activities, project are undertaken to develop standards.
In order to gain operating experience with wind turbines and to verify the economic and technical feasibility of WECS for the most promising applications, about 10 technical demonstration projects are being carried out.

Commercial developments

By the end of 1982 the Netherlands counted about 75 wind turbines with diameters varying from 10 to 16 meters and rated power varying from 10 kW to 60 kW.
All these machines are connected to the low voltage grid by means of induction generators and d.c./a.c. inverters; the latter system being applied increasingly frequently. The d.c./a.c. inverter systems allows the WECS to operate at a variable r.p.m. which leads to a lighter drive train. Further advantages are a better power factor, a better over-all efficiency, a smoother power output and the possibility for load characteristic control. The d.c./a.c. inverter systems are being used by both manufacturers of passive-pitch controlled wind turbines (figures 1, 4 and 8) and stall regulated wind turbines (figures 2, 3, 5 and 6). Practice shows that the above mentioned advantages do compensate the disadvantages of higher cost and harmonic distorsion.

The present cost of wind turbines are Dfl 700.-/m^2 swept rotor area for 16m diameter machines and Dfl 900,-/m^2 swept rotor area for 10m diameter machines, excluding VAT, but including foundation on stable soil,

installation and connection to the grid.
The relatively high interest rates and the tendency of oil and gas
prices to stabilize impose uncertainties on the short term market
expectations.

The application of heat generating WECS still is very limited, because
at present natural gas-generated heat is about twice as cheap as wind-
generated heat. The expectation that very soon wind generated heat
would become cheaper than gas-generated haeat, became not true because
wind turbines became more expensive and the price of natural gas tends
to stabilize instead of to increase. A similar argument applies to water
pumping systems for polder drainage.

A market analysis of small electricity producing wind turbines for
stock- and dairy farms, green houses, mushroom nurseries, cold stores,
drying houses, polder drainage and water treatment learns that the
total potential is about 25,000. This is on the basis of present prices
of fuel, electricity and wind turbines, the energy consumption of the
enterprises and the wind speed.
The actual potential decreases if the investment criteria of the enterprises and subjective aspects such as the attitude towards wind energy,
environmental spects, the effects of negative experiences of other WECS
users and poor consultancy are taken into account.
It is a general feeling that the actual total market potential can be
increased considerably if further research and development is directed
towards the following aims:
1. Decrease the total investment cost of wind turbines by a factor 2
2. Generate more experiences on WECS and more easily accessible know-
   how. The quality of consultancy should be improved; in this respect
   output prediction models, calculation of the self-supply degree, environmental aspects such as safety and acoustic noise, electrical
   connection standards of the utilities and the estimation of average
   wind speeds from isotachs taking into account terrain roughness and
   the effect of obstacles should get extra attention.

References

1. Standpoint of the Netherlands Government concerning wind energy and
   energy storage. Ministry of Economic Affairs, The Hague.

2. Wind Energy in the Netherlands
   Project Office for Energy Research (BEOP)-ECN
   P.O.Box 1, 1755 ZG Petten, the Netherlands

   BEOP-8, Petten, January 1982.

3. J. Dekker, F. Lekkerkerk, G. Looijestein, G. Valter
   Operating experience, control and measurements made on a 25m horizontal axis wind turbine

   Fourth International Symposium on Wind Energy Systems, Stockholm
   Sept. 21 - 24, 1982. Proceedings Vol. 2pp. 161-180.

4. H. Beurskens. Practical aspects of wind energy systems (in Dutch)
   Landbouwmechanisatie, Vol. 33, nr. 11, November 1982. pp. 1020-1031.

COMMERCIALLY AVAILABLE DUTCH SMALL WINDTURBINES

Fig.1 Lagerwey van de Loenhorst.
Rotordiameter 10.6m. Rated power 10 kW - 15 kW.
Induction generator or inverter system.

Fig.2 Bouma
Rotor diameter 16m. Rated power 55 kW.
Induction generator.

Fig.3 Polenko
Rotor diameter 16m. Rated power 15 / 55 kW.
Induction generator or inverter system

Fig.4 Windpaq
Rotordiameter 11m. Rated power 17.5 kW.
Induction generator

COMMERCIALLY AVAILABLE DUTCH SMALL WINDTURBINES

Fig.5 Multimetaal
Rotordiameter 11m. Rated power up to 22 kW.
Induction generator.

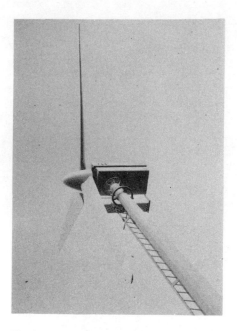

Fig.6 H-Energie Systemen
Rotordiameter 10m. Rated power 10 kW.
Induction generator or inverter system.

Fig.7 Windvang
Rotordiameter 5m. Rated power 4 kW.
Induction or synchronous generator.

Fig.8 Lagerwey van de Loenhorst
Twin. Rotordiameter 2 x 10.6m. Rated power
2 x 20 kW or 2 x 30 kW.
One inverter system.

Fig.9   Polymarin Darrieus rotor
Rotordiameter 15m. Rated power 100 kW.
Inverter system.

Fig.10 Electron-Fokker Darrieus rotor under test
in the DNW wind tunnel. The turbine has been installed
at Cilaut Eureun, West Java, Indonesia.
Rotordiameter 5.7m. Rated power 8 kW.
Inverter system.

Fig.11 Experimental Windturbine at the ECN Research
Institute, Petten.
Rotordiameter 25m. Rated power 300 kW.
Inverter system.

Fig.12 Test station for small and medium
size windturbines at the ECN Research
Institute, Petten.

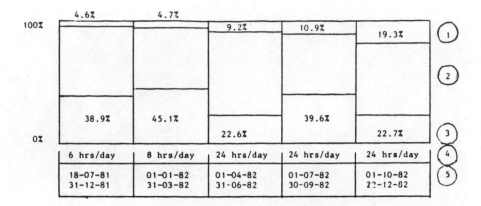

Fig. 13 Operation experience 25m HAT 18-07-81 to 22-12-82

(1) Wind turbine out op operation because of repairs, maintenance, computer or soft ware problems. In the period 01-10-82/22-12-82, 230 hrs were lost because of lightning, damaging electronics of data- and control systems.

(2) + (3)   Wind turbine operational

(3) Wind turbine produced power

(4) Number of hours per day that the wind turbine was allowed to run. Only these hours are taken in calculating the percentages of (1), (2) and (3).

(5) Operating period

The total number of rotating hours on 22-12-82 was : 2328.

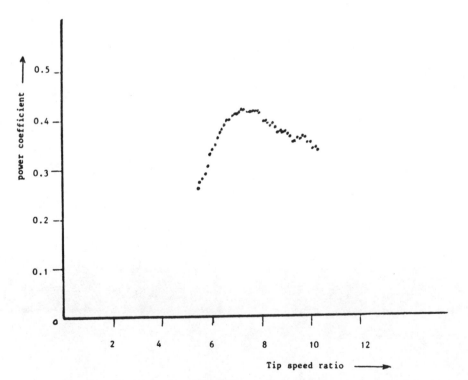

Fig. 14   The measured performance curve of the 25m HAT. The curve was determined with a constant r.p.m. control strategy.

THE WIND ENERGY PROGRAMMES OF THE FRENCH AGENCY
FOR THE MANAGEMENT OF ENERGY (AFME)

by Michel BREMONT (AFME)
s/c Louis DROUOT - Solar and Geothermal Department

1- INTRODUCTION

France is currently taking a fresh interest in wind energy.
The pumping windmills (of the multibladed type) remain fairly
unaltered, but there are changes in the field of aerogenerators.
The E.D.F. (French Electricity Supply Board) had run a series
of tests in the fifties, the result of which were such prototypes as:

- the NEYRPIC aerogenerator (1 MW, 35m in diameter)

- the BEST-ROMANI aerogenerator (0,8 MW, 32m in diameter)

These tests were conclusive as to the technical feasibility,
but not as to their economic efficiency, on account of the low
oil prices in those days.

Currently, though, given the new economic conditions, the
prospective evolution in the cost of non-renewable energies,
and after analysing the French wind potential (more particularly) that of the French Dominions and Overseas Territories),
together with the industrial prospects of a world-wide market,
and the new existing and available French high-level scientific and technical potentials, the A.F.M.E. has been prompted
to develop wind energy in a significant way since 1982.

2- THE A.F.M.E.'S OPTIONS TO CONCENTRATE ITS ACTIONS ALONG SEVERAL LINES

2.1- Low power professional aerogenerators

With this type of machine usefulness and reliability come
foremost at design stage. Wind energy can meet such needs
as powering beacons and lighthouses, or radio and T.V.
repeaters and we now have 20 years experience with several
hundreds of such machines.

2.2- Low power aerogenerators for non-professional use

Cost comes first at design stage. It is difficult to find
a satisfactory compromise between quality, which alone
paves the way to reliability, and the overall cost. Many
manufacturers will often suggest an answer to this, while
using insufficient scientific or technical means.
We have therefore created the National Wind Test Site in
LANNION (the CNEEL), in order to select the best performing

items.

The definitely significant development of wind energy in
Northern Europe and in the United States is now mostly
based on that type of machines.

Nevertheless, on account of the accidents that have already
taken place (especially during some recent storms), it
seems to us that particular attention ought to be given to
the elaborate design and fabrication of the aerogenerators.

2.3- Medium power aerogenerators

Ranging between 100 and 500 kW, these might prove interesting to utilities and therefore they must designed so as to
offer a high reliability and operating ability, and the
cost of the kWh must compare favourably with that of oil or
coal.

We do not think that small manufacturers working individually are able to reach such results with this size of
machine, if they aim at making aerogenerators with at least
a 20 year's life expectations.

When working out the dynamic calculation of the machines,
one must consider the vibrations due to the often overlooked
unsteady aerodynamics, causing frequent destruction, and one
must also be able to use the results of research in aeronautics, especially those concerning rotors (in helicopters).

Thanks to improvements in technology and materials, as well
as to a better knowledge of the stresses these machines
have to face, we ought to manage significantly lower costs.

3- THE A.F.M.E.'S STEPS IN WIND ENERGY

3.1- Budget

The Wind Energy Development Programme was a recent move.
The increased budget allowed by the A.F.M.E. is a token of
our determination to help setting up and developing a
French industry in that field.

It amounted to:

- 2.5 million FF in 1981
- 7.5 million FF in 1982
-13   million FF in 1983 Plus the demonstration
                                      operations

We have planned to keep up at that pace in 1984.

3.2- OUR ACTION NOW

In 1982, we undertook the following:

3.2a- Bidding for tenders in research

This has allowed us mainly to coordinate and give a

backbone to research unsteady aerodynamics that was
carried out by laboratories under the ONERA's supervision (National Office for the Study and Research in
aerodynamics).

The results achieved by aeronautics as regards aerodynamics and structures will therefore be used.

3.2b- <u>launching an industrial consultation for the
development low power machines</u>

The aim of that consultation was to:

- help the manufacturers that had been able to build
and sell machines in improving the already existing
ones, and developing new aerogenerators.

- pick out from among the new manufacturers those
that might become industrial partners for the A.F.M.E.
Our action then consists in helping French manufacturers
demonstrate the quality of their products by supporting a series of tests on the LANNION National Test
Site, and taking into careful account the site's technical assessments.

- set up a component policy, in order to provide manufacturers with the more elaborate parts of the machines,
those requiring heavier investments such as:

- <u>blades</u>: it will be the job of research centers
to define their optimum profile low
account of aerodynamics, stress,
vibrations, and we will give financial
support to the manufacturers making
those blades.
Such costs cannot be repeatedly faced
by each of our manufacturers separately,
and we hope that we'll soon have a whole
range of blades and hubs, meeting the
needs of all manufacturers, and also
greater safety requirements for the
users.

- <u>various systems</u>: e.g. electronic switch-boxes
for automatic coupling to the grid,
heating, pumping, etc......

3.2c- <u>Getting together the French Scientific and
Community</u>

Contrary to other countries, there had never been any
significant get together specialists in wind energy
before. That first meeting took place in Sophia-
Antipolis, VALBONNE, in November 82 and enabled us to
make basic assessments reports on our development
policy and initiate profitable contracts.

3.2d- <u>Launching a consultation on the feasibility study for a medium power aerogenerator</u>

That consultation is supposed to help us devise quickly enough two prototypes to be set up on the CNEEL.
The evaluation of the bidding will be made public in May 83, but we can say here and now that, on account of the competence of the people that have tendered, and the ONERA's participation to the work, we are definitely heading for success.

3.2e- <u>France's presence in the E.E.C.</u>

The E.E.C. grants a sizeable part of the renewable energies to the development of wind power: 13% figure which betokens a will to bring a significant contribution to that type of energy, in order to improve the Community's energetic resources.

3.2f- <u>Creating the C.N.E.E.L.</u>

This Test Site was planned in 1981 by the AFME and set up in 1982.
An organisation independent from manufacturers, it is to provide technical views and assessments, and has actually started operating, with 5 machines already installed.

3.2g- <u>Planning for the C.N.E.S.E.N.</u>

In order to make the CNEEL's work more complete, as far as wind systems are concerned, and make a national use of the French scientific and technical potential.
The AFME has gone forward and created the CNESEN (Northern Wind Energy Test Center).

The activities of both centers will be coordinated with:

- the CNEEL studying the dynamic behaviour of the machines on a test site;

- the CNESEN simulating the running of the shaft in a laboratory and studying the behaviour of the systems placed after the main shaft.

4- <u>CONCLUSION</u>

The means used in the AFME's Wind Energy Development Programme should enable us to:

- lower the cost of low power machines, as we improve their efficiency.

- catch up quickly with other European countries in the medium power range (40m in diameter), and offer high performing products.

WIND ENERGY PROJECTS

IN AUSTRIA

A. Szeless

Österreichische Elektrizitätswirtschafts AG,

Vienna , Austria

Abstract

In a country where due to its topography the wind energy
potential is quite moderate compared to some other European
countries the expenditures for research and development of
WECS have been accordingly modest. From 1977 until the end of
1983 the Federal Ministry of Science and Research will have
spent approx. 27 Mio AS of public money on wind energy research.
Figures on financial spendings by industries and electricity
utilities are not available but it can be assumed that their
share surely surpasses that of the Federal Ministry.

From the outset research and development efforts of WECS in
Austria were concentrated on the development of WECS series
in the power range up to 60 kW to be followed by the demonstration of competitive and economical operation towards the
final step of commercialization of these WECS. Thereby average
capital costs of 20000 to 30000 AS per kW installed should
not be exceeded.

Very early in the development of WECS is turned out, partly
because of the country's relatively low wind energy potential,
that it is advantageous, both from the economical and assured energy
supply point of view, to integrate the WECS as a component
into the overall energy supply system. On this basis one is
optimistic to be able to show that even in areas where the
utilization of wind energy alone is not expected to be competitive, combined systems (wind energy,solar energy, heat
pumps etc.) will turn out to yield satisfactory economical
results.

The start of operation of the experimental station at Leobersdorf
for the testing of WECS and combined systems (including WECS)
is a milestone in Austria's wind energy programme. A particularly interesting application of WECS is judged to be in the
agricultural sector for irrigation, hot water generation
(service water and heating) and greenhouse and stable heating.
At present 4 WECS are being tested in long duration tests to

verify predicted performance by actual operation. Altough some of these WECS are already commercially available the target capital costs per installed kW have generally not been reached yet. Austrian WECS manufacturers hope that eventually they will be able to sell their products also on foreign markets.

Introduction

The Austrian wind research and development programme has been presented on several occasions (1,2). Therefore, it suffices to only repeat some of the highlights of Austria's wind energy projects. Public funding via the Federal Ministry of Science and Research of wind energy research and development dates back to 1977. This Ministry also coordinates and finances the country's participation in the IEA wind energy project aimed at the exploitation of wind energy by means of wind energy converter systems (WECS) in the power range of 10-150 kW.

For the time period of 1980 to 1983 a total amount of 12 Mio AS has been allocated as public funding to the continuation of wind energy research and development.

Of course, research and development of wind energy systems is not only restricted to projects supported by public money.

Quite a lot of WECS research and development activities are going on in industry, some electricity utilities and in universities. However, it is impossible to quote financial expenditures on wind energy projects by these institutions. In some instances projects are being executed based on joint financing partly through public money partly through the industry. The ultimate goal of industry, of course, is to develop and commercially sell WECS in the power range of about 1 to 60 kW.

Wind energy projects

10-kW WECS at the Research Centre in Seibersdorf

The first 10 kW horizontal axis wind energy converter which has been designed by VOEST-ALPINE Co and the Institute for Precision Engineering, Technical University of Vienna was installed on a 35 m high tower at the Research Centre in Seibersdorf. In the first stage the turbine was equipped with a two-bladed rotor of 10,5 m diameter and a 10 kW asynchronous generator, with rated power of 10 kW at a windspeed of 7,5 m/sec, cut-in windspeed 4 m/sec, cut-out windspeed 30 m/sec, blade material was glass reinforced polyester, hydraulic pitch regulation and rotor speed 130 rpm.

After successful completion of the first test programm in 1980 it was decided[1] to carry on with the development of a WECS

1) Federal Ministry of Science and Research, VOEST-ALPINE Co, Institute of Precision Engineering

in the power range of 22-30 kW, e.g. essentially an extrapolation in size of the first 10 kW WECS. Since about 2 years
continuous tests are being carried out with the 22 kW wind
energy generator, of similar design as the original 10 kW
wind turbine. The test results have been quite satisfactory[1],
despite the fact that the site of the WECS is not optimal
with respect to the wind conditions.

1-kW mini WECS at the harbour entrance in Cuxhaven

SIEMENS Co., Austria[2] developed a small 1 kW WECS, which has
been installed as energy supply system of a light buoy at the
harbour entrance in Cuxhaven, Germany. The technical specifications are: two-bladed rotor of 2 m diameter, 1 kW at rated
windspeed of 8 m/sec, cut-in windspeed 3 m/sec, cut-out windspeed 20 m/sec, battery storage, fully automated. The operational results are quite encouraging, as almost 2 years of
testing have been successfully completed including 2 winter
seasons without any major interruptions.

30 kW-alpine WECS for the mountain refuge "Adamek-Hütte", elevation 2200 m

As has been reported earlier (3) Austria's first alpine wind
energy generator is operating and supplying electrical energy
since Sept. 1981 to the Adamek-Hütte, an alpine refuge at the
foot of the Gosau-Glacier at an altitute of 2200 m above sea
level. This project was jointly financed and carried out by
several partners[3]. The main technical specifications of the
30 kW alpine WECS are: 30 kW rated power at design windspeed
of 10 m/sec, cut-in windspeed 4,5 m/sec, cut-out windspeed
30 m/sec, two-bladed fibreglass reinforced rotor, 15 m rotor
diameter, 85 rpm rotor speed, hydraulic blade pitch control,
15 m high tower (guyed).

This wind energy system is a valid example for an isolated
consumer to achieve to a very high degree energy independence
from other energy sources. Previously, a diesel generator has
provided the necessary electric energy without existing
battery storage. Also the fuel had to be transported by
helicopter which at least doubled its price.

So far the 30 kW alpine WECS was in operation during 1 summer

1) 20-25% overall conversion efficiency
2) together with SIEMENS Co., Germany and Institute of Precision
   Engineering
3) Federal Ministry of Science and Research, Institute of Precision
   Engineering, VOEST-ALPINE Co., Regional Electricity Utility,
   Austrian Alpine Club

season[1] and in addition 2 winter test programmes have been run. Some difficulties have been encountered in particular during winter time. Especially the icing problem should be mentioned here, which made operation sometimes difficult or even impossible. So far no satisfactory solution has been found for prevention of icing of the rotor blades. Right now the rotor is back at the factory for inspection and again a number of fissures have been detected, their occurrence most likely to be attributable to the extreme environmental conditions with artic temperatures, icing and gusty winds of hurricane force.

Nevertheless, an overall saving of 80% of the diesel fuel could be achieved since the wind energy generator started its operation, with at the same time an increase of the level of comfort.

Still one must hold a realistic view on the possibilities of wide spread introduction of this type of alpine WECS. It is a question of costs and only in remote alpine regions which are not connected to the public electricity supply system and where the transportation of fuel (diesel fuel, coal, wood, etc.) becomes tedious and excessively expensive wind energy supply systems of this typ will be installed in the future. However, if new buildings (refuges) are to be constructed in alpine regions a combination of modern heat insulation techniques and the use of non-conventional energy sources (e.g. WECS in combination with solar energy) will often yield the most economical solution (minimum costs).

One such project is the reconstruction of the refuge Ober-walderhütte, elevation 3000 m above sea level near the Großglockner, Austria's highest mountain. In this case the whole energy supply system based on wind and solar energy is optimized with respect to the variations of the available energy sources and the energy demand accounting also for increased heat insulation measures.

The project for a WECS at the peak of the Sonnblick mountain[2], elevation 3100 m above sea level, has been abondoned because of the extreme climatic conditions and the lack of space to position the WECS and the photovoltaic solar cells.

1) including parts of spring and autumn
2) to supply energy to the meteorological station, radio link station and refuge

Experimental station at Leobersdorf for testing of alternative energy systems

In November 1982 the experimental station for testing of alternative energy systems located in Leobersdorf, Lower Austria started its routine operation. This project has been financed and is being carried out by several partners[1]. Right now 4 WECS in the power range of 1,5 - 22 kW are being tested.

Especially the grid-coupling tests of the 22 kW WECS[2] which will start shortly will be of great interest to the utilities.

An other test which is being carried out is the irrigation of vinyards and orchards with the small 1,5 kW WECS working in combination with photovoltaic solar cells. Using a recently developed irrigation system (only one third of the water consumption of conventional systems) vinyards and orchards for high quality crop are irrigated by the afore mentioned combined energy system. Aim of this project is the proof of economical feasibility of this irrigation system and at the same time increasing the quantity and quality of the crop.

Furthermore, a test is going on with a combined energy system based on a 22 kW wind energy generator, photovoltaic solar panels and a heat pump to supply energy to farms for hot water generation for service and heating purposes including greenhouses and stables.

All these tests of course aim at the demonstration that pure and combined wind energy systems can, when cleverly designed, operate economically under special circumstances and at specific sites. It is intended to develop these systems to the stage of commercial marketing. The market in Austria is restricted and it is hoped that the bigger share of the WECS manufactured in the country will go into the export.

1) Federal Ministry of Science and Research, Institute of Precision Engineering of the Technical University of Vienna, VOEST ALPINE Co., Provincial Government of Lower Austria, Federal Electricity Board

2) design by VOEST ALPINE Co.

Outlook

The research and development goal of WECS with rated power below 100 kW (mini and small WECS) which has been defined based on market and energy scenario surveys in 1977[1]) has turned out to be a good choice. This has been confirmed by many recent supply inquiries.

Moreover, very early in the development of WECS the point of view of integration of the WECS, as a competent, into the overall energy system has been adopted. Especially, in areas where the utilization of wind energy alone is not expected to yield a high availability the combined wind/(non-)conventional energy system has turned out to give advantagenous economical results. This is also the reason why since 1980 to a great extent system specific technical considerations have been included into the development efforts.

The start of operation of the experimental station at Leobersdorf for the testing of WECS and combined systems (including WECS) is a milestone in Austria's wind energy programme towards the step of commercialization. The market in Austria, however, is restricted and it is judged that one of the more promising applications of WECS will be in the agricultural sector for irrigation, hot water generation (service water and heating) and heating of grennhouses and stables.

Although WECS are already commercially available the target capitel costs of 20 000 to 30 000 AS per kW installed have generally not been reached yet. Austrian WECS manufacturers hope that they will be able to sell their products also on foreign markets.

1) by the project team responsible for the first Austrian 10 kW wind energy generator at the Research Centre in Seibersdorf

References

1.   Szeless, A.: "Outline of Austria's wind energy
     programme". Proceedings of the Fourth
     BWEA Wind Energy Conference, Cranfield,
     March 24-26, 1982

2.   Detter, H., Oszuszky, F. and Szeless, A.: "Wind
     energy research in Austria, pilot WECS
     projects in agriculture and extreme alpine
     regions". Fourth International Symposium
     on Wind Energy Systems, Stockholm,
     September 21-24, 1982

3.   Szeless, A.: "The Austrian alpine wind energy
     converter". Ad-hoc contribution at the
     Fourth BWEA Wind Energy Conference,
     Cranfield, March 24-26, 1982

ITALIAN WIND ENERGY ACTIVITIES

G. Gaudiosi   ENEA, Rome

Italy is heavily dependent on imported energy, so since the 1973 oil crisis interest in renewable energy sources has been increasing.

In May 1982 legislation was passed providing 1350 million dollars to subsidise energy saving and the introduction of renewable energy sources for residential, industrial and agricultural applications.

Wind energy has been considered viable in Italy by the National Energy Plan and by the above legislation. Wind regimes in Italy are comparable with most european countries. There are large areas of Appennines, Alps, Sardinia, Sicily, other southern regions and small islands with high enough wind potential to warrant wind generator technology development for internal and external markets.

Activities were initiated in 1979 with programmes involving ENEL and CNR. The following public Agencies are involved today in wind energy programmes:

ENEA (National Committee for Development of Nuclear and Alternative energies) was given responsibility in 1981 to develop renewable energy sources and provide the national activity in that technology. Its Wind programme is centered on wind resources, siting and development and testing of small intermediate and large size wind machines.

ENEL (National Electric Utility) set up in 1979 a six years wind energy programme (VELE) focused mainly on siting, wind generator technology development, a 500 kW wind power station, site applications and development of large size wind generator with ENEA and Industry.

CNR-PFE (National Research Council - Finalised Energy Project) is mostly involved in wind resource exploration and innovative wind systems and component development.

The following public industries have set up wind energy activities:

ENI Group is acting through two companies AGIP Nucleare, responsible for the renewable energies development, and TEMA specialised in energy system development.
FINMECCANICA Group is acting through two companies AERITALIA, aviation and space leader industry, and CESEN.

AERITALIA is developing small, intermediate and large size wind generators.

The following private Industry has wind activities:

FIAT GROUP through the SES, FIAT TTG, FIAT AVIATION and FIAT Engineering. FIAT TTG is involved in small and medium size wind generator development, while FIAT Aviation is collaborating with AERITALIA for the large size wind generator development. Other small industries develop small wind generators or pumps.

The Italian activities on wind energy can be classified in the following four areas: wind resources and siting, small intermediate, large size wind machine development and testing.

The relevant aspect of the present situation are given below for each area.

1. Wind resources and siting

    Historic wind data has been collected by the Military Aviation Meteorological Service and analised. A data bank has been set up by the IFA (Atmospheric Physics Institute) of the CNR. A preliminary wind map, based on 48 meteorological stations data, was published in 1981 by CNR PFE in collaboration with ENEL, AERITALIA and FIAT.

    AGIP NUCLEARE in collaboration with IFA has set up a climatic data Bank. ENEA will collaborate with CNR to improve the wind data bank.

    The present anemometric network of the Military Aviation Meteorological Service will be completed by CNR PFE in collaboration with ENEA and ENEL.

    Site evaluation has been performed by ENEL mainly in Sardinia, by ENEA in Puglia, Sardinia, Abruzzi, Sicily, by AGIP Nucleare in Puglia, Toscana, by AERITALIA and FIAT in other regions.

    Anemometric instrumentation evaluation will be done by ENEA.

2. Small wind machine development and testing

    This area refers to the development and testing on site of small wind generators and to the organisation of a National Test Station.

    Wind machines, horizontal and vertical axis, have been developed by FIAT, CESEN, AERITALIA, RIVA CALZONI and by some other small industries.

An original system has been developed by TEMA in collaboration with Reading University. The wind system consists of a Diesel engine coupled mechanically to a vertical axis wind generator (Musgrove type). Below are listed the wind generators under test on site in 1983 in Italy.

Under ENEA Support:

4 kW V.A. Darrieus type Wind Generator of CESEN at P. del Turolino (Genova)

3 kW H.A. 3 blades Wind generator of FIAT TTG for isolated site (Savona)

3 kW H.A. 3 blades Wind generator of FIAT TTG for isolated site (South Italy)

3 kW H.A. 1 blade Wind generator of RIVA CALZONI for battery charging (ENEA, Cassaccia Centre)

20 kW H.A. 2 blades Wind generator of AERITALIA for isolated community (South Italy)

Under ENEL Support:

68 kW H.A. 3 blades Wind generator of FIAT CRF, grid connection, Santa Caterina Test Centre

20 kW H.A. 2 blades Wind generator of AERITALIA grid connection, Santa Caterina Test Centre

15 kW H.A. 3 blades Wind generator of GRUMMAN/ AERITALIA, grid connection, Santa Caterina Test Centre

15 kW H.A. 3 blades Wind generator of GRUMMAN/ AERITALIA, grid connection, (Salina)

Under AGIP NUCLEARE Support:

9 kW V.A. Musgrove type coupled with Diesel, grid connection, Segezia

50 kW H.A. 2 blades Wind generator of FIAT TTG, grid connection, Segezia

Multi bladed wind pumps are tested by ENEA, innovative pumping system will be developed by ENEA and AGIP NUCLEARE.

A test centre has been operated by ENEL since 1981 with a maximum capacity of six wind machine positions.

During 1983 a National Test Station with 10-20 test
positions will be set up by ENEA probably in collaboration with other partners. Consideration will be
given to the ENEL Santa Caterina Test Centre, or any
other site in Sardina, Puglia or Sicily.

3. Intermediate size Wind Generators (Power < 500 kW)
   Development.

The area refers to the ALTA NURRA 500 kW Wind power
station of ENEL and to the prototype of 200 or 300 kW
of ENEA.

The Alta Nurra Wind Power Station consists of 10 unit
FIAT TTG of 50 kW each, and will be operated by ENEL.
The first unit is in operation, the other 9 units
will be installed during 1983. Wind cluster experience will be gained. Two preliminary designs have
been carried out under ENEA Contracts one for a
horizontal axis unit, 200 kW by AERITALIA, the other
for a vertical axis (Darrieus) unit 150 kW. A decision
will be taken before June 1983 for the construction of
a 200 or 300 kW wind generator prototype by ENEA and
Industry to be operated within 1984. AERITALIA has
presented a proposal based on the preliminary design.
ENEL is interested in the test campaign. Other industry
seems interested in this size, probably RIVA CALZONI,
and AUGUSTA (a helicopter company) in collaboration with
AGIP NUCLEARE.

An operational experience will be gained by ENEA on wind
generators of this size by the test campaign programmed
on the 100 kW wind generator of EN.EO (a private compressor company) sited in Bereito (North Appennines).
The EN.EO technology is Italian, though the parent
company is Swiss.

4. Large size Wind Generator (Multimegawatt) Development.

A preliminary design is in progress on a multimegawatt
horizontal axis unit. At the end of 1983 a decision
will be made on the construction of a prototype, which
could be installed by 1986 on a site to be selected,
probably in Sardinia.

The first contract for the preliminary design has been
awarded by ENEA and ENEL to an Industrial Consortium
(AERITALIA, FIAT AVIO in collaboration with ANSALDO
MOTORI).

ENEA with Industry will start the technology development of blades and of some critical components. CNR
will support some basic research.

Aerodynamical and structural computer codes are under
development by the ENEA support.

THE ECONOMICS OF EXISTING WIND TURBINES IN THE SIZE RANGE 10 TO 100 METRES DIAMETER

P.J. Musgrove   Reading University

Abstract

Present wind turbine costs are reviewed and it is concluded that wind turbines in the size range 15 to 30 metres diameter offer the most attractive economics, at least in the immediate future. The cost of electricity from such wind turbines is approximately 2.2 pence/kWh, competitive with electricity from coal, though long wind turbine life and low Operating and Maintenance Costs remain to be demonstrated. For private users in rural areas small wind turbines (10 to 20 metres diameter) operated in parallel with the grid currently offer payback periods of 6 to 9 years, exclusive of any subsidies or tax incentives. The installation of wind farms has proceeded very rapidly in the U.S.A. in the past two years, and it is expected that such installations will soon be legal in the U.K. In many parts of the U.K. the potential revenue to farmers from wind farms is several times greater than their income from present crops. Once the economic viability of wind farms has been demonstrated once can therefore expect the concept to spread rapidly.

Introduction

The assumption that large wind turbines (multimegawatt rated) will provide energy at a lower cost than small and medium sized wind turbines has influenced much of the development that has taken place in the wind energy field since 1973. As a result of this assumption most national wind energy programmes have concentrated on the development of the large machines. However data is now becoming available on the costs of wind turbines ranging in size from less than 10 metres diameter, to more than 100 metres diameter, so that the effect of size on wind turbine costs - and hence on wind turbine economics - can now be critically examined.

Wind Turbine Costs

As part of a study undertaken on behalf of the Directorate General for Energy, DGXVII, of the Commission of the European Communities, in the summer of 1982 cost data was collated for all the principal medium and large wind turbines now operating in Europe and the U.S.A. Cost data was mostly obtained directly from manufacturers, supplemented in the case of the larger American wind turbines by Divone's cost review, see ref.1. This data is summarised in figure 1, which indicates how the cost per unit rotor area varies with turbine diameter. Cost per unit rotor area, £/m^2, is shown rather than cost per rated kilowatt, £/kW, as the latter can be very misleading; since the rated power is proportional to the cube of the rated wind speed a small change in the latter can significantly alter the calculated £/kW, but have little effect on the annual energy output, and hence on the delivered cost of energy. Good modern wind turbines have comparable efficiencies and their energy output is primarily a function of their rotor swept area. For this reason the cost per unit rotor area is a more reliable parameter for comparing different wind turbine costs.

Most of the costs quoted in figure 1 include the tower and erection on site, but exclude foundations and grid connection (though these costs are included for the larger American machines). Though foundation and connection costs can be significant (of the order 15 to 20% of the total) their contribution is small by comparison with the scatter evident in the figure. It should be noted that costs relate to mid-1982 and currency conversions are at the rates prevailing in October 1982. Between October 1982 and March 1983 the pound has weakened by about 14% relative to other major currencies, including particularly the U.S.A., Denmark and the Netherlands, and if this persists most of the costs indicated would need to be revised upwards by 14%. As indicated by the key, most of the wind turbines have only been made, so far, in small numbers.

Though the data in the figure shows considerable scatter there is a clear trend towards increasing cost, per unit area, as the diameter increases. This could perhaps be expected since the mass, and hence the cost, of structures tends to increase at a rate intermediate between (size)^2 and (size)^3. However it is the cost of energy that is the crucial parameter, and one must allow for the fact that larger wind turbines have their rotors at greater heights above ground level, where wind speeds are higher. If one assumes that the velocity variation with height is given by the typical one-seventh power law then the wind turbine's power and energy outputs will be proportional to the tower height H raised to the power $(H^{1/7})^3$, i.e. to $H^{0.43}$. A measure of the cost per unit annual energy output is then provided by the parameter

$$C_{10} = \frac{Cost}{Area} \cdot \left[\frac{10}{H}\right]^{0.43} \qquad \ldots\ldots\ldots\ldots 1$$

where H is the tower height in metres. C_{10} can be interpreted as the equivalent cost, per unit rotor area, relative to the wind speeds experienced at a height of 10 metres (the normal meteorological measuring height). The variation of this cost parameter with diameter is shown in figure 2, which clearly indicates the beneficial effects, for large machines, of allowing for their greater hub heights. Since $5^{0.43} = 2.0$, a fivefold increase in hub height, e.g. from 15 metres to

75 metres, increases the wind power density by a factor of 2.
This means that for comparable energy cost one can afford to
pay twice as much, per square metre, for a large turbine with
a 75 metre hub height, by comparison with a small wind turbine having a 15 metre hub height. Despite the allowance for
the effect of height on the wind power density it is apparent
from figure 2 that the achieved and estimated costs for large
wind turbines do not show any significant improvement on the
achieved costs for smaller wind turbines: there is - as yet -
no evidence that large wind turbines will be significantly
more economic than smaller ones. Indeed the evidence so far
suggests that the smaller wind turbines, e.g. in the range
15 to 30 metres diameter, may well be more economic, at least
for land based applications.

Wind Energy Costs

The cost of energy depends on the lifetime of the wind energy
system, its operating and maintenance costs, and the required
rate of return, on one's capital investment. The concept of
an annual charge rate c is a very useful one, and c is related to r by the equation

$$c = \frac{r}{1 - (1 + r)^{-n}}$$

where r is the required real rate of return (i.e. after allowing for inflation) and n is the lifetime in years; n equal
annual repayments of cI will repay an initial investment I
and provide a real rate of return r on the capital invested.

A 5% rate of return, in real money terms, (as is usually
specified in the U.K. for public sector investments, see ref.
2) together with a lifetime of 20 years, i.e. r = 0.05, n =
20, corresponds to an annual charge rate of 8%. To this must
be added the annual costs associated with Operating and
Maintenance (O & M). Preliminary assessment of wind turbine
operating experience in the U.S.A., ref.3, indicates that for
large and medium sized machines annual O & M costs can be
expected to be about 2% of the capital cost; for smaller wind
turbines the O & M costs can be expected to be higher, but
there is - as yet - no reliable data on O & M costs for small
machines.

Adding a 2% allowance for O & M costs to the initial 8% charge
rate discussed above gives an overall charge rate - inclusive
of O & M costs - of 10%. Figure 2 indicates that a cost that
is reasonably representative of the more competitive current
wind turbine designs is £125/m^2 (normalised to 10 metres
height): with a 10% overall charge rate the annual cost of the
wind turbine is therefore £12.5/m^2. But what is the corresponding energy output?

If the annual wind velocity duration distribution is represented by the Rayleigh distribution (as is usual when estimat-

ing wind turbine energy outputs, see ref.4), the annual average power in the wind can be shown to be equal to $0.95 \rho V_m^3$, where V_m is the annual average wind speed and ρ is the air density. Published calculations for a variety of modern wind turbines indicate that an average power output of $P_A = 0.25 \rho V_m^3$ can be achieved (corresponding to an overall average power coefficient $C_p = 0.26$). Ref.4 recommends use of the standard density $\rho = 1.225 kg/m^3$, and with $V_m = 6$ m/s the value of P_A is $66W/m^2$, corresponding to an annual energy output of $580 kWh/m^2$. With an annual cost of £12.5/m² for a good modern wind turbine, and an annual energy output of $580 kWh/m^2$ (for $V_m = 6$ m/s at 10 metre height) the cost of electricity delivered by the wind turbine is consequently <u>2.15 pence/kWh</u>*.

Systems integration studies have shown that the value of wind turbines to a utility system, such as that of the CEGB, is primarily the value of the fuel saved. Since 80% of electricity in the U.K. is generated from coal, the cost of wind energy (for utility applications) must be compared with the fuel cost of generating electricity from coal fired power stations. The CEGB yearbooks for 1980-81 and 1981-82 indicate that this fuel cost is approximately in the range 2.0 to 2.4 pence/kWh, so given the assumptions stated with respect to lifetime and O & M costs, wind energy costs are competitive with the cost of electricity generated from coal.

It must be emphasised that further work is needed to demonstrate that a long life and a low level of O & M costs are consistent with a normalised capital cost of about £125/m². From figure 2 the lowest wind turbine costs at present correspond to wind turbines of about 15 metres diameter. However O & M costs for these smaller machines can be expected to be somewhat higher than for medium and larger scale machines, which have fewer components per unit energy output. It therefore seems probable that wind turbines in the size range of about 20 to 30 metres diameter will be able to offer superior economics, and it is significant that many manufacturers in Europe are now developing commercial wind turbines in this size range. Medium sized wind turbines have the advantage, by comparison with megawatt scale machines, that they can be developed (and their design subsequently refined) much more rapidly and economically; and their production and purchase involves significantly lower levels of financial risk, to manufacturers and purchasers respectively.

Recent estimates, both for Britain and the European Community (refs.5 and 6) have made clear that wind turbines could make a substantial contribution to present electricity needs. And the potential market for economic machines, in Europe alone, is many thousands. The public reaction to the deployment of wind turbines in large numbers is still uncertain, but with

---

*This energy cost corresponds – as stated – to a wind turbine cost of £125/m², when normalised to a height of 10 metres; this is equivalent to £185/m² for a wind turbine with a hub height of 25 metres (see equation 1), and to £300/m² for a wind turbine with a hub height of 75 metres.

good design and attention to aesthetics, (e.g. tubular rather than lattice towers), the modern wind turbine can be elegant. Here again medium scale machines may have a distinct advantage, by comparison with land-based megawatt scale wind turbines, since their scale is more in harmony with the rural landscape, and their appearance will consequently be less intrusive.

Wind Turbine Economics for Private Users

The preceding section reviewed the economics of wind turbines using criteria consistent with those used by utilities in evaluating the economics of coal or nuclear power stations. The private purchaser of a wind turbine has a different economic perspective, and requires an earlier return on his investment. In Europe most prospective purchasers will already be connected to the utility grid system, and will consider purchase of a wind turbine as a means of reducing their energy costs. Based on experience so far (mainly in Denmark and the Netherlands) the main market is for smaller wind turbines in the size range 10 to 20 metres diameter, and grid connected via an induction generator. The purchaser, typically a farmer, a small industrialist in a rural area, or the owner of a large country home, will take electricity from his wind turbine as and when the wind blows, so reducing his purchases from the utility. As such every kilowatt-hour he avoids purchasing has a value to him equal to the price at which the utility sells electricity to consumers, which is typically about double the direct, power station fuel cost. During periods of high wind speeds the wind turbine owner will frequently produce more electricity than he can use. The surplus is then sold to the utility, at a buy-back rate which is typically slightly less than the fuel saving value to the utility. Overall about half of the value of the wind turbine results from the reduced electricity purchases from the utility, and about half results from the sale of electricity to the utility. (For wind turbine owners who use oil for heating it may well be more economic to use surplus wind generated electricity as heat, and so reduce their oil consumption.)

In Denmark wind turbine sales at present total about 200 per year, and most new sales are for machines about 15 metres diameter. (In the last three years the most popular size has gradually increased from about 10 metres diameter to about 15 metres diameter, as the larger machines have become commercially available at more competitive prices.) Average costs for currently available Danish wind turbines are indicated in Table 1; these costs include delivery and on site erection, but exclude the cost of foundations and grid connection.

Table 1: Average Costs for Small Danish Windmills

Diameter	Cost per unit rotor area	Cost per peak watt
10m	$\simeq$ £160/m^2	$\simeq$ £0.56/W
15m	$\simeq$ £115/m^2	$\simeq$ £0.36/W

Based on Danish electricity prices and buy-back rates the corresponding pay-back periods are as indicated in Table 2. The pay back-period is calculated simply as the ratio of the initial cost to the annual value - to the purchaser - of the electricity output (without any allowance for the prospect of increasing energy prices over the lifetime of the installation.)

Table 2: Pay-back Periods for Small Danish Windmills

Annual Average Windspeed	Windmill Diameter	Payback Period (Years)	
		Without Subsidy	With 30% Subsidy
4.7 m/s	10m	11.6	8.1
	15m	8.3	5.8
5.5 m/s	10m	8.4	5.9
	15m	6.1	4.3

Small wind turbines which meet standards specified by the Riso test centre qualify for a government subsidy of 30%, and the effect of this on the pay-back period is also indicated in Table 2. For the larger wind turbines, in the windier locations, pay-back periods are attractively short - just over 6 years without the subsidy, and just over 4 years if the subsidy is included. And there are large areas of the U.K., and Europe, where average wind speeds are 5.5 m/s or higher.

Although the costs in Table 1, and hence the pay-back periods in Table 2, exclude the site specific costs of grid connection and the wind turbine's foundation, a more significant omission is that Table 2 does not include any allowance for O & M costs. Table 3 therefore shows how the pay-back period varies with the magnitude of the O & M costs, and shows how important it is that these costs be kept to a low level. For the more economic 15 metre machines, in locations where V_m=5.5 m/s, a 2% (of capital cost) annual allowance for O & M costs increases the pay-back period from 6.1 to 6.9 years (excluding any allowance for the government subsidy.) However if annual O & M costs are 5% of the capital cost, the pay-back period is extended to 8.7 years.

Table 3: Pay-back Periods for Small Danish Windmills with Operating and Maintenance Costs Included

Annual Average Windspeed	Windmill Diameter	Payback Period Years (excluding Government subsidy)			
		with O. & M. costs*			
		0%	2%	5%	10%
4.7 m/s	10m	11.6	15.0	27.3	∞
	15m	8.3	10.0	14.3	50.0
5.5 m/s	10m	8.4	10.1	14.4	51.9
	15m	6.1	6.9	8.7	15.4

*Annual O. & M. cost included as a percentage of initial capital cost

As operating experience is acquired and designs are refined, low O & M costs can be expected. And since even the most popular Danish machines are at present manufactured in relatively small numbers (less than one per week) there is considerable scope for cost reductions as the market expands and the production volume increases. However the effect of taxation on the benefit to the purchaser may well be even more important. A grid-connected electricity generating wind turbine directly reduces the purchaser's expenditure of after tax income on electricity (and, for some applications, oil). For a high rate tax payer this can substantially enhance the value of a wind turbine, by comparison with alternative investments yielding taxable income.

Ref.5 estimates that small wind turbines could contribute up to 10% of U.K. electricity needs, and similar estimates for other European countries indicate that the total market for economic machines in this size range totals many tens of thousands.

Wind Farms

The growth of wind farms in the U.S.A., and especially California, has been particularly noteworthy in the last two years, with major installations in a number of different locations. The wind farm concept was pioneered by U.S. Wind Power, and their installation in the Altamont Pass, east of San Francisco, comprises more than 200 wind turbines, each with a diameter of 17 metres and a rated output of 50 kW.

The basic concept is that the wind farm developer puts together a package for investors, which provides for the purchase, installation and maintenance of an array of wind turbines on a leased site in a location where average wind speeds
are high. An essential part of the package is the connection
to the local utility grid (or some large user of electricity)
plus their purchase, at an agreed rate, of all the electricity
produced. By law, in the U.S.A., utilities must buy back
electricity, from wind farms or individual producers, at a
rate which fully reflects the utility's marginal generation
cost. Individual investors in a wind farm typically contribute several thousand dollars, and several hundred such
investors are formed into a partnership to own and operate the
wind farm. Each investor is able to claim a Federal Tax
Credit equal to 25% of his share of the overall wind turbine
costs, plus a similar California State Tax Credit, and can
then write off the cost of the wind turbine investment against
State and Federal tax liabilities over the next three to five
years. The net result is that provided the investor is a high
rate tax payer (paying tax at the top rate of 50%) most of his
investment is returned in the first year, and almost all is
returned within five years, virtually regardless of how much
energy the wind turbines produce. Given good, reliable
machines the potential return on the investment is a very
attractive one. Not surprisingly, there are plans to instal
several thousand more small wind turbines, (plus larger
machines also) over the next two to three years.

It remains to be seen whether government/taxation incentives
will encourage similar entrepreneurial wind farm developments
in the U.K. or elsewhere in Europe. Although such installations would at present be illegal in the U.K., the Energy
Bill now before Parliament will - if passed - permit such
developments. And it is worth noting that farmers in the
windier parts of the U.K. could well profit from wind farm
developments. The Annual Review of Agriculture (ref.7) for
1982 indicates that the four million hectares on which cereal
crops are grown provide the farmer with a gross income of
approximately £500/ha. (i.e. £200 per acre). In locations
where wind speeds average 5.5 m/s (and large areas of the U.K.
experience wind speeds of this magnitude) wind turbines spaced
a nominal 8 diameters apart would give an annual electricity
output of 65000 kWh/ha. Valued at 2 pence/kWh, via sales to
the local utility, this would be worth £1300/ha. (i.e. £530/
acre); valued at 4 pence/kWh, via sales to local farm or
industrial users, the electricity output becomes worth £2600/
ha. (i.e. £1050/acre). And this income is, of course, additional to the farmer's continuing income from his crops. Even
though individual farmers may not have the capital to finance
the installation of wind turbines on their land, U.S. experience suggests that they can still benefit by leasing to wind
farm promoters the right to instal and operate wind turbines.

Given potential revenues of the magnitudes indicated above it may be expected that the use of wind turbines in rural areas could expand very rapidly, once the economic viability of such installations has been successfully demonstrated.

Conclusions

Despite the initial expectation that megawatt scale wind turbines would offer energy at a lower cost than smaller machines, present evidence suggests that wind turbines in the size range 15 to 30 metres diameter will offer the most attractive economics, at least in the immediate future. Wind turbines in this size range will deliver electricity at a cost of about 2.2 pence/kWh (competitive with electricity from coal fired power stations), given the requirement for a 5% rate of return in real money terms (i.e. net of inflation) on the capital invested in their construction. It is presumed that a 20 year life and low Operating and Maintenance Costs can be achieved, but this remains to be demonstrated.

For private users in rural areas small wind turbines (10 to 20 metres diameter) operated in parallel with the grid currently offer pay-back periods of the order 6 to 9 years, excluding any government subsidy or tax incentives.

The recent development of wind farms in the U.S.A. is particularly notable, and is responsible for the very rapid installation of large numbers of smaller wind turbines (10 to 20 metres diameter) in California and some other parts of the U.S.A. Though wind farms are at present illegal in the U.K. the Energy Bill now before Parliament proposes to make such developments legal. In many parts of the U.K. the potential revenue to farmers from wind turbines is several times greater than their gross revenue from present crops. Once the economic viability of wind farms has been demonstrated one can therefore expect the concept to spread rapidly.

Acknowledgement

The author wishes to thank the Directorate General for Energy, DGXVII, of the Commission of the European Community for permission to publish this assessment of wind turbine economics, most of which was initially undertaken on behalf of the CEC and reported to the Commission in November 1982. The views expressed are those of the author.

References

1. Divone L. and Luther E. 'Technical and Economic Progress in the Development of Wind Power'. Proc. 4th Int.Symp. on Wind Energy Systems, 119-130, Stockholm. Publ. by BHRA, 1982.

2. Pooley D. 'Wind Energy Development in the U.K.'. Proc. 4th BWEA Wind Energy Conference, 1-11, Cranfield. Publ. by BHRA 1982.

3. Vachon W.A. 'Large Wind Turbine Generator Performance Assessment. Technology Status Report No.2'. Arthur D. Little Inc. Report AP-1641, 1980.

4. Frandsen S., Trenka A. and Pedersen B. 'Recommended Practices for Wind Turbines Testing I. Power Performance Testing. I.E.A. Expert Group Study. 1982.

5. 'Strategic Review of the Renewable Energy Technologies' ETSU Report R13, HMSO, 1982.

6. Schnell W. 'The Wind Energy R and D Programme of the European Communities'. 4th Int. Symp. on Wind Energy Systems, Stockholm, BHRA, 1982.

7. Annual Review of Agriculture 1982. Cmnd 8491 HMSO 1982.

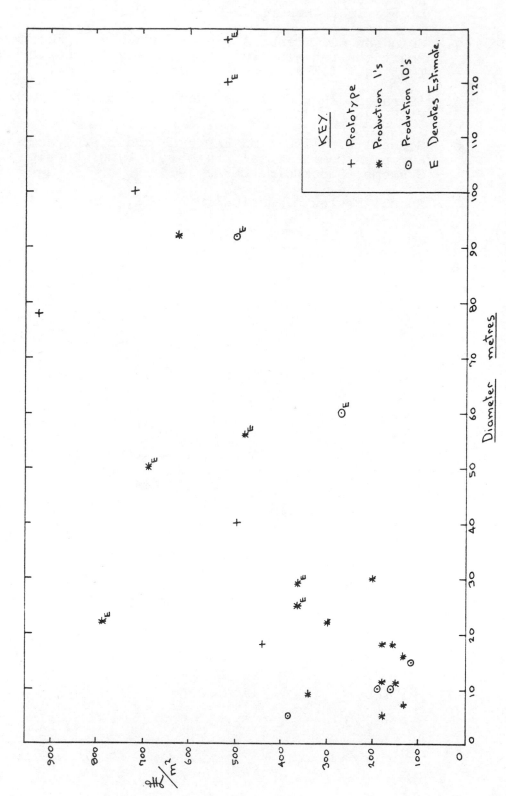

Fig. 1   Cost per unit rotor area versus wind turbine diameter

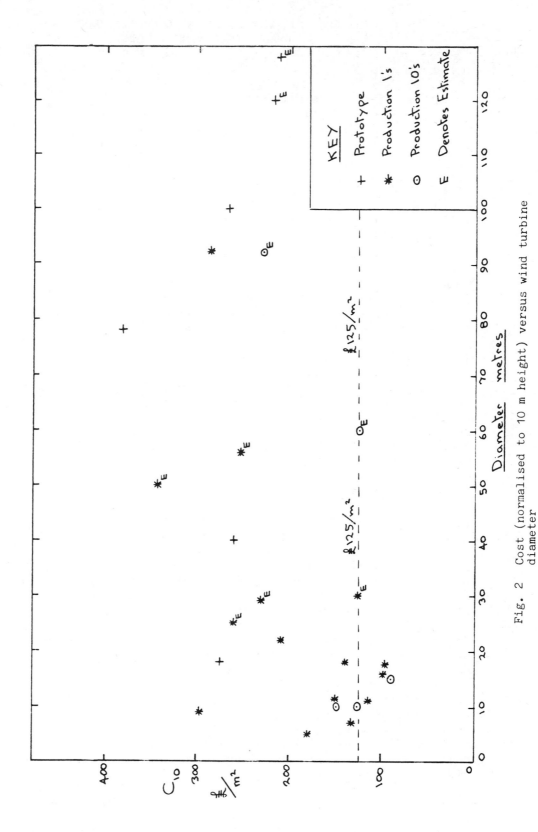

Fig. 2 Cost (normalised to 10 m height) versus wind turbine
diameter

THE ECONOMIC VIABILITY AND COMPETITIVENESS OF SMALL SCALE WIND SYSTEMS

P L Fraenkel and J P Kenna
Intermediate Technology Power Ltd
Mortimer Hill, Reading, Berks RG7 3PG

Abstract

I T Power Ltd has recently completed a worldwide survey of commercial windpumps and wind-generators, with rotor diameters of up to 15m. This paper indicates manufacturers' responses on prices and performance, and comments on these. The data collected was used in an economic model to compare energy costs of windpumps and wind-generators with those for equivalent solar and diesel powered systems. The principal results of this study and their implications are given.

Introduction

The authors have completed a number of economic assessments of small wind and solar energy systems (in comparison with conventional technology such as diesel power), e.g. (1), (2), (3), and in the process have developed and refined a computer-based economic modelling technique. The analysis described in this paper is based primarily on data collected as part of a study, by I T Power (4).

Current status of small-scale wind energy technology

Fig. 1 shows the costs in US $/m^2 for 94 models of windpump and Fig. 2 gives the equivalent results for 50 models of wind generator. Wherever possible the nearest tower size to 10 m was chosen for comparison, and fob prices were used. Care is obviously needed in comparing products from different countries, due to exchange rate fluctuation; for this reason, the products are identified by country, but significant scatter is present for products from one country such as the USA.

In the case of windpumps, the scatter was so large that we also looked at mass versus rotor size, and hence at cost/kg in order to investigate any machines that may be improbably lightly built to provide adequate durability; e.g. the mass of 4.9 m (16 ft) rotor windpumps appears to vary from as little as 150 kg (for a self-build design in India) to up to 1500 kg for UK, US and Australian factory-made farm windpumps of traditional design. The mass-specific cost similarly varies from little more than $1.5/kg for certain Australian windpumps, to more typically $3-6/kg for most products, with, at the other extreme, a few prices greater than $10/kg.

So far as wind-generators were concerned, the main scatter arose from manufacturers' performance claims. As would be expected, there is general correlation between size and efficiency, (Fig. 3), but a number of models appear to have implausibly high or low efficiencies. This problem may partially be due to the variety of ways in which manufacturers describe the performance of their systems; we had to try and standardise and attempted to rerate all systems to 10 m/s windspeed in order to obtain a better comparison. On the basis of this, the specific cost in $/W can be determined, and these are given in Fig. 4.

Fig. 2 indicates that with present small volumes of production, wind-generators in the 1 to 3 m diameter range which are appropriate for small battery charging installations, typically cost from about US $400/$m^2$ to $800/$m^2$. We used a value of $625/$m^2$ for our comparative economic analysis relating to small battery charging systems. However, Fig. 2 also indicates that we believe it should be possible to mass-produce small wind-generators for $200/$m^2$ of rotor area. This estimate is based on a price of $5/kg for finished mass-produced products ranging from ceiling fans (equivalent to very small wind-chargers) to motor-cars (possibly similar to production-line produced larger wind turbines). Combining this with a typical efficiency assumption allows a probable "least cost with mass production" to be derived in Fig. 4.

Economic viability of small wind systems

For the economic study we concentrated on quite small systems to satisfy demands in the 1 to 5 kWh/day range. We used typical present day costs for the windpump and windgenerator analysis (and the same for solar photovoltaic systems).

While the performance claims of manufacturers are variable in their credibility, such data as rotor size or the mass of a system are relatively reliable. Hence, although we are interested in manufacturers' performance claims, we compared systems on the basis of cost per unit area of rotor on the assumption that most good systems would produce a comparable output per square meter of rotor area in a given wind regime. We attempted to determine the actual energy output costs to be expected from small windpumps and wind-generators (the latter in conjunction with a battery storage). We also established comparable solar photovoltaic and diesel engine powered models for comparison.

The basis of the modelling consisted of defining a plausible energy demand pattern, wind regime and solar regime. From this the wind and solar systems could be sized such that they could satisfy the required demand in the most critical month; i.e. the month in which the mean wind or solar energy availability is lowest in relation to the demand. This means that in most cases, for much of the year, the wind or solar powered systems are required to have an excess capacity.

The lifecycle cashflows could then be determined on the basis of firstly, capital costs (including site overheads, installation, and the costs of auxilliary components) and secondly, of O & M costs, including replacement of components and accessories at predefined intervals, plus fuel in the case of diesel engines.

The lifecycle costs over a long period (30 years) were discounted to the present, at a differential rate of 10% and an annual equivalent cost was determined. From this, the unit energy costs could be determined by dividing the equivalent annual cost by the annual energy production.

Numerous parameters were required for the models; some key ones are as follows:

1. Windmills:
cut in windspeed	v_c	2.5 m/s
rated windspeed	v_r	10 m/s
furling windspeed	v_F	12.5 m/s
windpump efficiency at	v_r	9 %
windgenerator efficiency at	v_r	20 %
windpump installed cost		290 $/m² of rotor
windgenerator installed cost		625 $/m² of rotor

2. Solar photovoltaic:
PV array overall efficiency	10%
solar pump system efficiency	5%
PV array installed cost	16 $/Wp

3. Diesel engine:
size	2.5 kW
diesel/pump efficiency (high case)	9 %
diesel/pump efficiency (low case)	15 %
diesel/gen. efficiency (high case)	6 %
diesel/gen. efficiency (low case)	10 %
diesel fuel price (low case)	0.38 $/litre
diesel pump installed at	850 $/kW (shaft)
diesel generator installed at	1000 $/kW (e)

Because there is a lack of reliable data on the true efficiencies and maintenance costs actually achieved in practice from small diesel systems, we considered a "diesel low" and "diesel high" case to represent a range of conditions for low and high O and M costs respectively.

The results of running the models with the above data, are indicated in Figs. 5 and 6 respectively, which show unit costs versus daily load. The pumping system unit costs were all derived for a static pumping head of 20 m.

Tables 1 and 2 go further in summarising some typical comparative system capital costs and unit output costs under unfavourable, marginal and adequate energy resource regimes.

General conclusions

Considerable variation in windpump and wind-generator system costs are apparent, but sufficient pattern emerges to permit an approximate "typical" price per unit area of rotor to be defined. The indications are that if the market were sufficiently large to justify mass-production, the smaller sizes of wind-generators could be produced at between one third and one half of present day costs, but the larger sizes of system studied are less likely to fall so much in price.

Despite their quite high present day costs, wind systems came out as the least-cost option for mean wind speeds exceeding 4 m/s in the case of windpumps and 4.5 m/s in the case of small wind generators, for the models assumed. This suggests that many of the more arid parts of the world, plus most coastal and marine wind regimes could be exploited cost-effectively for small-scale applications using windpumps or wind-generators. However, wind system economics are extremely sensitive to mean windspeed assumptions, so much so that at under 2 m/s in the case of windpumps, and under 3 m/s in the case of wind-generators, wind becomes the highest cost option with the parameter values chosen.

This emphasises the considerable importance of accurate windspeed monit- oring. Inaccurate monitoring seems more likely to result in wind systems not being used where they could have been, rather than vice-versa. Errors in windspeed measurement seem more likely to be underestimates rather than overestimates due to badly sited anemometers or mechanically "sticky" anemometers.

We also conducted sensitivity analysis to determine the most critical parameters; these indicated a number of further points:

1   The mean windspeed in the "critical" month is more important than the annual mean windspeed for a site since it determines the system size for a given demand pattern.

2   The demand pattern has a profound effect on wind and solar power system economics; the less peaky the demand pattern (i.e. the ratio of peak demand to average demand) the more cost-effective are the renewable systems likely to be.

3   Energy costs from wind generators are very sensitive to assumptions on maximum periods of calm, as these influence battery costs consid- erably. Fig. 6 illustrates significant differences in output costs of energy from systems with 36h and 72h storage capacity in a mean windspeed of 4.5 m/s

4   Financial factors such as the choice of discount rate have a profound effect. We used a "commercial" discount rate and a diesel fuel cost of 38 ¢/litre (£1.14/gal) for the low case diesel. In practice, in remote areas, fuel costs could be much higher thereby further favour- ing wind or solar powered systems.

Acknowledgements

This paper is based mainly on work completed for the World Bank under UNDP Global Project GLO/80/003.

References

1. Fraenkel, P.L., "The relative economics of windpumps compared with engine driven pumps", Proc. BWEA Conf., Cranfield, 1981

2. Fraenkel, P.L., "The relative economics of windpumps compared with diesel engine and solar photovoltaic powered pumps on boreholes in Kenya", I T Power Paper, Ref 82034 for I T Industrial Services, Reading, May 1982.

3. Kenna, J.P., et al., "The Energy Connection", Development Forum, UN Information Committee, Geneva, January, 1983

4. I T Power Ltd., "Wind Technology Assessment Study", for UNDP/ World Bank, February, 1983.

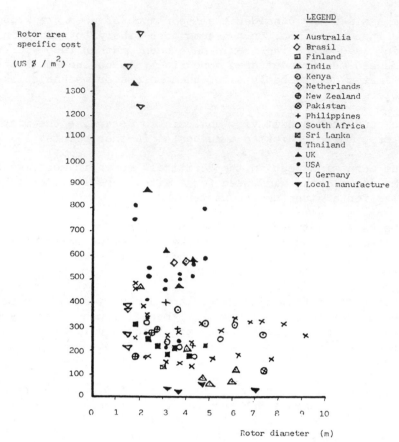

Rotor area
specific cost

(US $ / m²)

LEGEND

× Australia
◇ Brasil
▣ Finland
△ India
⊗ Kenya
◇ Netherlands
⊕ New Zealand
⊗ Pakistan
+ Philippines
○ South Africa
▨ Sri Lanka
◼ Thailand
▲ UK
● USA
▽ W Germany
▼ Local manufacture

Rotor diameter (m)

Figure 1: WIND-PUMP ROTOR AREA SPECIFIC COST v ROTOR DIAMETER

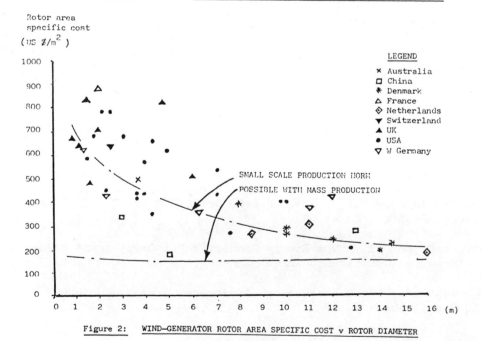

Rotor area
specific cost

(US $/m²)

LEGEND

× Australia
□ China
✳ Denmark
△ France
◇ Netherlands
▼ Switzerland
▲ UK
● USA
▽ W Germany

SMALL SCALE PRODUCTION NORM

POSSIBLE WITH MASS PRODUCTION

Figure 2: WIND-GENERATOR ROTOR AREA SPECIFIC COST v ROTOR DIAMETER

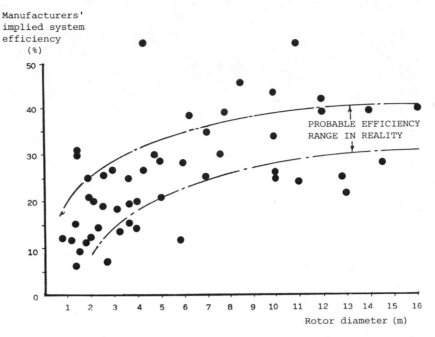

FIGURE 3.   WIND-GENERATORS - MANUFACTURERS' IMPLIED
            EFFICIENCY v ROTOR SIZE

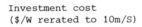

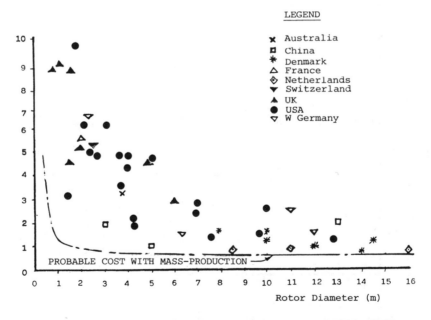

FIGURE 4.   WIND-GENERATOR INVESTMENT COST v ROTOR SIZE

Unit water cost

(US $ / m³)

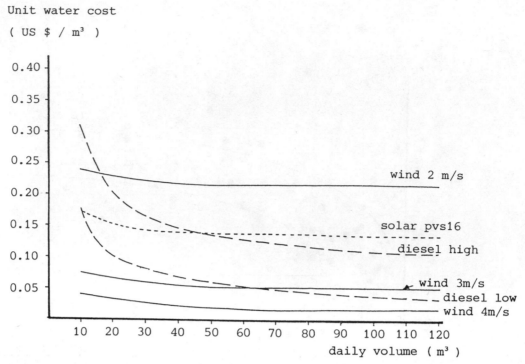

Fig. 5. Unit water cost as a function of daily water demand

Energy output cost

(US % / kWh)

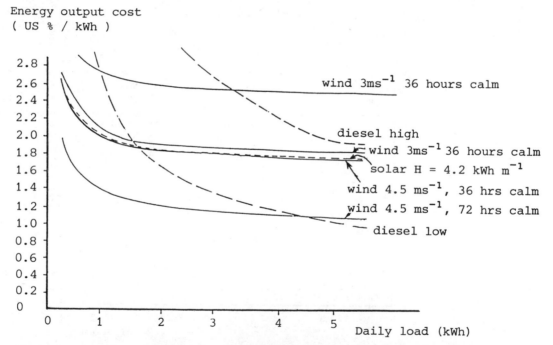

Fig. 6. Energy cost as a function of daily load.

POWER SOURCE	DESIGN MONTH MET CONDITIONS	POWER SOURCE SIZE	CAPITAL COST $	ENERGY COST $ per hydraulic kWh
SOLAR	2.8 kWh/m²	2450 Wp (1)	38,150	3.5
	4.2 kWh/m²	1650 Wp	27,500	2.7
	5.6 kWh/m²	1225 Wp	21,000	1.9
WIND	2 m/s	14.2 m (2)	41,000	4.2
	3 m/s	6.7 m	15,000	1.1
	4 m/s	4.4 m	5,000	0.5
DIESEL low	-	2.5 kW (3)	2,500	0.9
DIESEL high	-	2.5 kW	2,500	2.4

SUMMARY OF COSTS FOR ALTERNATIVE PUMPING DEVICES PROVIDING 60 m³ PER DAY

TABLE 1   THROUGH 20 M. HEAD

Key Assumptions: 10% discount rate; 30 year period of analysis
                   solar and wind;   50 maintenance per year
                   Diesel - low; $200 maintenance per year; $477 fuel per year
                   Diesel - high; $400 maintenance per year; $1430 fuel per year

Notes: (1) Electrical peak watts
                   (2) Rotor diameter
                   (3) Rated shaft power

POWER SOURCE	DESIGN MONTH MET CONDITIONS	POWER SOURCE SIZE	CAPITAL COST $	ENERGY COST $ per electrical kWh
SOLAR	2.8 kWh/m² (1)	1050 Wp (2)	21,000	2.4
	4.2 kWh/m²	700 Wp	15,000	1.8
	5.6 kWh/m²	525 Wp	12,000	1.5
WIND	3 m/s 36 hours calm	4 m (3)	13,000	1.8
	3 m/s 72 hours calm	4 m	16,000	2.5
	4.5 m/s 36 hours calm	2.3 m	6,500	1.1
DIESEL low	-	2.5 kW (4)	3,000	1.0
DIESEL high	-	2.5 kW	3,000	2.0

SUMMARY OF COSTS FOR GENERATING SYSTEMS PROVIDING A DAILY ENERGY OUTPUT OF 3 kWh

Key Assumptions: 10% discount rate; 30 years period of analysis
                   solar and wind;   50 maintenance per year
                   Diesel low; $200 maintenance per year; $263 fuel per year
TABLE 2   Diesel high; $ 400 maintenance per year; $708 fuel per year

Notes: (1) Assuming three days of battery storage
                   (2) Electrical peak watts
                   (3) Rotor diameter
                   (4) Rated shaft power

The Cost of Electricity from Wind Turbines
The Through Life Cost Concept

J. C. Riddell
J. C. Riddell & Associates
Wetherby, Yorkshire

Abstract

The Operational Cost is an important input into the design of a Wind Turbine. This paper looks at these costs from an Engineering standpoint and shows how the power produced between overhauls can be used both as a design objective and as an effective basis for evaluating unit costs of Electricity generated. In this way, the sales price for such equipment may be indicated.

The method of investigation is also used to suggest the necessary conditions where the integration of a Wind Turbine and Diesel Electric Generator will be cost effective. Guidelines for such installations are suggested.

Introduction

The recent fall in the price of Oil from $34 to below $30 a barrel following a fall in World demand has shown that Oil and Energy is a price sensitive commodity. It was the rise in Crude Oil prices in 1973 that did much to stimulate the renewed interest in Wind Energy. It is useful now to examine, from an Engineering point of view, the basis on which the cost of Wind Energy may be determined. We should also consider if this will encourage its use as a power source.

Basis of Cost

It is said (Ref 1, 7) that costs should reflect reality so that effort expended on one activity can be compared precisely with another. However there are many external influences on the capital cost of a piece of Plant, such as a Wind Turbine, that are outside the influence of the designer (Ref 4, 5). For example:

1. Government Grants that meet a proportion of the installed cost of the machine.

2. Tax Incentives that encourage specific machinery purchases.

3. Loan charges from Banks.

4. Inflation Rate on money during the life of the machine.

It is unfortunate that notions of cost are often based upon a twelve month period for this is time over which taxes are levied (Ref 5). However machinery usage does not always fit precisely within the same period. For a Wind Turbine the power produced in one location may well be very different from that which the same machine will generate in a location with a lesser mean wind speed. It is clear that if the same annual charges are levied on both machines distortions in the true cost of electricity from the wind will occur.

Through Life Cost Concept

Being a power producer, Wind Turbines are only of use when producing power. If the basis of cost suggests that a Wind Turbine will cost more per kilowatt hour than a grid supply or a diesel electric generator, it follows that it will be rejected in favour of the cheapest alternative. 'The Through Life Cost Concept' takes into account the costs of maintenance and repair of a piece of plant over its useful life and those of supervision and establishment that are inevitable in any installation.

We can express it so:

Price per kilowatt of Electricity =

$$\frac{\text{Purchase Price Installed + Total Maintenance Costs}}{\text{Total Kilowatt Hours Produced}}$$

This arrangement has a very important advantage - it takes out the time dimension. The annual time period of measurement can confuse totally (Ref 7) and bring about circumstances that ensure the wrong machine is bought for the wrong reason and located in the wrong place. Wind Energy may never recover from the frustrated expectations of the customer that will result if this occurs.

It is unrealistic to suppose that the Wind Turbine will run without servicing or overhaul for its useful life. It is realistic to define the amount of power that it can produce before certain levels of attention and overhaul are necessary. The example is analogous to the Time Between Overhauls (TBO) of the aircraft engine or helicopter rotor, or the 'lifted' components of many types of machinery: i.e.

Light Aircraft Engine	2,000 hours
Helicopter Rotor	500 hours
Electric Generator	5,000 hours

In a Wind turbine it becomes the P.B.O. (Power Between Overhauls).

The Importance of Overhaul Life

The concept of Overhaul Life brings together, in a convenient form, the objectives for Design, Service Requirements and Sales Price.

1. Design: Overhaul life defines the level of performance that the Machine must satisfy.

2. Service: Overhaul Life defines the cost of the operation of the machine.

3. Sales Price: From Overhaul Life the purchase price of the Machine that the Market will accept in open competition, can be derived.

The Designer is required to consider the loads imposed upon the machine in operation. These loads are clearly not going to be uniform in all parts of the machine and calculation will not predict precisely all conditions of operation. Those parts that show signs of distress can be removed - at the point of the overhaul. It follows that as confidence is built up of the operation of the machine in service, then the overhaul life can be extended. The fatigue life of a component - the product of level of stress and the number of times it is applied - should be a multiple of the overhaul life.

Service Experience

The existence of an overhaul life implies that parts will be changed at that time, and thus a level of expenditure is implied. It is often assumed erroneously that machines will work effortlessly all the time. In the case of a power source, the machine must be repaired as soon as possible after breakdown. The advantage of an overhaul life is that there can be a high level of confidence that the machine will not cease to function in the course of the period between overhauls.

Repairs in the field are always much more expensive than in the factory, and therefore effort expended in the design stage to ease the work of exchange and rectification in the field is always well rewarded in service.

It will be appreciated that when a machine is fully developed and properly maintained, it will remain in service for many years although component parts may have been replaced many times in the lifetime of the machine.

The Price the Market will bear

It matters little to the ultimate users what method is used to generate the electricity he buys. He or she is entirely concerned with cost. The price of diesel generated electricity can vary widely depending on location. Prices as high as 75p per kilowatt have been mentioned. However, uncertain wind turbine systems will not be purchased unless there is a substantial saving in the generating cost of electricity when wind is used. The additional capital cost of the Wind Turbine will have to be recovered by the cost savings produced by its use.

Additional assumptions will be made before we can relate in an example the overhaul life and capital cost for a given unit cost of a Kwh.

They are:

1.   Machine maintenance costs are 2% per annum of installed first cost.

2.   Overhaul Costs are 25% of installed first cost.

3.   The Power Output is 250,000 kwh between overhauls.

4.   The annual kwh/output output achieved is taken at 28% of the rated output. (Ref 7).

5. The useful life of the machine is five overhaul periods before obsolescence ensures scrapping.

6. The price per kilowatt of electricity from the Wind Turbine is 5p.

So from the formulae for Capital Cost X we get:

$$\text{Price 5p} = \frac{0.02X \text{ (overhaul life in years} - 1) + 0.25X + 0.2X}{\text{Energy generated between overhauls}}$$

Table 1.

Rated Capacity	15 kw	20 kw	50 kw
Overhaul Period	Price	Price	Price
2 yrs	£ 7,978	£10,640	£26,595
3 yrs	£11,320	£15,306	£38,265
4 yrs	£14,706	£19,608	£49,000
5 yrs	£17,700	£23,585	£59,000

It is also interesting to see how the prices of machines of different rated capacities are effected by the duty required of them.

Table 2.

Annual Power Output	Price	Equivalent Rated Cap.	Overhaul Period
125,000 kwh p.a.	£26,595	50 kw	2 years
84,000 kwh p.a.	£25,510	33 kw	3 years
62,500 kwh p.a.	£24,500	25 kw	4 years
50,000 kwh p.a.	£23,585	20 kw	5 years

It is clear from these two tables that there is little advantage in buying a small machine and working it very hard for it will be worn out very quickly. It is more economical to buy a larger machine and de-rate it with a smaller rotor.

The interchange between overhaul life cost of operation and the first cost that can be paid for the same unit cost of electricity is also very clear. Too often the installed cost has been the basis of buying choice, and this has led to disappointment. The machines purchased were not strong enough to withstand the demands made upon them.

Market Opportunities

It can also be seen that where electrical generation costs are high then Wind Turbines can provide a useful source of low cost power. Clearly there is a real financial advantage in linking Wind Turbines to Electrical grids where the power is provided by diesel electric generators. The first acceptance of Wind Power will be in isolated locations. The need for long periods of trouble free running between overhauls is of first importance.

The cost of diesel fuel is not controlled entirely by the well head price, for transportation costs can be very high. In difficult country the cost of driving a light cross country vehicle can exceed £1 per kilometre, and thus fuel to remote stations becomes very expensive indeed.

Influence of Fiscal and Other Measures

Investment Grants

In the U.K. Investment Grants may be paid by Government on Capital Plant Purchased for approved purposes. The rate will vary but is currently between 25% and 50% of installed cost. The award of a grant has the effect of lowering the point when there is advantage in buying the subsidised machine. However the maintenance costs are related to the unsubsidised installed cost.

Tax Incentives

Tax Incentives can reduce the effective cost of Wind Energy very considerably, but they are charges against the individual or company tax assessment. So to be of value, it is necessary to have a lot of tax to pay.

However, Tax Incentives can be effective stimulants to a market as we have seen in Denmark and the United States. There is a danger that when they are removed or reduced the customer demand falls significantly so that the balance of the manufacturing industry is upset. Companies report operating losses leading to Company failure until the balance between supply and demand is restored.

Inflation and Loans

Attempts have been made to establish the effect of inflation in the life of the investment (Ref 1, 4, 5), using the well known compound interest formulae. The result has been to encourage users to buy now, in the belief that it will cost more in future. Inflation works on both income and expenditure, and therefore the argument is not sound. A buying decision is a comparison between alternatives at a precise moment, so inflation has little influence. All alternatives are equally effected by it.

There may be some benefit in buying through the use of a Bank Loan or Extended Credit arrangement, for it may be that there is an implicit discount on the cost of the equipment in such arrangements. However, it is always better to buy for cash and negotiate on that basis, for that will ensure that the plant has the lowest installed cost. For a bank loan purchase to be successful, the benefit from the Wind Turbine must be more than the extra cost of the Loan Interest and servicing charges. The effect is to raise the threshold at which the purchase of a Wind Turbine is economic.

Wind Diesel Integration

The application of the "Through Life Cost" concept to the integration of Wind Turbines and Diesel Generators shows us clearly when this arrangement should be used for power generation.

Ref 5, 6 demonstrated that there is a link between the cost of wind and diesel power and wind penetration, that is defined

Wind penetration (%) = $\dfrac{\text{Power in kwh supplied by the wind}}{\text{System Power required in same period}}$ x 100

It can be seen that there is a minimum amount of power that the Wind Turbine must supply at a lower rate before the additional cost of the Wind Turbine is paid back (S/G). This amount is determined by the total cost of the system electrical supply (G), and it is this ratio (S/G) that decides how much wind energy is needed before the presence of the Wind Turbine is economic — that is the minimum wind penetration of the system.

A non dimentional plot has been drawn up for differing values of the costs of Unit Wind Electricity (W) and unit diesel electricity (D), against wind penetration for break even at selected S/G ratios of saving and system cost.

For a prospective integrated installation, the ratios can be worked out to ensure that the installation parameters are above the line where there will be economic advantage in an integrated system.

Conclusions

The value of the Energy Produced between overhauls is a useful measure to overcome the difficulties met when additional annual charges are found to be a distorting factor. It is also an effective method to describe the price and the objectives that must be met by both the designer and the service requirements for the Wind Turbine to become an effective power producer. The "Through Life Concept" requires that equal importance is given to the installed cost and the subsequent operating costs in the purchase decision.

This investigation suggests that it is likely that the power between overhaul values of 250,000 kwh to 750,000 kwh depending on size are realistic for current kilowatt range of Wind Turbines. This indicates that an overhaul period of between four and five years will be necessary. The purchase price of about £1,000 per kilowatt installed of rated capacity is thus justified for operating costs of 5p per unit of electricity generated.

The effect of Fiscal measures has also been discussed, but as they are independent of design considerations, they were not part of the thrust of the argument. They have importance when relating the cost of operation and purchase of a Wind Turbine to a specific location and market.

When an engineering comparison is made, Wind in a suitable location can be compared favourably with other forms of power generation given that the quality of design will provide the reliability in service and the power production between overhauls that are necessary to satisfy the market.

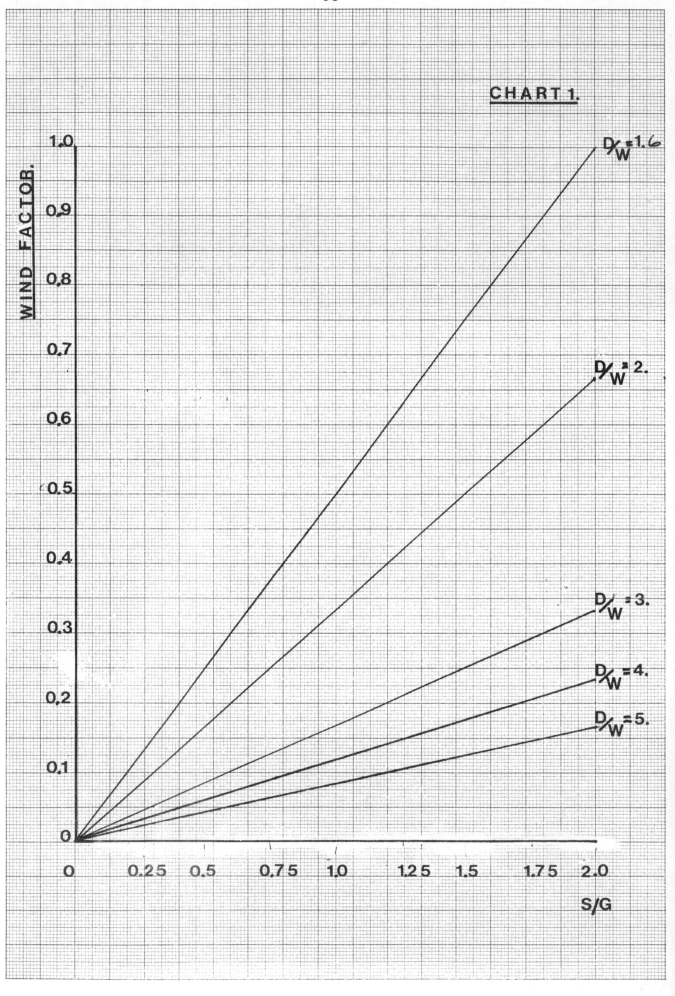

CHART 1.

References

1. A Cost Benefit Analysis of Wind Power Generators, by C. Nottingham. Accountancy June 1978.

2. Markets for Wind Energy Systems - Where and at What Price? Johnanson, Goldenblatt, Marshall, Tennis. AIAA/SERI Wind Energy Conference, Boulder Colorado, April 1980.

3. Economic Incentives to Wind Systems Commercialisation, Michael Lotker, The Synetics Group Inc., AIAA/SERI Wind Energy Conference, Boulder Colorado April 1980.

4. Economics of Selected W.E.C.S. Dispersed Applications. Stella Krawiec, Solar Energy Research Institute, Golden Colorado, AIAA/SERI Wind Energy Conference, Boulder Colorado April 1980.

5. Integration of Small Wind Turbines with Diesel Engines and Battery Storage. G. Stack, N. Lipman, P. Musgrove. Reading University, B.W.E.A. Conference, Cranfield 1982.

6. An appreciation of the 10 metre Windmatic Aerogenerator operating on Orkney. W. M. Somervlle, N.E.I. Ltd., W.G. Stevenson N.O.S.H.E.B. B.W.E.A. Wind Energy Conference April 1981.

7. Technical and Economic Progress in the Development of Wind Power. L. V. Divone, 4th International Symposium on Wind Energy Systems, Stockholm 1982.

RECENT DEVELOPMENTS AND RESULTS OF THE
READING/RAL GRID SIMULATION MODEL

E.A. Bossanyi[*] and J.A. Halliday[**]
[*] Department of Engineering
University of Reading, U.K.
[**] Rutherford Appleton Laboratory
Chilton, Didcot, Oxon., U.K.

Abstract

A detailed description is given of the latest version of the electricity grid simulation model, with which the integration of renewable energy sources and the use of storage can be investigated in some detail. Results are given for the integration of wind energy up to large penetrations for a case based on the expected CEGB system in about 1985.

Introduction

This paper describes the latest state of an electricity grid simulation model currently being used at Reading University and the Rutherford Appleton Laboratory. The model was originally developed by G. Whittle and a description of the original model and some of the results from it can be found in refs. 1, 2, 3.

Grid system operation is frequently studied using probabilistic simulation models (see ref. 4), in which probability distributions are assigned to various grid system variables (i.e. loads, plant availabilities, etc.), so that the performance of the system can be assessed statistically, e.g. by calculation of loss-of-load probabilities. Such models require relatively small amounts of computer time, but do not take into account time-dependent effects, for example in wind energy studies the predictability of wind power up to a day in advance is quite important, as it takes some hours to start up large fossil-fired generating stations. Also the scheduling of pumped storage plant is a time-dependent effect which cannot easily be modelled using probabilistic models.

It is possible to extend such models to include some time-dependent effects (ref. 5) but to treat these effects fully it is necessary to use a time-step simulation model.

The Whittle model uses a time-step of one hour, i.e. it uses hourly load and wind power data and simulates the operation of the whole grid system hour by hour during the whole simulation period, which is usually one year but can be varied.

The main disadvantages of such an approach are that it requires a large amount of computer time, and also that it is difficult to derive any results relating to system reliability - this depends on events of low probability such as unforeseen failures of large generating sets, or a sudden drop in wind power coinciding with a period of peak electricity demand. To estimate these results reliably would mean running the model through many years of simulation, for which suitable hourly input data are in any case not available.

Nevertheless, the model is still very useful for studying the integration of renewable power sources such as wind power. Detailed estimates of fuel savings can be derived from, say, a year's simulation, along with information about optimum grid control strategies, such as the way in which spinning reserve and pumped storage should be scheduled. In addition, the operating regime experienced by generating plant units can be studied, for example the load factors and numbers of hot and cold starts for steam turbine units.

Types of Generating Plant Modelled

Conventional generating plant is divided into nuclear, fossil-fired steam turbine, gas turbine, and pumped storage hydro plant.

Nuclear stations are assumed to give a constant output throughout the simulation, i.e. they cannot be used for load-following.

In the absence of any renewable sources, the model then attempts to meet the remainder of the load using steam turbines. If this is not possible for any reason then pumped storage (if present and not drained) is used, and then gas turbines as a last resort. Unlike steam turbines, pumped storage and gas turbines are assumed to be able to respond instantaneously to changes in demand - in practice it might take several minutes to start up a gas turbine, but this is short compared to the hourly time step used in the model.

Thus nuclear and gas turbine plant can be modelled very simply, while steam turbine and pumped storage require more complex operation strategies involving forward planning. This is because steam turbines have a long start-up time, and pumped storage can be used partly to level out the daily load cycle. These considerations are detailed below. If the level of nuclear plant output exceeded minimum system demand, it would be necessary to model their part-load behaviour in some way, but this has not been necessary to date. It is in any case far from clear how nuclear stations would respond to this kind of operation.

Modelling of Steam Turbine Plant

In the following description the figures in parentheses indicate the values of particular variables used to obtain the results described later in this paper.

It would perhaps be desirable to model separately each individual steam turbine unit for a given system, since their characteristics can vary considerably, but in the interests of simplicity and generality this has not been done.

The steam turbines are modelled as a given number of units, all of the same size (500 MW) and with a given start-up time (eight hours). They can, however, be arranged in a merit order, specifying the full load efficiency as decreasing linearly with position in the merit order. The full load efficiency of the first unit is specified (37.5%, based on nett calorific value) and also the decrease per unit (0.085%). These values were selected to give approximate agreement with the CEGB system.

Once a unit is generating, it can be operated at any level from full load down to a specified part-load limit (50%). It is assumed that any change in output level can occur in less than the hourly time-step. Part-load efficiency is defined by a Willans line: at x% load the fuel consumption is 0.15 + 0.85 (x/100) times the full load fuel use. Fuel use is also specified during the start-up sequence as 0.102 times the full-load consumption. In addition the unit can be held on standby at any level of readiness between cold and hot; on n-hour standby, the resulting fuel use is 0.01875 (8 - n) times full-load consumption. Thus a unit which has been cooling for, say, three hours is assumed to require three hours to start up again; if it is required in four hours' time, it must be kept on three-hour standby for an hour. The model keeps a record of the state of readiness of each unit which is updated every hour.

If a unit is not required in a given hour but is likely to be needed again within the next four hours, it is kept running at the part-load limit to avoid switching it off and on again. This evidently means that other units will have to be run at less than full load in the meantime, to avoid surplus generation (if the part-load limit is greater than zero). In addition, if a unit is not needed for four hours but will be required again within 24 hours, it is kept on one-hour standby to avoid excessive cycling.

If one or more units are running part-loaded in any hour and the pumped storage reservoir is not full, the loading is increased and the surplus power used to replenish the reservoir.

Operating Strategy

Because of the long start-up time of steam turbines, it is necessary to predict up to eight hours ahead how many units will need to be operating in that future hour. This is done by predicting the demand for that hour, subtracting the output of the nuclear stations and of any renewable sources (including a prediction of wind power), adding any spinning reserve requirements, and, if pumped storage is to be used for load-levelling, making an adjustment for this. All these aspects are explained below.

The eight-hour forecast is repeated each hour, to ensure that sufficient steam turbines are started up in time. When each hour actually arrives, the actual demand may be different from that predicted. This can mean that there is insufficient steam turbine plant available to meet the actual demand. Since additional steam turbines cannot be brought on line at such short notice, power from the pumped storage system is used if available. If there is still a shortfall then gas turbines are used. If there are insufficient gas turbines installed on the system, the model records a "loss-of-load event".

If the forecasting error is such that too much power is available (even after using as much as possible of the surplus to replenish the pumped storage reservoir) the output of steam turbines is reduced. If a renewable such as wind power is present in sufficient quantity it is possible for there to be surplus power even after all steam turbines have been taken down to the part-load limit, in which case the surplus power is dumped notionally by shutting down some of the wind turbines.

Electricity Demand and Load Forecasting

The model reads in from a file the electricity demand on the system for each hour of the simulation period. A ten-year magnetic tape of half-hourly load data for the CEGB system was used to create suitable datafiles for each calendar year.

This data is used to provide a forecast of future demand for any hour, used for scheduling steam turbines and for planning load-levelling with the pumped storage plant. When the hour in question actually "arrives" this forecast demand is multiplied by a randomising factor to give an actual demand for the hour which is slightly different from the predicted demand. The factor used is a random number taken from a normal distribution with a mean of 1.0 and a standard deviation of 0.015. In this way an element of demand uncertainty is introduced, which is of a similar magnitude to the uncertainties on the CEGB system over a similar timescale.

Wind Power Data and Predictions

Wind power available during each hour of the simulation period is also read in from a datafile. Hourly wind speed data covering a number of years for 14 U.K. sites were obtained from the Meteorological Office and these can be converted to wind power using a given wind turbine characteristic. For the results presented in this paper the wind speeds were extrapolated to the appropriate hub height (61 m) from the height of measurement using a power-law exponent (0.14), and wind power data was then created using a Mod-2 characteristic (ref. 6) but scaled down to have a rated wind speed of 1.5 times the ten-year mean wind speed for the site.

Wind forecasting is much less reliable than demand forecasting. Consequently the wind power data is used as the actual available wind power for any hour, while forecasts of wind power are done each hour using the persistence method, a very simple method which works reasonably well (ref. 7). Each hour, the wind power available in that hour is used as a forecast of wind power available in future hours, up to eight hours ahead (the start-up time for steam turbines). Evidently there can be quite large discrepancies between forecast and actual wind power.

One system uncertainty which the model does not take into account is the possibility of unexpected failure of large generating units. As already mentioned, this aspect is better dealt with by a probabilistic model (ref. 4), owing to the difficulty of running an hour-by-hour simulation model for a sufficiently large number of years.

Tidal Power

In addition to wind, the model has also been used with tidal power. Although variable, this source is highly predictable and is treated in a similar way to nuclear plant output by the model - see ref. 8.

Spinning Reserve

Spinning reserve results from the operation of steam turbines at part load, and is the amount of additional power which would be available if all steam plant which are running part-loaded at a given hour were turned up to full power. Steam turbine start-ups are planned in such a way that a certain amount of spinning reserve would be available in any future hour if demand and wind forecasts proved accurate for that hour. Thus if forecasting inaccuracies result in a shortfall of power in a given hour, this reserve can be used at short notice (shorter than the hourly time-step) and is used in preference to pumped storage or gas turbines.

Three control parameters have to be specified when running the model which determine how much spinning reserve is to be allowed for in any future hour when scheduling steam turbine start-ups. The first, SR1, specifies an amount of spinning reserve proportional to the predicted system load, to cover demand uncertainties. The second, SR2, specifies an amount of spinning reserve proportional to the predicted available wind power to cover wind power forecasting errors. SR3 is a fixed minimum level of spinning reserve. Thus the spinning reserve planned for a future hour is given by SR1 . D + SR2 . W , or by SR3 if it is greater, where D is the predicted demand for that hour and W is the predicted wind power (equal to the current wind power).

These control parameters are fixed at the start of the simulation. The model can thus be run several times with different values until a roughly optimal combination is found (i.e. one which minimises system costs). In real life, of course, spinning reserve strategy may be varied by the system operators according to season, hour of day, scheduled television programmes, etc., and can therefore be "tuned" more finely than is possible with this model.

In practice, the optimum value of SR1 is sometimes slightly negative, i.e. enough steam turbines are started up to meet slightly less than predicted demand, so storage and gas turbine use will be slightly higher. If a fixed minimum level of spinning reserve is not required, SR3 is set to a large negative number.

Load Levelling

An optional load-levelling strategy, more sophisticated than that used in ref. 3, has now been built into the model which ensures that the storage reservoir (if present) is replenished during the night-time demand trough so that it can be used to help meet the subsequent afternoon/ evening peak.

This has the advantage that generation from low-merit (i.e. less efficient) stations at the peak is replaced by generation from higher-merit stations at the trough. However, the inefficiency of the storage system (for present purposes the efficiency has been assumed to be 88% in each direction, or 77.44% overall) must be taken into account, so this advantage is lost if the difference in efficiency between the low and high merit stations is not great, as is generally the case. Nevertheless the availability of storage for levelling off the peak can mean that several extra steam turbine units may not be needed at all and can be left cold, thus reducing start-up and standby losses, and this can make the exercise worthwhile. Also if some of the annual peak demand can be met from storage, the total installed capacity can be reduced. With the cost and efficiency figures used in this paper, the use of storage for load-levelling does in fact appear to be worthwhile.

The disadvantage is that the availability of the storage as a fast-response substitute for spinning reserve and gas turbines is slightly reduced. Therefore an additional control parameter has been introduced, QSF, which specifies a fraction of the reservoir capacity which is to be used in planning load-levelling, i.e. the load-levelling algorithm assumes a reservoir capacity of only QSF times the actual capacity. Once again, an optimum value of QSF can be found by trial and error.

The algorithm is invoked at the start of each day and uses the predicted load profile for the day to calculate a new "levelized" profile, such that the difference between the profiles can be accounted for by operation of the pumped storage plant (using no more than QSF times the reservoir capacity). The algorithm has the following limiting conditions:-
(a) in any hour, the amount of pumping or generating cannot exceed the rating of the pumped storage plant;
(b) the same amount of energy, Q, is added to the store during the trough as is drained from it at the following peak;
(c) Q must be less than or equal to QSF times the reservoir capacity.

Under normal circumstances the daily peak is smoothed to a flat plateau, the level of which is limited by condition (a) to the peak demand minus the pumped storage rating, as illustrated in fig. 1. If this does not use all the allowed reservoir capacity, it is possible to shave more off the hours adjacent to the peak, (subject to (a) above). However, this is not done if it means that steam turbine power is displaced by power from the storage, because those steam turbine units will need to be running in the peak hour of the day anyway, and as explained above, the inefficiency of the storage plant does not justify load-levelling unless steam turbine units can be kept switched off altogether. However such additional shaving of near-peak hours is permitted if gas turbine generation is thereby saved, which only happens if the installed steam turbine capacity is insufficient to meet the peak. There is no corresponding requirement for the trough to have a flat bottom.

A simple iterative procedure calculates the amount of levelling at peak and trough according to the above rules, and this is done at the start of each day. The resulting "levelized" load profile is used instead of the predicted load in scheduling steam turbine start-ups eight hours in advance. It is then not normally used again; the storage is simply used in whatever way is best in any given hour. This usually results in load-levelling more-or-less as planned, but with the storage still available to cope with forecasting errors in any hour.

However the levelized load profile is used again in the case of a day in which the installed steam turbines are insufficient to meet the peak demand and the pumped storage capacity is insufficient to eliminate the need for gas turbines at the peak altogether. This means that the pumped storage (which is of higher merit than gas turbines) will be needed earlier than was envisaged by the load-levelling planning algorithm, and the reservoir may already be empty before the peak hour is reached. The peak would then have to be met by gas turbines. It is more sensible in this case to use gas turbines earlier and save some pumped storage for the actual peak. The model achieves this by restricting pumped storage output in near-peak hours to no more than the amount calculated by load-levelling algorithm. Although this does not affect fuel costs much (it merely changes the timing of the output of gas turbines, which all have the same efficiency) it does reduce the amount of gas turbine capacity required.

The disadvantage is that during those few hours the availability of storage for meeting unforeseen variations is reduced.

Outputs available from the Simulation Model

The model produces a number of outputs which can be used to understand the operation of the system. It produces annual summations of the system load, the available wind power and the wind power accepted into the system, the output and fuel usage of steam turbines and gas turbines, the energy flow into and out of the pumped storage reservoir, and the output of nuclear plant. It can also produce daily and hourly print-outs of these values. In addition the number of hot and cold starts are recorded for each steam turbine unit, as well as the total for all the units. Also printed out are statistics about the number and magnitude of any loss-of-load events and the minimum available plant reserve during the simulation period. Finally the fuel, fixed and capital costs for the simulation are printed out.

The results presented in this paper are solely concerned with the fossil fuel savings resulting from the addition of wind turbines to an existing CEGB system, so only the fossil fuel costs are relevant. In a study such as that described in ref. 12 where a long-term view is taken, the number of steam and gas turbines is varied to find the optimum plant mix, so the capital and fixed costs for these plant types are required.

The Model Applied to the UK: Wind Energy Integration in the Near-Term

The plant mix of the CEGB system as expected in 1985 was used as the basis of the research reported in this paper. This plant mix was calculated from the existing plant mix, knowledge of stations under construction, and the plant lifetimes given in ref. 9. For the simulation model the steam turbine capacity required is in fact that presumed to be available for operation at the time of peak annual demand. Thus the steam turbine capacity expected in 1985 was multiplied by the winter peak availability factor of 0.86 (from ref. 9). The steam turbine full load efficiencies are based on CEGB figures for 1981-2 (ref. 10). The fossil fuel costs used are in March 1981 prices. Full details of all these and the other inputs and assumptions used in this paper are given in Table 1. As explained above, capital and fixed costs are not required for present purposes.

Wind speed data from three different sites were used simultaneously to give some geographical diversity, the three sites selected being Plymouth, Dungeness and Ronaldsway. In view of the considerable inter-annual variation of wind speed (ref. 11), simulations were carried out using data for three different years: 1978 (a near-average year), 1976 (a year with a low annual mean wind speed) and 1974 (which had a high mean wind speed). Table 2 shows the mean wind speeds for the three sites and the combined means for each year. Wind power data for each site was created as explained above, and the three resulting datafiles were combined in equal proportions. The CEGB load data for 1974 and 1976 was normalised to give the same total demand as in 1978, to allow a more meaningful comparison of the model results for the three years.

Optimisation Procedure

Optimisation consists of varying the three control parameters SR1 (load-dependent spinning reserve component), SR2 (wind power-dependent spinning reserve component) and QSF (fraction of pumped storage reservoir used for load-levelling) until a combination is found which minimises the fossil fuel cost of the simulation run, subject to the requirement that no loss-of-load events occur. The parameter SR2 is of course not used in the cases where no wind turbines are installed. For each of the three years, wind penetrations of 0, 5, 10, 20 and 40 GW were used. This therefore required three 2-parameter optimisations (the no-wind cases) and twelve 3-parameter optimisations, which used a large amount of computer time. A standard "simplex" minimising routine was used to do the optimisations automatically, although at high wind penetrations the minimum was rather flat, and could only be located precisely by doing some additional runs on a trial and error basis.

Results

To illustrate the operation of the model, Figure 2 shows how the system copes on a rather bad day, when the wind power available drops sharply to zero just before the day's peak demand hour. In hour 17, spinning reserve on steam plant is insufficient to meet the shortfall. The pumped storage plant is therefore brought in at its maximum capacity, but gas turbine plant is still required. Wind power picks up again in hour 19; gas turbines are no longer required, and there is some excess steam turbine power which is used to replenish the pumped storage.

Detailed results of all fifteen optimised simulation runs are presented in Table 3. It can be seen that the optimum value of SR2 increases with wind penetration. Wind power increases the number of steam turbine cold starts, but has little effect on hot starts. Note that gas turbines generate only a very small proportion of the total annual demand of about 218600 GWh.

Figure 3 shows the principal results for the three different years. In Figure 3(a) the savings in fossil fuel cost on the system are shown. The "ideal" saving is defined as the available wind energy for the year multiplied by the average fuel cost per kWh for fossil-generated electricity in the no-wind case.

The difference between the ideal and actual savings is the "total penalty" plotted in Figure 3(b) as a percentage of the ideal saving. Figure 3(b) shows that at low penetrations the penalty is small and is mostly attributable to the "operating penalty", which is caused mainly by the reduction in efficiency of the steam turbine plants due to increased part-loading and the more frequent starting and stopping. The slightly increased use of gas turbines also makes a small contribution to this "operating penalty". However, at high penetrations (above about 15 GW) there is no further increase in operating penalty, but a large increase in total penalty due to wind power having to be discarded. Figure 3(c) gives the proportion of annual electricity demand which is supplied by the wind, and shows that at the very highest penetration considered (40 GW), although nearly half the available wind energy is discarded, it still supplies between 28.8% and 33.6% of total electricity demand (depending on the year).

Wind energy is discarded because of errors in forecasting: the very simple persistence method used in the model gives no advance warning at all of any changes in wind conditions, whereas proper meteorological forecasts are able to predict the arrival of fronts and weather systems, although their predicted time of arrival may be wrong by several hours (ref. 14). Nevertheless, the use of such forecasts should reduce considerably the amount of wind power discarded at high wind penetrations, and correspondingly increase the saving of fossil fuel.

Further improvement would result from reducing the steam turbine part-load limit. This would involve burning some oil along with the coal in coal-fired stations, but at high wind penetrations this might be economically worth while. At high penetrations a considerable amount of wind power is discarded because nuclear plant cannot be downloaded. This loss is attributable to the inflexibility of nuclear and not to the unpredictability of the wind.

Figure 3(a) shows that there is a difference in fossil fuel cost savings of the order of 15% between good and bad wind years, but the use of data from a near-average wind year such as 1978 does appear to give representative results.

Conclusions

The simulation model has been described in detail and results presented for integration of wind energy onto the CEGB grid. The results demonstrate that it is important to have wind speed data for a number of years, so that a near-average year can be selected. It is shown that wind power could supply about 20% of annual demand without any major difficulties. With 10 GW of wind capacity, supplying 14.3% of demand in a typical year, the fossil fuel cost savings amount to £515m (1981 prices), only 6.8% less than the ideal savings, while with 20 GW of wind capacity, supplying 24.6% of demand, the savings are £845m, implying a penalty of 23.6%. It is suggested that the savings could be improved by using a less simplistic wind forecasting method, particularly at high penetrations.

Acknowledgements

The authors would like to express their thanks firstly to Mr. G. Whittle who wrote the initial versions of the Simulation Model, and also to Dr. R. H. Taylor and his colleagues of the CEGB Planning Department for their useful comments and information. Thanks are also due to the Science and Engineering Research Council, who funded the research project, to the other members of the project team, Dr. P. J. Musgrove, Professor N. H. Lipman and Professor P. D. Dunn, and to D. Infield.

References

1. Whittle, G. E. et al. "A Simulation Model of an Electricity Generating System Incorporating Wind Turbine Plant" Proc. 3rd Int. Symp. on Wind Energy Systems, Copenhagen, BHRA, 1980.
2. Whittle, G. E. "Effects of Wind Power and Pumped Storage in an Electricity Generating System". Proc. 3rd BWEA Conference, Cranfield, 1981.

3. Whittle, G. E. "Wind Power System Simulation - Optimal Mix". Proc. 4th BWEA Conference, Cranfield, 1982.

4. Rockingham, A. P. "The Economics of Windmills for Large Electricity Grids". Ph.D. Thesis, Mech. Eng. Dept., Imperial College, London, 1982.

5. Janssen, A. J. "A Frequency and Duration Method for the Evaluation of Wind Integration". Wind Engineering 6, 1, 1982.

6. Solar Energy Research Institute. "The Mod 2 Wind Turbine Development Project". SERI/SP-732-728, U.S.A., Oct. 1980, p.8.

7. Bossanyi, E. A. et al. "The Predictability of Wind Turbine Output". Proc. 2nd BWEA Conference, Cranfield, 1980.

8. Bossanyi, E. A. "Wind and Tidal Energy Integration into an Electricity Network". Proc. 4th BWEA Conference, Cranfield, 1982.

9. CEGB "Sizewell 'B' Power Station Public Enquiry: CEGB Statement of Case". April 1982.

10. CEGB "Statistical Yearbook 1981-2".

11. Halliday, J. A. and Lipman, N. H. "Wind Speed Statistics of 14 Widely Dispersed UK Meteorological Stations". Proc. 4th BWEA Conference, Cranfield, 1982.

12. Bossanyi, E. A. "Use of a Grid Simulation Model for Longer-Term Assessment of Wind Energy Integration", to be published 1983.

13. Halliday, J. A. "A Study of Wind Speed Statistics of 14 Widely Dispersed UK Meteorological Stations, with Special Regard to Wind Energy". Rutherford Appleton Laboratory Report, 1983.

14. Taylor, R. H. et al. "Integration of Wind Power onto an Electricity Supply System". Proc. 1st BWEA Conference, Cranfield, 1979.

Table 1 - Model Description and Input Parameters

Time Step:	1 hour
Load Data:	Demand on CEGB grid
Normalising Factor	1974 1.057996 1976 1.0507002 1978 1.0000000
Max. Demand (MW)	43162 43655 44758
Min. Demand (MW)	9640 9474 9907
Total Demand (GWh)	218590 218600 218597
Uncertainty Factor	Normally distributed around 1.00 with standard deviation of 0.015
Nuclear Plant capacity:	8170 MW
Output	Constant 5310 MW (i.e. 65% load factor)
Steam turbine plant:	42500 MW, or 85 units of 500 MW each
Start-up time	8 hours
Part-load limit	50%
Full-load efficiencies	37.5% for 1st unit decreasing linearly to 30.4% for the 85th unit
Fuel use, relative to	(0.15 + 0.85 (x/100) at x% load
full load	(0.102 during start-up
	(0.01875 (8 - n) at n-hour standby
Gas turbine plant:	Effectively instant start-up
Rating	3450 MW
Pumped storage plant:	Based on Dinorwic plus existing schemes
Rating	1860 MW
Storage capacity	10200 MWh
Efficiency	88% each way, thus 77.44% overall
Wind speed data	Combination of Plymouth, Dungeness & Ronaldsway
Mean speed at 10 m	1974 6.573 m/s 1976 5.480 m/s 1978 6.013 m/s
Height correction	to hub height of 61 m using power law exponent of 0.14
Wind turbine plant	Based on USA Mod 2, scaled to V_r = 1.5 x mean
Cut-in speeds	Pl'th 5.17 m/s D'ness 6.25 m/s R'way 6.01 m/s
Rated speeds (V_r)	10.08 m/s 12.19 m/s 11.71 m/s
Furling speeds	16.47 m/s 19.92 m/s 19.13 m/s
Generation costs (March 1981 prices)	
Fuel costs	Steam turbine 0.6081 p/kWh (thermal)
	Gas turbine 5.0428 p/kWh (electrical)

Table 2 – Mean Wind Speeds

Site	1970	1971	1972	1973	1974	1975	1976	1977	1978	1979	ten-year mean
Plymouth	5.50	4.66	5.67	4.94	5.98	4.90	4.62	5.36	5.22	5.33	5.19
Dungeness	**	5.44*	6.70	5.89	7.35	6.32	5.66	6.98*	6.47*	6.12*	6.31
Ronaldsway	5.89	5.10	6.04	5.77	6.39	5.99	6.16	6.84	6.35	6.36	6.06
3 site mean	-	-	6.136	5.533	6.573	5.736	5.480	-	6.013	-	5.853

Table of mean wind speeds (m/s) at an effective height of 10 m (from reference 13).

(* signifies an incomplete record for that year;   ** signifies total absence of data for that year)

Table 3 – Simulation Model Results

Year	Wind Capacity (GW)	Operating Strategy			Fossil Fuel Cost (£m)		Wind Power (GWh)		Outputs (GWh)		Maximum Gas turbine Demand (GW)	Average steam turbine efficiency %	Steam turbine starts			
													Total		Max. per unit	
		SR1	SR2	QSF	Steam turbine	Gas turbine	Available	Used	Steam turbine	Gas turbine			Hot	Cold	Hot	Cold
1974	0	-0.008	-	0.95	3007	5.1	0	0	172658	101	3044	34.91	6944	714	257	31
	5	-0.014	0.37	0.80	2698	9.2	17922	17903	154505	182	2860	34.82	6879	1309	267	62
	10	-0.013	0.61	1.00	2434	6.1	35845	35136	137079	120	3291	34.25	6841	1476	270	63
	20	-0.022	0.83	1.00	2100	5.4	71692	58428	113713	106	3291	32.92	6550	1216	257	56
	40	-0.004	0.80	1.00	1884	4.5	143380	73371	98752	88	3291	31.88	6435	1331	253	62
1976	0	-0.004	-	0.95	3014	3.6	0	0	172647	71	1569	34.84	6998	760	204	26
	5	-0.006	0.31	0.80	2777	7.0	13636	13631	158819	137	3182	34.78	7081	1306	233	51
	10	-0.019	0.75	1.00	2577	7.3	27273	26843	145426	144	3065	34.32	6891	1129	222	41
	20	-0.010	0.77	1.00	2267	5.0	54547	47610	124607	98	3411	33.42	6798	1424	227	48
	40	-0.020	0.86	0.68	2047	6.7	109091	62938	109249	133	3273	32.46	6769	1214	222	45
1978	0	-0.005	-	0.90	3016	3.9	0	0	172679	76	1528	34.81	7330	823	241	31
	5	-0.008	0.26	0.55	2742	6.2	15811	15806	156650	122	3052	34.74	7648	1458	273	63
	10	-0.013	0.55	0.70	2499	5.3	31622	31202	141098	105	3350	34.33	7335	1607	261	67
	20	-0.002	0.71	1.00	2171	3.5	63246	53712	118472	69	3446	33.18	6988	1732	256	73
	40	-0.013	0.85	0.90	1977	3.2	126488	67915	104270	62	3183	32.06	6873	1223	246	57

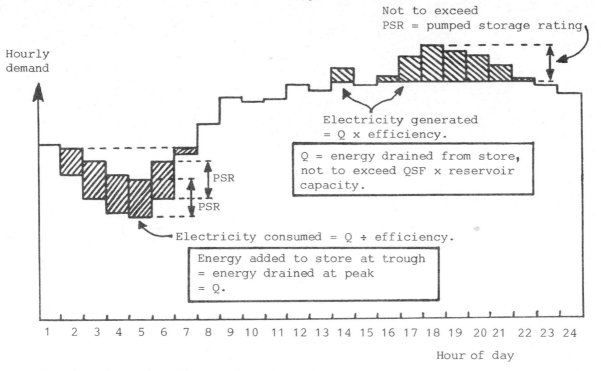

Fig. 1   Schematic Illustration of Load-levelling Strategy

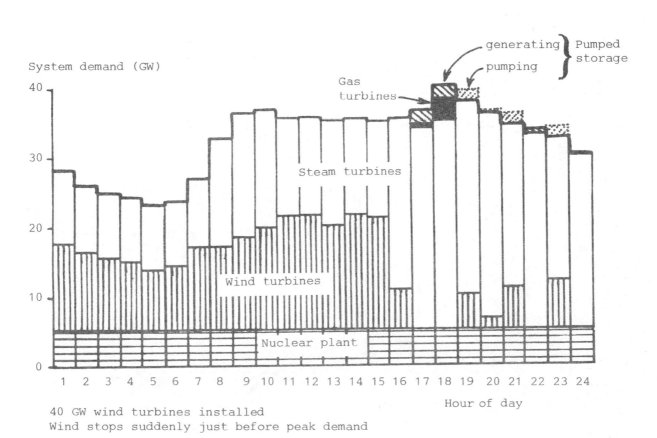

40 GW wind turbines installed
Wind stops suddenly just before peak demand

Fig. 2   System Operation on Day of Maximum Gas Turbine Demand· (for 1978)

Fig. 3 Wind Power Availability and System Savings for Three Different Years

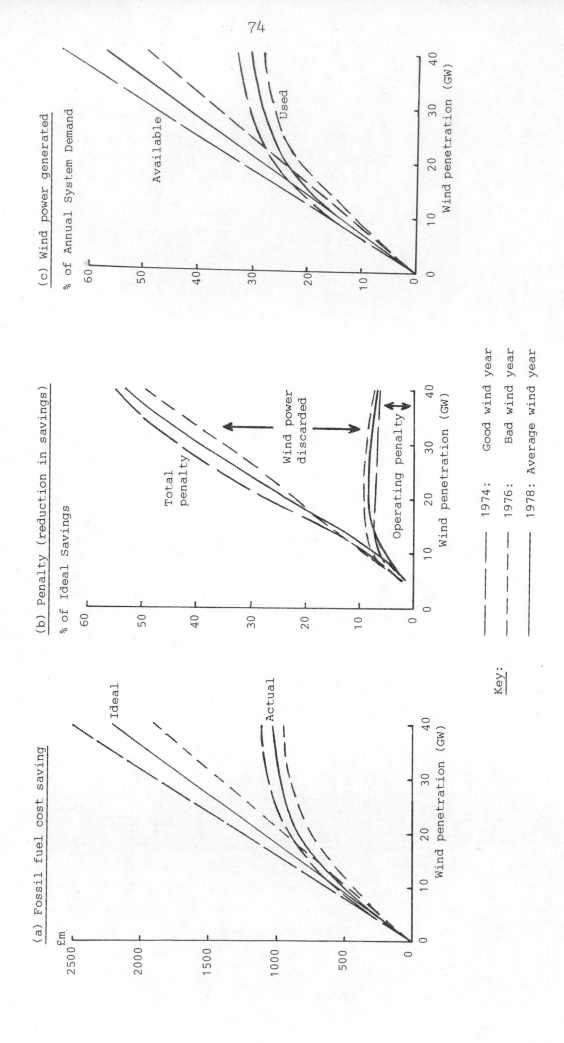

(a) Fossil fuel cost saving

£m

Ideal

Actual

Wind penetration (GW)

(b) Penalty (reduction in savings)

% of Ideal Savings

Total penalty

Wind power discarded

Operating penalty

Wind penetration (GW)

(c) Wind power generated

% of Annual System Demand

Available

Used

Wind penetration (GW)

Key:

1974:	Good wind year
1976:	Bad wind year
1978:	Average wind year

OPERATING RESERVE, UTILITY SIZE AND POWER VARIANCE

J C Dixon and R J Lowe

Faculty of Technology, The Open University, UK.

Abstract

The fuel saving value of wind generated electricity is reduced because of the need for utility operating reserve to service wind power variance, a loss generally estimated at about 10%. It is here shown that this fractional cost is essentially proportional to the wind power penetration with a ramp slope that is a strong function of utility size, and array size or number of sites. Parallel construction at several sites appears to be markedly superior to concentrated developments, and it appears desirable to use 300 MW rated-power arrays in preference to, say, 1 GW.

Introduction

The inclusion of wind generated electricity in a utility system changes the demand on conventional plant. A convenient way to deal with this is to treat the wind power as negative load. Thus the wind power reduces the remaining mean load, but variation of the wind power available effectively adds variance to the conventional plant load. To service these power changes, some increased operating reserve capability is required. This has an associated cost, which reduces the value of the wind power. Rockingham (1) has indicated a conservative (ie high) estimate of about 14% of fuel saving value, which would result from complete coverage of the wind power by part loaded fossil-plant. Rockingham and Taylor (2), extending work by Farmer et al (3), later estimated an 8% penalty where the wind power deviation may be covered by a combination of part-loaded steam plant, gas turbines and pumped storage. A number of computer model studies of utility operation incorporating wind power have included consideration of operating reserve requirements (eg Marsh (4), Johanson and Goldenblatt (5), Whittle et al (6), Whittle (7), Bossanyi (8)). For example Whittle et al (6) quote a 12% penalty, and later Whittle (7) quotes substantially smaller values from an improved model incorporating existing load variance.

One purpose of this paper is to present a parametric analysis, as opposed to a numerical simulation, which makes the influence of variables such as utility size and wind power penetration rather clearer. This approach is not intended to replace numerical simulations but rather, by aiding comprehension, to improve their development and interpretation.

The effect of load variance

The utility load, neglecting wind, has a deviation (dispersion) σ_L for a planning period t (eg despatch period of one hour). The wind power contribution has a deviation σ_W. Provided that the wind and load deviations are uncorrelated (conservatively, since some favourable correlation is likely to be present over longer time periods because demand is related to weather), the total deviation is σ_T, and the extra deviation, which will govern the extra incurred costs, is $\sigma_T - \sigma_L$.

$$\sigma_T^2 = \sigma_L^2 + \sigma_W^2 \qquad (W^2)$$

$$\sigma_T = \sigma_L (1 + (\sigma_W/\sigma_L)^2)^{0.5} \approx \sigma_L (1 + 0.5(\sigma_W/\sigma_L)^2)$$

This is a conservative approximation; it will be good provided that $\sigma_W \ll \sigma_L$, ie provided that the penetration is not large ($I \ll 1$). The added deviation is:

$$\sigma_A = \sigma_T - \sigma_L = 0.5 \ \sigma_W^2/\sigma_L$$

The added system deviation, expressed as a fraction of the actual added wind deviation, is the deviation factor F; ie

$$F = \sigma_A/\sigma_W = 0.5\ \sigma_W/\sigma_L$$

Hence the existing load deviation σ_L can effectively mask the wind deviation σ_W, and greatly reduce the incurred cost. The parameter F is a direct measure, indeed is defined as, the ratio of system added deviation to the wind power deviation. To express F in different terms, where a is the load specific deviation, b is the wind specific deviation, P is the mean wind power and L is the mean load power;

$$a = \sigma_L/L \quad ; \quad b = \sigma_W/P$$

$$F = 0.5\ \sigma_W/\sigma_L \quad = 0.5\ bP/aL \quad = 0.5\ (b/a)\ (P/L) \quad = 0.5\ k\ I$$

where $k = b/a$ and $I = P/L$. Hence the degree of masking depends upon the relative specific deviation k and the wind power penetration I.

If the actual cost of the provision of reserve is modelled as proportional to the deviation to be serviced, at the rate c p/h.kW (ie per kW of deviation), then the actual cost per hour is:

$$C = c\ \sigma_A$$

This cost is a loss of value of the delivered wind power P, so the reduction in value of energy (in p/kWh), or operating reserve cost, is

$$R = C/P = c\ \sigma_A/P \quad = c\ 0.5\ \sigma_W^2/(\sigma_L\ P) \quad = 0.5\ c\ b^2\ P^2/aLP \quad = 0.5\ c\ (b^2/a)\ (P/L)$$

$$R = 0.5\ kI\ cb \quad = F\ cb$$

Notionally, if there were zero load deviation and the wind deviation in its entirety had to be serviced directly, ie full coverage, then the hourly cost would be $c\ \sigma_W$, and the lost energy value would be

$$R' = c\ \sigma_W/P = cb$$

The cost is therefore reduced by the pre-existing load deviation in the ratio $R/R' = F$, so the deviation factor F ($= \sigma_A/\sigma_W$) is also a measure of the cost reduction due to existing load deviation.

For a given utility, with given load deviation $\sigma_L = aL$, if the wind deviation is modelled as proportional to installed mean wind power ($\sigma_W = bP$ with constant b) then $k = b/a$ will be independent of P. (This is the usual modelling assumption adopted, often implicitly, - we shall reconsider it later.) The deviation factor (and cost factor) $F = 0.5kI$ will simply be proportional to penetration I. Hence the reduction in value of energy (R) will be proportional to the penetration, and hence to the wind power. To be quite clear on this point, we emphasize: for a moderate penetration the total lost value of the wind energy (£/hr) is proportional to the installed wind power squared; the mean lost value per unit of energy (p/kWh) is proportional to the installed wind power; the marginal lost value for the next kW installed is also proportional to the previously installed wind power.

This simple but important prediction cannot yet be tested against 'reality' in any meaningful sense, since it would require extensive statistics of utility operation with various wind penetrations. However it is interesting to compare it with results from a numerical model of the NEGEA utility described by Johanson and Goldenblatt (5), Figure 1, who give operating reserve costs (in M$/y) and net load duration curves (9) as functions of wind power. From these it is possible to express the costs in terms of reduced value of energy (ie p/kWh), shown in Figure 1. The predicted linearity holds good to large penetrations. The slope is about 0.05 p/kWh (at 1980 prices, with various inflation assumptions), and it is evident that, for penetrations of any normal level, the extra cost is very small.

The remainder of the paper, then, constitutes an evaluation of the slope of the cost ramp, which will depend on both the utility and the wind, ie on both a and b.

Utility load variance and size

Different utilities will, in general, have different load specific deviations, a. In fact, a typical utility load can plausibly be modelled as many independent loads whose unpredicted deviations are not correlated (notwithstanding diurnal cycles that a utility anticipates and adjusts for). In such a utility model, the load variance will be proportional to the load, and the load deviation will be proportional to the square root of the load. Hence a larger utility would have a smaller specific load deviation, and will be less able to cover wind deviation at a given penetration (although this will correspond to a greater actual wind power). This neglects the utility's conventional generation uncertainty, which further reduces the effective σ_A. Previously we have written $\sigma_L = a L$. Now we write $\sigma_L = d L^{0.5}$, with the implication that d will be relatively constant amongst various utilities. For the NEGEA analysis L = 600 MW, and $\sigma_L = 140$ MW, so $d = 5.72$ $MW^{0.5}$ (24 hour prediction). On this model, a utility of 25 GW mean load, about CEGB size, would have load and specific deviations:

$$\sigma_L = 5.72 \times 25000^{0.5} = 900 \text{ MW} ; a = \sigma_L/L = 0.036$$

Hence, for a given wind specific variance, R would rise some 6.45 ($\sqrt{(25000/600)}$) times more rapidly with penetration. Nevertheless, at 20% penetration, the lost value is still under 0.07 p/kWh (about 3% of fuel saving value).

A single graph of the mean reduction in value, R, for different mean power utilities, will result from choosing the appropriate non-dimensional independent variable instead of I. Either kI or 0.5kI (ie F) would be appropriate. Also, it would be natural to non-dimensionalise the wind deviation σ_W directly against σ_L, giving σ_W/σ_L as a variable. In fact these are equivalent, since

$$\sigma_W/\sigma_L = bP/aL = kI = 2F.$$

The reduction in value can also be non-dimensionalised naturally against fuel saving value, V, although this may lead to some complications where different fuels suffer different escalation rates. However, from Figure 1, $R/V \approx 0.05$ I, giving Figure 2.

To correlate this result with other estimates, we may note that when the penetration is such as to give 0.5kI = 1.0, the R value will equal the particular notional value R', corresponding to full coverage. This will occur at I = 2/k (on the extrapolated linear approximation) and hence at different penetrations for different utilities (1.75 for 600 MW mean load (NEGEA), 0.27 for 25 GW). However the corresponding lost value R (p/kWh) is then a function only of the wind deviation, since the load deviation has, in effect, been set to zero. This linear theory gives R = 9% for kI = 2 (ie 0.5kI = 1) from Figure 2, in encouraging agreement with the previously mentioned estimates (1,2,6) from zero-load-variance models.

Wind power variance

The variance of wind generated power will depend upon wind behaviour, turbine characteristics and turbine locations. Figure 3 shows the form of the wind spectral density (10,11). Wind behaviour is generally characterised by relatively narrow variance peaks at annual and diurnal frequencies, a broad synoptic peak centred on a period of about 4 days, and a turbulence region. The variance of power changes over a lead time t is proportional, approximately, to the integral of the spectrum over the frequency range n > 0.1/t. Thus in estimating the uncertainty of wind power output at a lead time of 1 hour (a representative dispatch period), it is necessary to consider the variance in a significant part of the tail of the synoptic peak of typical wind spectra in addition to the turbulence component.

The synoptic contribution to wind power variance

Figure 4 shows an example of the variance (ie deviation squared) of power changes over time t for a single wind turbine similar to the Mod-2, based on hourly wind speeds at St Mawgan in the UK (Met Office data). The deviation has here been non-dimensionalised by the mean power and so is a generalisation of the specific standard deviation b, expressed as a function of lead time, ie b(t). A specific standard deviation based on hourly average data can be fitted by the empirical equation, modified from Farmer et al (3),

$$b(t)^2 \approx 2.0\ b_0^2(1 - e^{-t/12})$$

where b_0 is the simple standard deviation of the normalised hourly wind power outputs. For turbines with a specific power p = 1.0, $b_0 \approx 1.0$. Hence, to a good approximation,

$$b(t)^2 \approx 2.0\ (1 - e^{-t/12})$$

We assume, conservatively, that at the synoptic frequencies all the turbines in any one array are fully correlated, so the same equation will also be applicable to each array.

For N arrays, the outputs changes will be partially correlated. Justus and Mikhail (17) have shown that

$$b_N^2 = e\ \bar{b}^2/N\ (1+(N-1)\bar{r})$$

where b_N is the specific deviation for the N individual arrays together, $\bar{b}$ is the mean specific deviation of the N arrays, and $\bar{r}$ is the mean cross correlation coefficient of all possible pairs of the N arrays. For a siting policy in which arrays having equal annual energy output are installed, then with turbine specific power close to 1.0, $e \approx 1.0$. Justus and Mikhail have also shown that for a siting policy in which arrays are spaced regularly, $\bar{x}$ is approximately one third of the maximum intersite distance. The mean cross correlation coefficient $\bar{r}$ will approximately equal the coefficient of a pair of arrays with this mean separation. For the UK, the maximum separation may be of the order of 400 km, giving $\bar{x} \approx 150$ km.

The cross correlation coefficients of power changes over time t, for a large number of UK wind power sites, using hourly average wind data, are together plotted against lead time and site separation in Fig 5. Empirically, with no implication of causal significance,

$$r(t,x) \approx (x/(2.6t - 1.6) + 1)^{-0.6}$$

giving the smooth surface of Figure 6. For very widely separated sites, the cross correlation coefficient is dominated by the diurnal fluctuation, as is indicated by the coherence function plot of Figure 7. This effect manifests itself by the appearance of a minimum in the plot of cross correlation vs. lead time, at a lead time of 24 hours; eg, Figure 8 shows the cross correlation coefficient for St Mawgan and Wick (site separation 690 km). The diurnal fluctuation of wind speed for a given site is likely to be a strong function of measurement height and season (Petersen, 12). In this situation, 10 m height data is likely to give poor estimates of cross correlation coefficients for hub heights of 80 m. Fortunately, the correlation arising from the diurnal fluctuation is generally small in absolute terms, so in the present context this is not a serious problem.

The resulting cross correlation coefficient for the UK at 1 hour lead time is 0.05. The effective diversity factor is sketched in Fig 9 for the case where $\bar{r} = 0.05$.

The turbulence contribution to wind power variance

The hourly average data used alone neglect the turbulent power changes, which occur on a time scale of of a few seconds to a few minutes. We have estimated the variance of the power changes in the power output of a single wind turbine in a wind with 20% turbulence (high, but possibly appropriate for arrays), using the turbine power characteristic of Fig 10. Cut out in high wind speeds is assumed to be instantaneous. The variance of turbine power as a function of wind speed is shown in Fig 11; it is dominated by power changes generated at the cut out wind speed. The specific deviation of turbine power is 0.74 at a specific power coefficient of 1.0, increasing to about 0.82 at 1.5. The diversity factor (defined as the ratio of the specific standard deviation for the array to the specific standard deviation for a single turbine) was calculated using an equation developed by Lipman et al (13), but with considerably longer length scales (300 m cross wind and 900 m longitudinally) based on Busch and Panofsky (14) and Pasquill (15), and assuming a turbine spacing of 1 km. The resulting diversity is approximated by $D^2 = 2/T$ where T is the number of turbines. The instantaneous cut out strategy is undesirable for several reasons, and the frequency of cut outs can be reduced an order of magnitude (Bossanyi, 16). The lower limit on turbulent power changes is given by assuming a strategy that makes the variance due to cut out negligible. In this case the specific standard deviation of a single turbine due to turbulence is 0.62.

The total wind variance

We can include the effects of geographical diversity by incoporating the diversity factor thus:

$$R = 0.5 \ c \ (b^2/a) \ D_N^2 \ I \qquad (\text{replacing } R = 0.5 \ c \ (b^2/a) \ I)$$

where b is the specific standard deviation of power output for a single array, and D_N is the diversity factor for power output power changes for N discrete arrays of turbines. To estimate R for a 1 hour lead time, we have combined the turbulent deviation of wind power power changes with variation of hourly mean wind power outputs, by adding their variances. The resulting specific standard deviation is 0.84. The diversity factor of this uncertainty is given by

$$D_N^2 \approx 1.55/T + 0.23 \ (1/N + (N-1)\bar{F}/N)$$

Implications for UK operating reserve costs

Two extreme siting policies for wind turbines are: (1) Start construction of a number of wind turbine arrays simultaneously; later additions to grid wind capacity come from increases in array installed capacity, (2) Install turbines in multiples of a standard array. In the first case the diversity factor D_N will be approximately constant, and wind operating reserve costs, as p/kWh, will at first rise linearly with penetration I, with a slope dependent on the number of sites, Figure 12. In the second case, operating reserve costs constitute a discrete variable that will rise slowly as the effect of the non-zero cross correlation increases. Figure 13 shows how the operating reserve cost depends on array size for various values of penetration. To derive these results we used the following values: wind power specific deviation $b_1 = 0.84$; load specific deviation a = 0.01; grid mean power output L = 25 GW; R'/V = 0.15; reserve = 4 standard deviations. With these values, the approximations start to break down at I ≈ 0.2.

Summary

The fuel saving value of wind generated electricity is reduced because of the variance of power output, a reduction previously estimated at typically 10% of the value. However, pre-existing load variance masks the wind variance, so that the operating reserve costs will, in practice, increase linearly with penetration, beginning at zero. The slope of this cost ramp depends upon several factors, including in particular the number of wind sites, and the array size, and the utility size. A better non-dimensional variable to reduce the curves is not the commonly used penetration, but σ_W/σ_L, or 0.5kI. The operating reserve costs are strongly dependent on array size, and in this respect it appears advantageous to use 100 MW mean-power (ie 500 MW rated) arrays rather than 500 MW. Parallel construction at several sites is significantly better than highly concentrated development.

References

1   Rockingham A P; 'System Economic Theory for WECS' BWEA 2, Cranfield, 1980.
2   Rockingham A P and Taylor R H; 'The Value of Wind Turbines to Large Electricity Utilities' IEE Future Energy Concepts Conference, London, 1981.
3   Farmer E D, Newmann V G, Ashmole P H; 'Economic and Operational Implications of a complex of wind-driven generators on a power system' IEE Proceedings, 127A, 5, June 1980.
4   Marsh W D; 'Requirements Assessment of Wind Power Plants in Electric Utility Systems' EPRI ER-1110-SR, DOE Conf 790352, July 1979.
5   Johanson E E and Goldenblatt M K; 'Wind Energy Systems Application to Regional Utilities' EPRI ER-1110-SR, DOE Conf 790352, July 1979.
6   Whittle G E et al; 'A Simulation model of an Electricity Generating System Incorporating Wind Turbine Plant' BHRA 3rd ISWES, 1980.
7   Whittle G 'E; Effects of Wind Power and Pumped Storage in an Electricity Generating System' BWEA 3, Cranfield, 1981.
8   Bossanyi E A et al; 'Wind Turbine Response and System Integration' IEE Future Energy Concepts Conference, London 1981.
9   See Ref 5; Operating reserve costs from Table 4 p144, LF from Figure 8.
10 Van der Hoven I; 'Power Spectrum of Horizontal Wind Speed' Journal of Meteorology, 14, 2, April 1957.
11 Oort A H and Taylor A; 'On the Kinetic Energy Spectrum Near the Ground' Monthly Weather Review, 97, 9, Sept 1969.
12 Petersen E L; 'On the kinetic energy spectrum of atmospheric motion in the planetary boundary layer' Riso report No 285, DAEC Jan 1975.
13 Lipman N H et al; 'Fluctuations in output from wind turbine clusters' Wind Eng 4, 1, Jan 1980.
14 Busch N E and Panofsky H A; 'Recent turbulence spectra'. Quart J Royal Met Soc. 94 pp132-148 1968.
15 Pasquill F; 'Atmospheric diffusion'. Ellis Horwood, Chichester, UK, 1974.
16 Bossanyi E A; 'The frequency of wind turbine shut downs' BWEA 3, Cranfield, UK, 1981.
17 Justus C G and Mikhail A; 'Energy statistics for large wind turbine arrays'. Wind Eng 2, 4, 1978.

Nomenclature

Functional dependence on t is sometimes left implicit, ie a for a(t).

$a(t)$	-	σ_L/L	specific deviation of load power changes over time t.
$b(t)$	-	σ_W/P	specific deviation of wind power changes over time t.
b_0	-	-	simple specific deviation of wind power.
c	£/y.kW	-	operating reserve cost (kW of deviation)
C	p/hr	-	operating reserve cost
d	$MW^{0.5}$	$\sigma_L/L^{0.5}$	utility load deviation constant
D	-	-	diversity factor (deviation ratio)
e	-	-	a constant
F	-	σ_A/σ_W	deviation factor
I	-	P_{mean}/L_{mean}	wind power penetration
k	-	b/a	relative specific deviation
L	MW	-	utility mean load
LF	-	-	wind system load factor
n	s^{-1}, h^{-1}		frequency
N	-	-	number of arrays
p	-	$P_{rat}/0.5\rho V_{mn}^3 S$	specific power
P	MW	-	mean wind power
r	-	-	correlation coefficient
$\bar{r}$	-	-	cross correlation coefficient averaged over all site pairs
R	p/kWh	-	reduction in value of energy
t	h	-	lead time
T	-	-	total number of turbines
V	p/kWh	-	value of energy (fuel saving)
x	km	-	distance (separation)
$\bar{x}$	km	-	mean separation over all site pairs
σ_A	MW	$\sigma_T - \sigma_L$	added deviation (of changes over lead time t)
σ_L	MW	-	utility load power deviation without wind power (")
σ_T	MW	$(\sigma_L^2 + \sigma_W^2)^{0.5}$	total deviation (")
σ_W	MW	-	wind power deviation (")

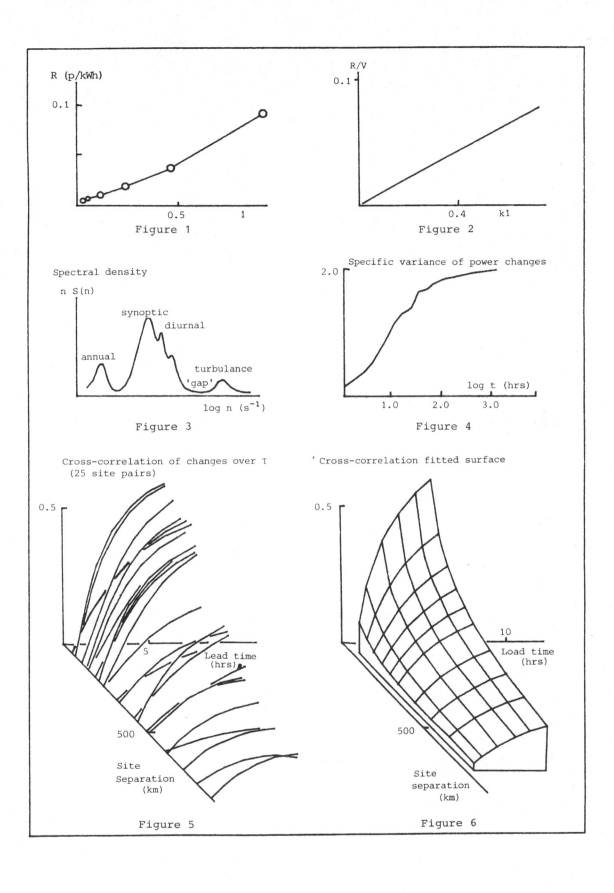

R (p/kWh)

0.1

0.5   1

Figure 1

R/V

0.1

0.4   k1

Figure 2

Spectral density

n S(n)

synoptic

diurnal

annual

turbulance
'gap'

log n (s⁻¹)

Figure 3

Specific variance of power changes

2.0

log t (hrs)

1.0   2.0   3.0

Figure 4

Cross-correlation of changes over τ
(25 site pairs)

0.5

5

Lead time
(hrs)

500

Site
Separation
(km)

Figure 5

'Cross-correlation fitted surface

0.5

10

Load time
(hrs)

500

Site
separation
(km)

Figure 6

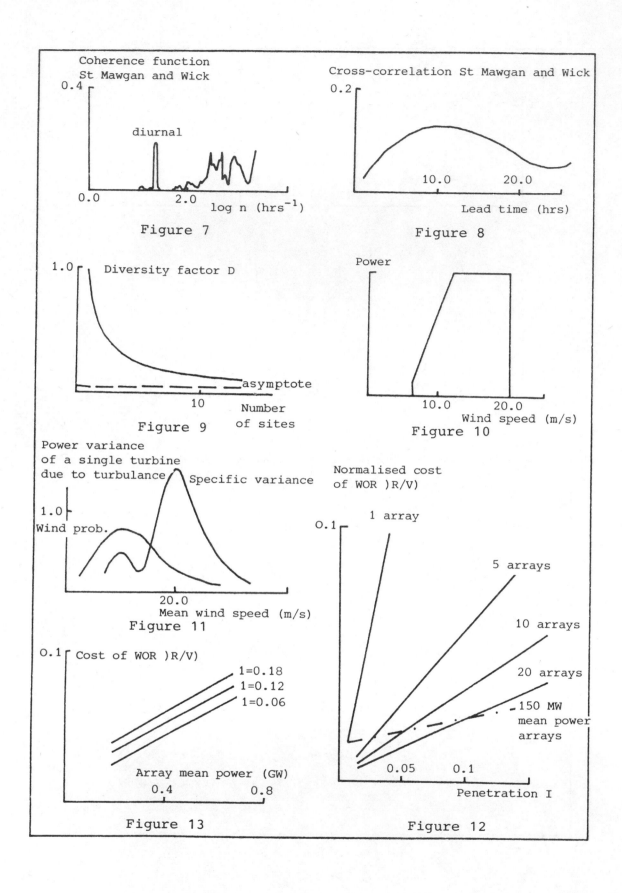

Coherence function
St Mawgan and Wick

0.4

diurnal

0.0   2.0

log n (hrs^{-1})

Figure 7

Cross-correlation St Mawgan and Wick

0.2

10.0   20.0

Lead time (hrs)

Figure 8

1.0   Diversity factor D

asymptote

10   Number
of sites

Figure 9

Power

10.0   20.0

Wind speed (m/s)

Figure 10

Power variance
of a single turbine
due to turbulance   Specific variance

1.0
Wind prob.

20.0
Mean wind speed (m/s)

Figure 11

Normalised cost
of WOR)R/V)

0.1   1 array

5 arrays

10 arrays

20 arrays

150 MW
mean power
arrays

0.05   0.1

Penetration I

Figure 12

0.1   Cost of WOR)R/V)

1=0.18
1=0.12
1=0.06

Array mean power (GW)

0.4   0.8

Figure 13

SIMPLIFIED DYNAMIC BEHAVIOUR OF A
STRAIGHT BLADED VERTICAL AXIS WIND TURBINE

M.S. Courtney
Department of Mechanical Engineering
The City University, London

Abstract

This paper presents a preliminary dynamic analysis of a straight bladed, fixed geometry VAWT. A two-degree-of-freedom model involving bending and twisting of the cross-arm is developed. The equations of motion are presented in a general form. The dynamic behaviour of the model is presented firstly by examining the uncoupled equations of motion. Coupling terms are then introduced and are seen to cause dynamic instability under certain conditions. Mitigating factors including twist damping and blade offset are discussed. Finally preliminary results from a Floquet stability analysis, involving all periodic and non-periodic terms, are presented.

Introduction

The emergence of the VAWT has given rise to an urgent need for a thorough dynamic analysis of this type. Existing work, e.g. Ref. (1), has concentrated on the natural frequencies of the stationary turbine with particular regard to the resonance problem. Rotational and aerodynamic phenomena alter certain of the natural frequencies and introduce the possibility of static and dynamic instability. It is therefore essential at some stage to include both these phenomena in a dynamic model.

The complexity of design and the unsteady nature of the aerodynamics will require a sophisticated computer model involving many degrees of freedom to provide useful results. However such a model provides little or no insight to the fundamental dynamic behaviour of the turbine. This can best be gained by studying simplified models. Algebraic constraints limit this to models of one or two degrees of freedom, to which analytical solutions may be obtained for the dynamic characteristics. This paper describes such a model. Previous single-degree-of-freedom models had identified extremely low levels of cross-arm twist damping. An instability observed on the Rutherford Laboratory VAWT appeared to involve a coupling of cross-arm twisting and bending. These modes are therefore selected for this analysis.

In all respects the analysis is kept as simple as possible. The principal implications of this philosophy in relation to the structural and aerodynamic models are as follows:-

(a)   Structural Model

(1)   Fixed geometry, symmetrical H configuration (see Fig. 1).
(2)   Two degrees of cross-arm freedom: bending (in the vertical plane) and twisting. The rotational axis of the turbine is fixed and the blades regarded as rigid bodies with distributed mass.
(3)   Assumed modes approach so that the oscillatory mode shapes may be described by simple polynomial or trigonometric functions.
(4)   Constant rotational speed Ω.

(b) Aerodynamic Model

(1) Single streamtube aerodynamic model with no wind shear. All points on
the blade at all azimuth angles experience the same disc wind speed v.
(2) Simple aerodynamic strip theory is employed, assuming a linear lift
curve of gradient a_1.
(3) No aerodynamic stalling is included.
(4) Blade aerodynamics only are considered. The aerodynamic effect of the
cross-arm and tower is neglected.

Many of these assumptions, particularly with regard to the aerodynamics, are
somewhat unrealistic or limiting. Even so, the value of the analysis lies
in identifying those parameters which have most effect on the dynamic
characteristics. Later studies with more sophisticated models may confirm
the results obtained or may identify limitations in the assumptions made.

Equations of Motion

These are obtained from the Lagrange equations which requires us to form
the kinetic and potential energies, T and V respectively, and the
generalised forces Q_1 and Q_2.   (Subscripts 1 and 2 refer throughout to the
bending and twisting modes respectively.)

The generalised co-ordinates q_1 and q_2 represent translational and angular
displacements at the blade end of the cross-arm. Each mode has a
corresponding function defining the mode shape. The displacement of any
point on the cross-arm may be expressed as

$$\overline{z} = f_1(\overline{x})q_1(t) \qquad\qquad 0 < \overline{x} \leqslant 1$$

$$\gamma = f_2(\overline{x})q_2(t) \qquad\qquad 0 < \overline{x} \leqslant 1$$

where γ is an angle of twist. The system $\overline{x}, \overline{y}, \overline{z}$ is a set of dimensionless
cross-arm co-ordinates normalised on the turbine radius R (see Fig. 1).

(a) Kinetic and Potential Energies

The basis of the analysis lies in forming an expression for the absolute
position vector $\underline{P}$ of a point $(\overline{x}, \overline{y}, \overline{z})$ on the blade or cross-arm. This is
best accomplished by describing the twisting and bending displacements as a
series of matrix co-ordinate transformations, giving $\underline{P}$ as

$$\underline{P} = \underline{T}_\Omega\left[\underline{A} + \underline{B}\,\underline{x}\right]$$

$\underline{T}_\Omega$ is the transformation matrix associated with the turbine rotation. $\underline{A}$ is
the translation vector containing terms associated with the position of the
co-ordinate system $\underline{x}$. These include terms relating to the horizontal and
vertical displacements due to the bending. $\underline{B}$ describes the angular
displacement of the co-ordinate system $\underline{x}$ due to both bending and twisting.

The kinetic energy is given by

$$T = \tfrac{1}{2}\int \underline{\dot{P}}\cdot\underline{\dot{P}}\ dm$$

where the dot indicates differentiation with respect to time. This
integration is carried out for the blade $(\overline{x} = 1)$ and for the cross-arm
$(0 < \overline{x} \leqslant 1)$ separately. First moment and product of inertia terms obtained

are disregarded on the basis of symmetry. Potential energy formulation presents no difficulties since the only contributions are cross-arm strain energies.

(b) Generalised Forces

These arise from the aerodynamic forces on the turbine blades, cross-arm and auxiliaries (struts, masts etc.). For present purposes blade forces only will be considered. Further assumptions are as given in the Introduction.

The generalised forces required for the Lagrange equation are defined by

$$Q_1 = \frac{\delta W}{\delta q_1} \qquad , \qquad Q_2 = \frac{\delta W}{\delta q_2}$$

where δW is the virtual work done during incremental displacements δq_1 and δq_2.

Elemental blade forces are resolved into normal and tangential components δF_n and δF_t. The generalised forces are then

$$Q_1 = C_\beta \int \delta F_n z - 2C_{S_B} q_1 R \int \delta F_n$$

$$Q_2 = - \int \delta F_t z$$

where C_β and C_{S_B} are functions of the bending mode shape $f_1(\overline{x})$ relating to the flapping angle of the blade and to the shortening of the radius of rotation respectively.

(c) Assembled Equations of Motion

After performing the required differentiations for the Lagrange formulation the equation of motion may be assembled. These are of the general form:

$$\begin{bmatrix} M_1 & 0 \\ 0 & M_2 \end{bmatrix} \ddot{\underline{q}} + \begin{bmatrix} D_{11}\,\Omega & D_{12}\,\Omega f(t) + G\overline{\Omega} \\ D_{21}\,\Omega f(t) - G\Omega & 0 \end{bmatrix} \dot{\underline{q}} +$$

$$+ \begin{bmatrix} k_1 + cf_1\Omega^2 + A_{11}\,\Omega^2 f(t) & A_{12}\,\Omega^2 \\ 0 & k_2 - cf_2\Omega^2 + A_{22}\,\Omega^2 f(t) \end{bmatrix} \underline{q} = 0$$

Terms D_{ij} and A_{ij} are associated with the aerodynamics and may be periodic functions of time $\left[f(t) = f\left(t + \frac{2\pi}{\Omega}\right) \right]$. Rotational effects introduce the gyro damping coupling term G and the centrifugal stiffness terms cf_1 and cf_2.

Analysis of Equations of Motion

This is a system of considerable complexity and there is some value in categorising the terms. The most obvious distinction is between terms which are periodic (functions of the cross-arm azimuth angle Ωt) and terms which are constant.

An important distinction can also be made between terms appearing on and off the leading diagonal. The former indicate parameters of the uncoupled oscillatory modes whilst the off diagonal terms are parameters coupling the two modes.

Whilst the overall behaviour can only be determined by analysing the complete equations, some insight can be gained by studying sub-systems of the equations.

The analysis proceeds in the following manner. Firstly the uncoupled equations are examined. Non-periodic coupling terms are then included. Finally the periodic terms are also included.

(a) Uncoupled Equations of Motion

Neglecting the aerodynamic damping term the natural frequencies of the uncoupled equations, ω_{1n} and ω_{2n}, are given by

$$\omega_{1n}^2 = \omega_{10}^2 + C_S\Omega^2$$

$$\omega_{2n}^2 = \omega_{20}^2 - \Omega^2$$

where ω_{10} and ω_{20} are the bending and twisting fundamental natural frequencies of the stationary turbine. Unless the **blade height is very** much greater than the turbine radius, the stiffening coefficient C_S is positive and the bending natural frequency will increase with rotational speed. This is because the projected length of the cross-arm shortens slightly as it bends. Since this tends to reduce the kinetic energy, centrifugal forces act to restore equilibrium. These **forces** supplement the structural stiffness resulting in a **higher** natural frequency. With the twisting mode the kinetic energy increases as the blade **is** displaced from the vertical since the effective radius of all points away from the centre of twist increases. The centrifugal forces therefore act against the structural stiffness, reducing the natural frequency.

When the rotational speed exceeds the stationary twisting natural frequency ω_{20}, the twisting mode becomes statically unstable. This is known as a "divergent" speed since the solution for the equation of motion includes an exponentially growing displacement. The turbine obviously cannot operate under this regime and so the natural frequency ω_{20} is the theoretical maximum operating speed.

Another important feature of the uncoupled equations of motion concerns the periodic twisting stiffness term. This introduces the possibility of "parametric" instability, a phenomenon associated with periodic coefficients in the equations of motion. The classical example of this is a strut with a sinusoidal end loading (**Ref. 2**) for which the equation of motion is

$$\ddot{q} + \omega_n^2(1 - \beta\cos\Omega t)q = 0$$

Although no complete solution is available the stability characteristics are well established both theoretically and experimentally. These are functions of the parameters β and Ω/ω_n. Re-casting the twisting equation of motion into this standard form, the controlling parameters are found to be

$$\beta = \frac{2C_a}{\eta} \frac{\Omega^2}{\omega_{20}^2 - \Omega^2} = \frac{2C_a}{\eta} \frac{x}{1 - x}$$

$$\frac{\Omega}{\omega_n} = \sqrt{\frac{\Omega^2}{\omega_{20}^2 - \Omega^2}} = \sqrt{\frac{x}{1 - x}}$$

where $x = \left(\frac{\Omega}{\omega_{20}}\right)^2$, $C_a = \frac{\rho a_1 CHR}{M_B}$ and $\eta = \frac{\Omega R}{V}$. Selecting a value for the constant C_a/η we may plot these parameters in the range $0 < \bar{x} < 1$ (Fig. 2). Inserted on the same axes is the boundary for the principal region of instability. (Other zones exist but are quelled by small amounts of damping.) As the rotation speed increases the plotted parameters pass into the unstable region. The value of the critical speed does not vary greatly with the parameter C_a/η, lying in the range 0.75-0.85 of the natural frequency ω_{20}. Decreasing the value of the constant C_a/η will result in a slightly higher critical speed.

A similar periodic term appears in the bending equation of motion. However, the values of the parameters β and Ω/ω_n in the range $0 < \bar{x} < 1$ are too low to come near the principal stability boundary. As far as the uncoupled equations are concerned this periodic term will not cause instability.

(b) Coupled Equations - Non-periodic Terms Only

By suppressing periodic terms, simple analytical methods may be used to examine the natural frequencies and the stability characteristics. The coupled natural frequencies may be reasonably accurately calculated by considering only the mechanical gyro coupling terms appearing in the damping matrix. These frequencies are shown in Fig. 3 for a typical case of $\omega_{10} < \omega_{20}$. The uncoupled frequencies are also inserted to illustrate the influence of the coupling. This is essentially to separate the two frequencies so that they no longer cross. However, the divergence at $x = 1$ ($\Omega = \omega_{20}$) remains and since there is no discontinuity the character of the lower natural frequency must change from purely bending at $\Omega = 0$ to purely twisting at $\Omega = \omega_{20}$.

Stability analysis requires us to include all the non-periodic terms. Substituting a solution $q = q_0 e^{\lambda t}$ yields the characteristic equation which is a quartic in λ, with complex roots of the form:

$$\lambda = \mu \pm i\omega$$

An unstable solution is indicated by a positive value of the real component μ. Assuming that such an unstable solution exists, manipulation of the coefficients of the characteristic equation provides a criterion:

$$\omega_*^2 > \Omega^2 + \omega_{20}^2$$

which must be satisfied for oscillatory stability to exist. (ω_* is the higher of the two natural frequencies applying at the rotational speed Ω.) By assuming that ω_* is changed only slightly when the aerodynamic terms are absent we can obtain an expression for the critical speed Ω_{crit}.

$$\frac{\Omega_{crit}^2}{\omega_{20}^2} = \frac{2(1 - C_\omega) - C_G}{C_G - 2(1 - C_s)}$$

where C_ω = standstill frequency separation parameter = $\left(\frac{\omega_{10}}{\omega_{20}}\right)^2$

C_G = gyro coupling coefficient = $\frac{G^2}{M_1 M_2}$

$$C_S = \text{bending stiffening coefficient} = \frac{cf_1}{M_1}$$

Fig. 4 shows a plot of $\frac{\Omega^2_{crit}}{\omega^2_{20}}$ vs. C_ω for typical values of the coefficients C_G and C_S. Numerical results show the instability to be rather weak. A small amount of twist damping is therefore sufficient to significantly move the stability boundary. Structural damping (1 or 2% critical) may well be sufficient to quell any instability at low values of x. Fig. 4 shows how the unstable zone contracts as twist damping increases.

Alternatively the critical speed may be raised from zero by offsetting the blade mass axis relative to the cross-arm flexural axis. This introduces additional coupling so that some of the strong aerodynamic bending damping is made available to the unstable twisting mode. The critical speed increases as the offset increases, the improvement being greater at higher values of the frequency separation parameter C_ω.

(c) Periodic Terms Included - Parametric Stability

Parametric stability has already been introduced in the analysis of the uncoupled equations of motion. These were cases which could be cast into a standard form so that published data on stability boundaries could be untilised.

Generally the stability characteristics can only be determined numerically using a digital computer. Floquet analysis is the most widely used method.

Preliminary results using Floquet analysis confirm the existence of the uncoupled parametric instability in the expected speed range. Beneath this critical speed the periodic terms appear to introduce an additional region of stability as shown in Fig. 5. However, the boundary in this region is very sensitive to slight changes in the dynamic model. Considering the crudity of the aerodynamic model it is not wise to place much faith in this stable region.

Concluding Remarks

The limitations of the model are fairly self-evident and lie mainly with the aerodynamics. A more sophisticated analysis should include aerodynamic stalling and an improved aerodynamic model. Such models are likely to be inaccessible to analysis, except by computer.

Thereafter improvements could be made to the mechanical aspects of the model. Variable geometry, if applicable, should be included at some stage. It may also be necessary to consider unsymmetrical configurations. However, the present analysis was never intended to be truly rigorous, indeed the intention has been primarily to provide insight. It is hoped however that the results presented in this paper will provide a basis from which to construct a stability investigation on more realistic and complex models of vertical axis wind turbines.

Acknowledgement

The author wishes to thank Prof. G.T.S. Done for his contribution to this work and for his generous advice in preparing this paper.

References

1. Ficenec I. etc., "Dynamic analysis of a variable geometry vertical axis wind turbine", Proc. of the 4th BWEA Wind Energy Conference, Cranfield, 1982.

2. Bolotin V.V., "The dynamic stability of elastic systems", Holden Day Inc., 1964.

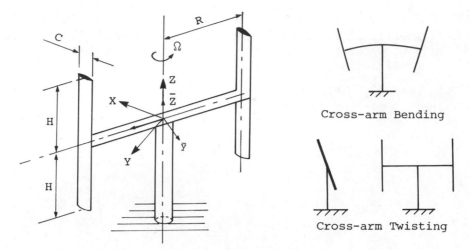

Fig. 1. Turbine Configuration, Co-ordinate Systems & Oscillatory Modes.

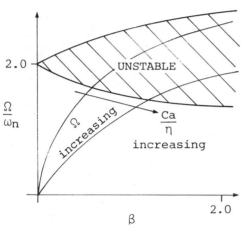

Fig. 2. Uncoupled Twisting
Parametric Stability Diagram

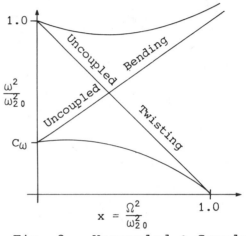

Fig. 3. Uncoupled & Coupled
Natural Frequencies.

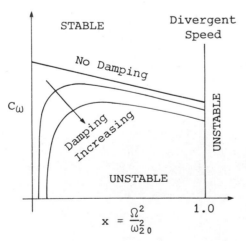

Fig. 4. Stability Diagram
Non-periodic Terms Only.

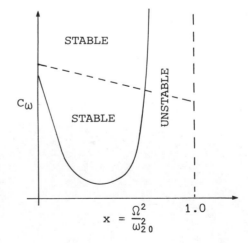

Fig. 5. Stability Diagram
Periodic Terms Included.

SOME EFFECTS OF SPEED ON THE ROTOR DYNAMICS OF THE RUTHERFORD 6 METRE VGVAWT

J H Webb, College of Aeronautics, Cranfield, A.J.Pretlove and L.W.Lack, Department of Engineering, University of Reading

SUMMARY

Both observations and experimental evidence indicate that the Rutherford Laboratory VGVAWT has severe vibration problems at certain speeds of rotation, producing substantial torsion and bending amplitudes of the crossarm/blade system. A simplified theoretical analysis of blade and crossarm dynamics, assuming a two degree of freedom representation, indicates that the effects of rotation on the crossarm bending and torsional frequencies is considerable on this particular machine, increasing the former and reducing the latter with increase in speed. The coupling between these motions, in conjunction with the cyclic forcing from the blade aerodynamic forces, is shown to produce crossarm dynamic bending strains which are generally in good agreement with those measured experimentally. Torsional strains in the crossarm are also in substantial agreement. It is concluded that the observed vibrations appear to be due primarily to coupled torsion bending modes of the crossarm, the frequencies of these being markedly speed dependent.

1. Experimental Results

Extensive experiments have been conducted on the Rutherford Laboratory VGVAWT to measure strains, deflections and accelerations in the rotor structure comprising blades and crossarms. The aim of the work in general has been (i) to understand the dynamics of the system, and (ii) to collect static and dynamic strain data which is intended to be useful generically for both the quasi-static and fatigue aspects of structural design. The results given here have been obtained when the machine was fitted with two fibreglass blades. Later experiments have been conducted on the machine converted to three fibreglass blades and the current configuration is with two metal blades. A theoretical analysis of the quasi-static loading together with corroborative experimental results has been given (1) together with some experimental measurements of dynamic strains. One aspect of the measured dynamics of the system has been a strong oscillation involving both bending and torsion of the crossarm at about 55 r.p.m. This has been the subject of analysis using a cine camera mounted on the central part of the rotor and rotating with it (2). The camera was pointed at the blade/crossarm joint and the resulting film analysed frame by frame. The purpose of this work was to determine whether the oscillation was a resonance or an instability. Unfortunately, the analysis did not positively identify one or the other. This may be because the analysis was based on an assumption of linear coupled motion whereas it is now known that the coupled equations are non-linear; perhaps this should have been expected because measured crossarm torsion is predominantly at the rotation rate (1p) whereas crossarm bending is predominantly at 2p, see Fig.1 for an example. The film has also shown that the approximate amplitudes of motion at the critical speed (55 r.p.m.) are (i) $\pm 5°$ torsion and (ii) ± 50 mm bending deflection at the crossarm tip.

In order to elucidate the nature of the motion the theoretical analysis in the following sections of this paper has been made. To validate the theory certain basic data for the Rutherford Laboratory machine is required and some of it is listed in ref.3. The salient data for the purpose of this paper is as follows:-

(i) Geometrical data (see also fig.2)

Crossarm semi-span s = 3.0m
Tie wire height h = 1.5m
Blade length = 4m
Blade length above the crossarm = 2.4m
Blade c.g. position above the crossarm = $\overline{R}$ = 0.43m
Blade and crossarm chord = 300mm
Shear centre position from leading edge = 119mm
Mass centre position from leading edge = 133mm

Blade and crossarm section   NACA 0015
(ii) Dynamic Data
Blade mass = 17.72 kg
Blade moment of inertia about crossarm axis = 24.1 kgm^2.
Crossarm static torsional stiffness = 1634 Nm/rad.
Cross arm bending stiffness (tip force/tip deflection) = 1.73 N/mm
Cross arm torsional natural frequency (machine stationary) = 1.21 Hz
Cross arm bending natural frequency (machine stationary) = 1.39 Hz

2.   Theoretical Investigation

The theoretical analyses of rotor dynamic behaviour given here are based upon models which are as simple as possible but which are capable of detailed correlation with experimental results. Use has been made of Lagrange's equations to derive dynamical equations from these models, and expressions for aerodynamic forces have been deduced from simple, linear blade element theory. Examination of the experimental results obtained from the Rutherford machine (1), suggested that over the speed range of interest, ie. 30-65 r.p.m., the blade/crossarm tip motion involved primarily just two degrees of freedom, vertical bending of the crossarm in its fundamental mode, and a torsion mode of the blade/crossarm. Whilst this indicated that an idealisation of the rotor embodying just two coupled freedoms might provide an adequate representation, when associated with aerodynamic forces, it was not clear how the torsional frequency might change with rotor speed, if at all. Previous work (4) had established the increase in bending frequency with rotor speed due to centrifugal stiffening of the crossarm, and the results from this suggested that a single degree of freedom idealisation involving torsional motion of the blade/crossarm might be of interest, before investigation of the two degree of freedom model. These two idealisations of the rotor are discussed below.

2.1 Uncoupled Torsion Motion

Details of the blade/crossarm configuration of the Rutherford machine are shown in Fig.2 and have been discussed above. At speeds less than about 70 r.p.m the blades are nominally vertical, furling occurring at higher speeds. The tie wire is held taut during rotation of the rotor by centrifugal forces, the centre of gravity of the blade being above the hinge connecting the blade to the crossarm. It may be observed that if elastic tip twisting of the crossarm is permitted, with no other elastic deflections, then the blade rotates about the axis AB. This axis is skewed with respect to the rotor rotation axis, and it may thus be expected that the torsional motion will be influenced by rotor rotation.

To determine the motion of this system, Lagrange's equation is used, i.e.

$$\frac{d}{dt}\left(\frac{\partial T}{\partial \dot{q}}\right) - \frac{\partial T}{\partial q} + \frac{\partial U}{\partial q} = Q$$

with

T = kinetic energy U = potential energy
q = generalised coordinate Q = generalised force

Angle ϕ is the rotation of the blade about the AB axis and is used as the generalised coordinate. The absolute velocity of a point on the blade at distance r above the hinge is given in terms of axes located in the mast, rotating with the rotor as (see fig.2)

$$\bar{V} = -r\left[\sin\phi \sin\theta\left(\dot{\phi}\cos\theta + \Omega\right)\right]\underline{i} + r\left[\cos\phi \sin\theta\left(\dot{\phi} + \Omega\cos\theta\right) + \Omega\left(\frac{s}{r} - \sin\theta\cos\theta\right)\right]\underline{j}$$
$$- r\left[\dot{\phi}\sin\phi \sin^2\theta\right]\underline{k}$$

Now the kinetic energy is given, for a narrow blade, by $T = \frac{1}{2}\int m\,\bar{V}\cdot\bar{V}\,dr$

and the potential energy, due to torsion of the crossarm, is given by $U = \frac{1}{2}k_\beta^2\phi^2\sin^2\theta$

where k_β is the static tip torsional stiffness of the crossarm. After some algebraic manipulation the frequency of free vibration in torsion is found as

$$\omega_\phi = \left[\omega_{\phi NR}^2 + {}^{93}\Omega^2 \left(\frac{\bar{R}h}{k^2} - 1 \right) \right]^{1/2}$$

where $\omega_{\phi NR}$ is the non rotating frequency, measured at 1.21 Hz, and k is the radius of gyration of the blade about the hinge. This result suggests that the torsional frequency may be markedly affected by rotor rotation. Rather similar effects to this are observed in propellers where blade twisting moments (so called "propeller" moments) are produced by rotation, and these are sometimes exploited to provide blade pitch control. In the present case, the behaviour is modified by the tie wire, offset c.g. location, and skewed torsion axis.

To test the validity of this result a simple model of the system was constructed using a rod of steel, 10mm diameter, 220mm long hinged at its upper end by a balljoint to an arm projecting from a vertical shaft (fig.3a, 3b). The steel bar, forming a compound pendulum, was restrained to the shaft by a string, providing a kinematic model of the rotor system. When the shaft was rotated there was satisfactory agreement between the calculated and observed rotational speeds at which divergence of the pendulum occurred (see fig.3b).

Application of the theory to the Rutherford machine is shown in Fig.4, which indicates a decrease in torsional frequency with increase in rotor speed. A near coincidence of torsion frequency with rotor 1p frequency at about 58 r.p.m. is noteworthy. For comparison the vertical bending frequency measured on the machine (2) is also shown, where again the near coincidence of the rotor 2p frequency with this at 55 r.p.m. is significant.

At about 70 r.p.m. the blade starts to furl, and the behaviour will then be substantially modified. These changes have been calculated using an extension of the present theory, and are shown in Fig.4. It is seen that there is a rise in torsion frequency when furling occurs and this is largely a result of the decrease in the effective moment of inertia of the blade.

2.2. Coupled Bending Torsion Motion

The inclusion of crossarm vertical plane bending freedom considerably extends the analysis, but provides a model which permits estimation of crossarm tip amplitudes for comparison with experimental results. The basis of the model used is shown in Fig.5 where the two degrees of freedom are w, the crossarm tip vertical displacement associated with the fundamental bending mode of the crossarm, and ϕ, the rotation of the blade about axis AB, as previously. The blade is assumed narrow, rigid, and initially vertical, with no furling action. Crossarm and tie wire extensions due to direct stresses are neglected, as are crossarm deflections in the plane of rotation. A quasi-static deflection of the crossarm tip, δ_0, due to gravity and modified by centrifugal stiffening of the crossarm, must be accounted for as it affects the initial geometry.

Application of Lagrange's equation to this model gives dimensional equations of motion (in SI units) for the Rutherford rotor geometry as:-

$$\frac{Q_\phi}{I} = 0.8\left(\ddot{\phi} + \omega_P^2 \phi\right) + 0.213\, w\ddot{\phi} + \dot{w}\dot{\phi}\left(0.213 - 0.064w\right) - 0.146\,\ddot{w}\phi - 0.715\,\Omega\dot{\phi}\phi$$
$$- 0.188\, w\dot{w} + \delta_0 \text{ terms}$$

$$\frac{F_w}{M} = \ddot{w} + \omega_w^2 w - \dot{\phi}^2\left(0.344 + 0.0345w\right) - 0.199\,\ddot{\phi}\phi + 0.111\,\dot{w}^2 w + 0.128\,\Omega\dot{\phi}w$$
$$+ \delta_0 \text{ terms}$$

These equations neglect squares and products of displacements. The symbols ω_ϕ, and ω_w represent the torsional and vertical natural frequencies shown in Fig.4. I is the moment of inertia of the blade about the hinge, and M is the blade mass. The remaining geometry is defined in Fig.2. The generalised moment Q_ϕ and force F_w arise from the aerodynamic forces acting on the rotor. They are variables dependent on wind speed and direction relative to the blade, the rotational speed of the rotor, the geometry of the rotor, and the displacements and velocities w, ϕ, $\dot{w}$ and $\dot{\phi}$. The moment Q_P is largely a result of aerodynamic force components parallel to the blade chord which produce a moment about the AB axis, although the tie wire drag force has an effect.

Neglecting any induced effects, the component velocities normal and parallel to the blade chord at a given blade azimuth location ψ are:-

$$U_p = V\sin\psi - r\sin\theta\cos\theta\;\dot\phi\phi - \left(\frac{w - \delta_0}{s}\right)\dot{w}$$

$$U_T = V\cos\psi + \Omega s + r\sin\theta\dot\phi$$

(Fig.6), where V is wind velocity, r blade station above the hinge, and θ the angle of the AB axis to the vertical. Assuming $U_p \ll U_T$, the moment of the local blade force chordwise about the AB axis is

$$M_T = \int \left\{ \tfrac{1}{2}\rho ac\, U_p^2 - \tfrac{1}{2}\rho c\delta\,(U_p^2 + U_T^2) \right\} r\sin\theta\, dr$$

the integration extending over the blade length. Here a is the lift curve slope (5.73/rad), c is the blade chord, ρ is air density, and δ is the blade profile drag coefficient (assumed to be 0.012 constant). This moment, together with the contribution from the tie wire, gives Q_ϕ.

The major contribution to F_w appears to arise from the aerodynamic forces acting on the crossarm. The torsion motion of the crossarm/blade produces twisting of the crossarm with consequent changes in the angle of attack. This results in another source of coupling between the torsion and bending degrees of freedom of the crossarm. The aerodynamic force produced by the crossarm depends upon its modal deflections and velocities in torsion and bending. These were taken to be a linear distribution of twist for torsion, and a bending mode deflection corresponding to a tip loaded cantilever for bending. Referring the aerodynamic forces on the crossarm to its tip gives the generalised force F_w.

3. Results from the Coupled Model

Investigation of the coupled model of the rotor has so far been confined to an analogue computer simulation. This is not entirely satisfactory since some of the smaller non-linear terms have been, of necessity, scantily included due to machine limitations. It is proposed at a later stage to replace this with a digital simulation. However, the results obtained so far compare very favourably with the experimental observations, both qualitatively and quantitatively.

Fig.7 shows the vertical response obtained from the simulation together with experimental points. The two peaks in the response correspond to (i) the 3p cross over with the bending frequency of the cross arm (Fig.4), and (ii) the 2p/1p bending and torsion cross over at approximately 57-58 r.p.m. At 30 r.p.m. the waveform obtained from the simulation for the vertical motion contains a large 3p component, whilst at 55 r.p.m. there is a strong 2p component. Both of these are in accordance with experimental observations. Fig.8 shows the simulated response at 55 r.p.m. The 1p torsional response which has also been found experimentally (see Fig.1) is evident. The wind speed here is taken as 4 m/s.

Comparison of experimental and simulated torsional response over the speed range 20-65 r.p.m. show a peak in the simulated response at about 35 r.p.m., whereas the experimental values appear to peak at about 50 r.p.m. This speed discrepancy is not totally resolved at present, but may be associated with the restricted theoretical model which is currently simulated.

The results obtained appear to confirm that an adequate model for investigating many dynamic phenomena on wind turbines of the Rutherford type may be relatively simple, containing quite limited degrees of freedom. The present work is still at an early stage, and it is hoped that a much more extensive investigation of the model will yield interesting and useful results.

4. References

1. A.J.Pretlove and D. Hess (1982). Strain Analysis of a Vertical Axis Wind Turbine. Proc. 4th BWEA Conference, Cranfield, March 1982.

2. A.J.Pretlove (1981). Crossarm Vibrations on the Rutherford Laboratory VGVAWT. University of Reading Energy Group Report 81/3.

3. D.Hess and M.D.Percival (1981). Further Structural Tests on the Rutherford Laboratory Aerogenerator. University of Reading Energy Group Report 81/4.

4. G.Stacey (1981). Excitation Frequencies on a Vertical Axis Wind Turbine. University of Reading Energy Group Report 81/5.

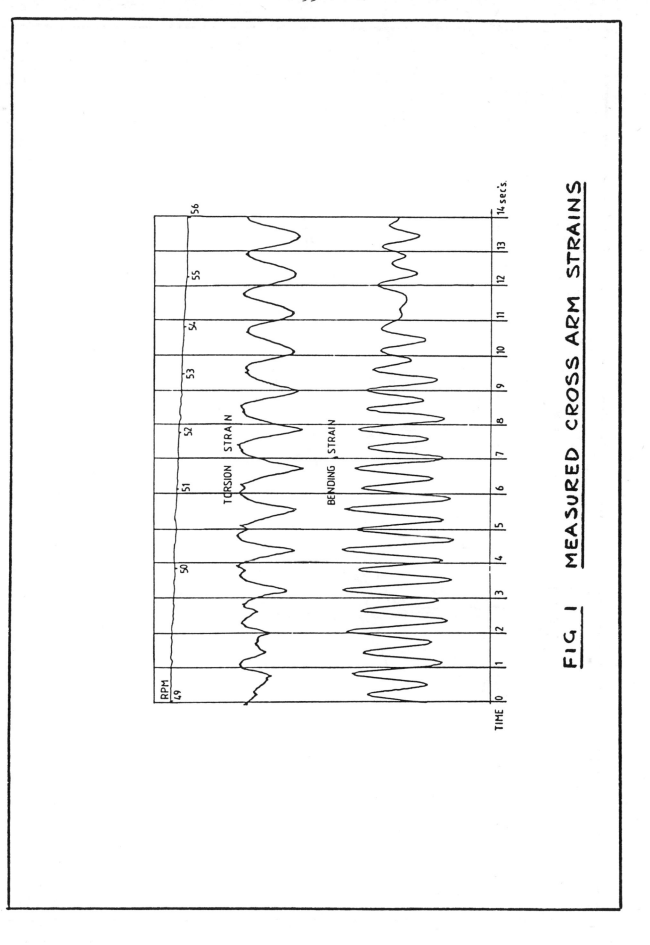

FIG 1   MEASURED CROSS ARM STRAINS

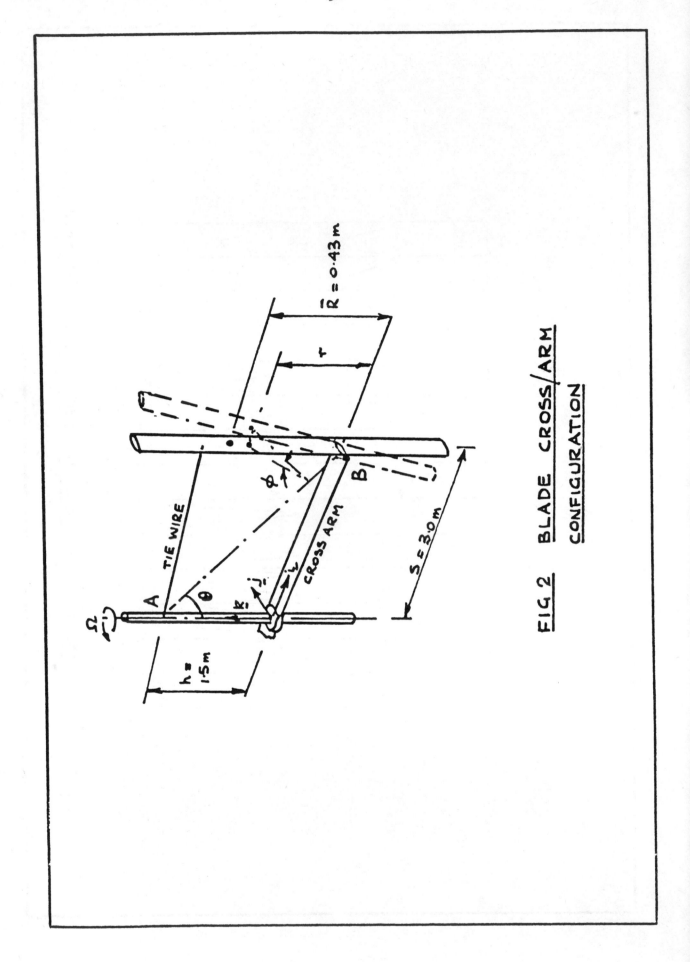

FIG 2 BLADE CROSS/ARM
CONFIGURATION

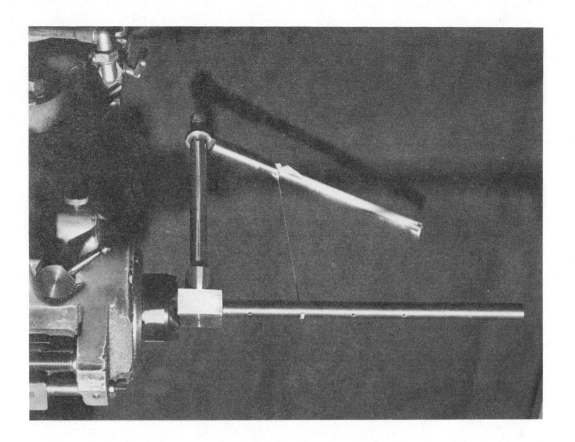

FIG. 3b : TEST MODEL AT 1.2 x
CRITICAL SPEED

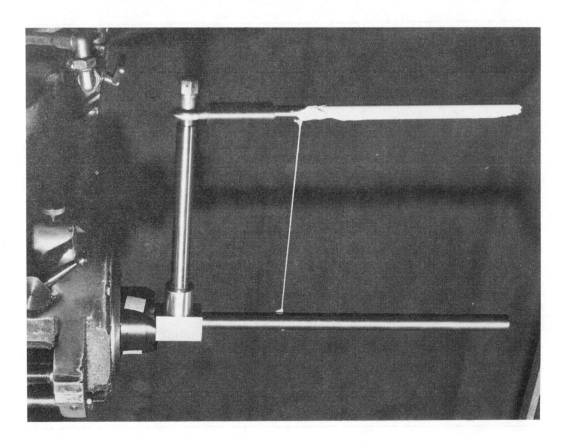

FIG. 3a: CRITICAL SPEED FOR
TEST MODEL

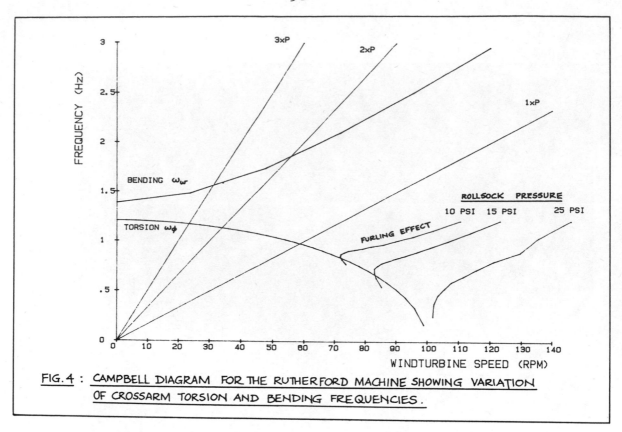

FIG.4 : CAMPBELL DIAGRAM FOR THE RUTHERFORD MACHINE SHOWING VARIATION OF CROSSARM TORSION AND BENDING FREQUENCIES.

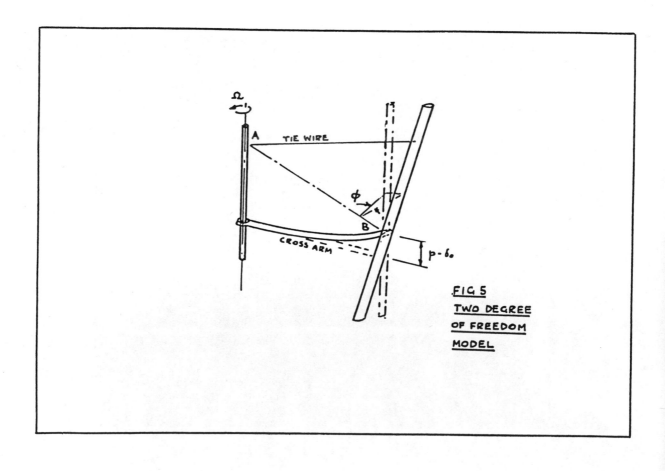

FIG 5
TWO DEGREE
OF FREEDOM
MODEL

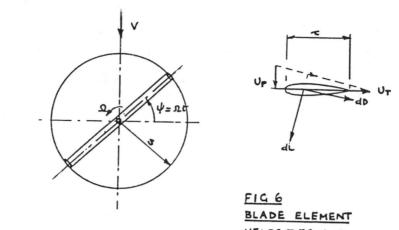

FIG 6
BLADE ELEMENT
VELOCITIES AND
FORCES

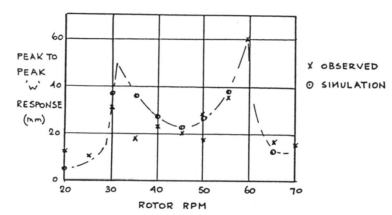

X OBSERVED
⊙ SIMULATION

FIG 7   COMPARISON OF SIMULATED
AND EXPERIMENTAL VERT. RESPONSE

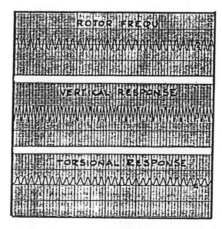

FIG 8   SIMULATED RESPONSE
AT 55 RPM

ASPECTS OF THE DYNAMICS OF A VERTICAL
AXIS WIND TURBINE

I.P.Ficenec and A.Saia
NEI Cranes Ltd.
Central Engineering Dept.,Leeds.

Abstract

The paper describes work carried out during an investigation into the dynamic behaviour of the 'H' type rotor of the vertical axis wind turbine. A two degree of freedom model representing the cross arm bending and torsional motions is developed and analysed with respect to the variation in system frequencies and stability with rotor speed. This model is extended by the introduction of blade aerodynamics giving an aeroelastic system whose behaviour is investigated.

Introduction

An examination of the salient features of this type of rotor (fig 1) suggested the possibility of interaction between the in-plane flap and out-of-plane propeller modes of the rotor blade motions. Although this intermodal coupling may be absent at zero rotor speed, coupling terms are introduced at non-zero rotor speeds either due to the rise of additional 'mechanical' terms or aerodynamic forces. The study was aimed at providing an understanding of this interaction and the background to a satisfactory practical design.

A model with two degrees of freedom was developed for the rotor system in vacuo, which represented the aforementioned types of motion and system equations were derived. These were then analysed with reference to the influence of rotational speed on system frequencies and stability. The characteristic equation was derived for the system matrices in general form and analysed in order to obtain an insight into the conditions required for different system stability conditions for the rotor. This led to the identification of the system parameters controlling the system behaviour.

Aerodynamic forces were introduced into the analysis and an aeroelastic model of the system developed. The effect of aerodynamic terms on the system behaviour is compared with the results derived from a purely mechanical model.

Model for rotor in vacuo

Initial considerations of the effects of rotation on simple dynamic systems led to the possible modifying influence of rotational inertia forces on the effective elastic parameters of the system. Through the analysis of various simple models it became clear that systems with displacements in a radial and tangential direction could lead to 'softening' and 'stiffening' phenomena whose magnitude were functions of rotational speed as well as the physical characteristics of the system. These effects would clearly render the system natural frequencies and hence possible forced response behaviour, to be functions of rotor speed for a given rotor design. In addition, certain softening terms could eventualy give rise to instabilities at certain critical rotational speeds or speed ranges, clearly a situation to be avoided by any practical design.

In order to study the basic dynamics of the rotor of the wind turbine a two degree of freedom model was developed to represent the rotor system. The two degrees of freedom chosen were those corresponding to the blade in-plane flap and out-of-plane propeller modes described by the cross arm tip vertical displacement or 'flap' deflection and the 'propeller' type rotation of the blade relative to the cross arm axis, see fig 2. As indicated earlier these two modes can be uncoupled at zero rotor speed. However, the rotation introduces additional terms which can give rise to coupling between these two modes. With the coupling present the blade movements are a combination of flap and propeller motions termed the 'plap' mode. The blades were assumed to be rigid with the cross arm assumed to provide the torsional and flexural stiffness.

The system equations of motion were derived using the Lagrange method following the approach outlined in Appendix I. The system stability and natural frequencies were established by evaluating the system eigensolutions from the matrix form of the equations for various sets of rotor parameters. For each set of rotor parameters the eigensolutions were plotted as functions of rotor speed, illustrating the variation of system natural frequencies, the presence of instabilities or instability regions and their severity at different rotational speeds. A typical example of such results is given in fig 3 and 4 showing the variation of the system natural frequencies (in the local rotating frame) and stability with rotational speed for a non-rotating flap mode frequency to propeller mode frequency ratio of 1.6. The dashed curves in fig 3 show the variation in uncoupled system frequencies. As the figures illustrate, the natural frequency plots consist of a rising curve and a lowering curve which eventualy falls to zero indicating the onset of instability with the level increasing with rotor speed. This instability appears as a divergence in the rotor frame of reference. Figures 5 and 6 show the variation of effective flap and propeller mode stiffnesses respectively with rotor speed. As fig.6 indicates the onset of instability coincides with the rotor speed for which a system stiffness becomes zero and subsequently goes negative which is also indicated by an analysis of the system equations. Clearly, any practical wind turbine design would ensure that the critical speed for the onset of instability is well above the operating speed range of the machine.

It became apparent that the system behaviour reflected in the form of the curves from these analyses could differ quite radicaly depending on the various relative magnitudes of the terms in the equations of motion and the assumptions on which the equations were derived. Figures 7 and 8 indicate the considerable change in the form of the results due to a simplification of one of the terms in the system stiffnesses.

The results presented in the figures discussed above were based on numerical solution techniques. In order to gain a better understanding of how the resulting dynamic behaviour was dependent on the various terms in the equations, an analytical investigation of the system characteristic equation was carried out. This took the form of decomposing the characteristic equation into groups of terms incorporated in a series of inequality expressions which govern the form of the roots of the characteristic equation and how they vary with rotor speed. For a given rotor design these groups are purely functions of rotational speed and by using these sets of algebraic inequalities one can investigate what form the eigensolution variation with rotational speed will take in particular speed ranges.

Figures 9 and 10 show typical examples of the variation with rotor speed of some of these groups which together with the inequality expressions provide a means to assess the types of roots of the characteristic equation which may occur in the various speed ranges together with the existence and type of instabilities. In addition this approach gives an indication of the rotor parameters which are required in order to avoid undesirable behaviour.

Model for rotor including aerodynamic forces

Using a simple single streamtube model, the previous system was extended to include aerodynamic forces generated by movements due to the two degrees of freedom in the model, as indicated in Appendix II. The additional aerodynamic terms which arise are functions of the rotational speed and consist of both constant and periodic coefficients. The previous methods of solution could still be employed for the extended system equations containing the constant coefficient aerodynamic terms only. Using the system parameters which produced figs. 3 and 4, the system stability with rotor speed is plotted in fig.11 for zero system damping, the variation in system natural frequencies remaining essentialy unchanged. The previous instability region remains basicaly unchanged but an additional low level instability range now exists due to the introduction of the constant coefficient aerodynamic terms. This region is relatively sensitive to the level of system damping and this is demonstrated in fig 12 showing an enlarged picture of the low speed end of these results indicating how the onset of the shallow instability region is pushed to higher rotational speeds by the introduction of even moderate amounts of system damping. In practice the damping, being of both a structural and aerodynamic nature, is likely to be significantly higher than the value used. Naturally, the aim of any practical design is to ensure that any instability is well above the operating range of the machine.

In order to solve the complete system equations including the periodic coefficient aerodynamic terms, the Floquet method was employed and the system stability characteristics (dashed lines) are shown in fig 13 together with those shown in fig 11. As this illustrates, the previous results remain essentialy unchanged.

Conclusion

A dynamic model for the rotor system has been developed coupling the two salient blade degrees of freedom and results produced indicate how the system frequencies and stability characteristics vary with rotor speed. Regions of instability have been identified and associated with the reduction of a system stiffness to zero. However, the study also indicated how to control the design parameters involved in order to ensure that the instabilities are maintained well above the operating speed range of a given machine.

The results have been obtained by numerical solution methods but in addition an analytical study of the system characteristic equation has been carried out leading to a set of inequality expressions which effectively characterise the possible types of eigensolutions for the system given a set of physical parameters and speeds.

This model has been developed further in order to introduce basic aerodynamic forces. Again solutions for the system frequencies and stability have been obtained indicating the introduction of a new region of instability in addition to those found previously, the system frequencies remaining essentialy the same. The onset of this shallow instability is shown to be sensitive to the amount of damping present and can be moved to higher rotational speeds by the introduction of even moderate amounts of system damping. The level of damping considered in this paper is believed to underestimate the damping actualy present for reasons mentioned earlier. Furthermore, additional measures such as the introduction of intermodal coupling can be utilised to control this instability.

In conclusion, the study gave an insight into some aspects of the dynamic behaviour of this type of rotor, provided associated theoretical models and indicated ways in which trouble free design can be achieved with confidence.

Acknowledgements

The authors would like to thank NEI Cranes Ltd. for permission to publish this paper and the Dept. of Energy for their support of the project associated with this work.

Symbols
----

z - cross arm tip deflection
β - blade propeller rotation
ω_F - non-rotating flap frequency
ω_P - non-rotating propeller frequency
ω - system natural frequency
α - real part of system eigenvalue $\lambda_s = \alpha + i\omega$
Ω - rotational speed
U_0 - wind speed
λ - tip speed ratio

Appendix I - Mechanical model

The system equations of motion were derived in terms of the generalised coordinates referenced to a local frame rotating with the rotor. Following the standard approach for such cases the absolute velocity of an elemental mass dm on the blade is,

$$\underline{v} = \frac{d\underline{r}}{dt} + \underline{\Omega} \times \underline{r}$$

where $\underline{r}$ is the position vector for the elemental mass referenced in the local rotating frame.

With the components of the absolute velocity $\underline{v}$ determined, the kinetic energy of the element can be calculated from,

$$dT = \frac{1}{2} dm\ \underline{v}^T . \underline{v}$$

The total kinetic energy for the blade can then be expressed as,

$$T = \int_{-l_0}^{l_0} dT(l)\ dl$$

where l_0 = half the blade length.

The potential energy U was derived entirely from the flexural and torsional stiffness of the cross-arm and can be written as,

$$U = \frac{1}{2} K_z z^2 + \frac{1}{2} K_\beta \beta^2$$

K_z = flap stiffness

K_β = propeller stiffness

Similar expressions can be written for the dissipation energy F and all three energies used in the Lagrange equation,

$$\frac{d}{dt}\left(\frac{\partial T}{\partial \dot{q}}\right) - \frac{\partial T}{\partial q} + \frac{\partial V}{\partial q} + \frac{\partial F}{\partial \dot{q}} = 0$$

q = generalised coordinate

from which the system equations may be derived.

A full derivation of these equations is rather lengthy to be included here. The resulting equations may be expressed in matrix form as,

$$\begin{bmatrix} M_{eff} & 0 \\ 0 & I_{eff} \end{bmatrix} \begin{Bmatrix} \ddot{z} \\ \ddot{\beta} \end{Bmatrix} + \begin{bmatrix} c_z & a\Omega \\ -a\Omega & c_\beta \end{bmatrix} \begin{Bmatrix} \dot{z} \\ \dot{\beta} \end{Bmatrix} +$$

$$\begin{bmatrix} K_z + b\Omega^2 & 0 \\ 0 & K_\beta + d\Omega^2 \end{bmatrix} \begin{Bmatrix} z \\ \beta \end{Bmatrix} = \begin{Bmatrix} 0 \\ 0 \end{Bmatrix}$$

where M_{eff} and I_{eff} are effective mass and mass moment of inertia of the blade, c_z and c_β are damping coefficients and a, b and d are system constants.

Appendix II - Aeroelastic model

The absolute velocity of an elemental strip of blade was expressed in the local rotating frame of reference and its components combined with the components of wind in order to yield the variation in elemental lift and drag. For the condition of tangential blade speed being greater than the wind speed, generalised forces Q_z and Q_β were derived and included in the previous 'in vacuo' system equations yielding the full aeroelastic equations. Again the derivation is too lengthy to include in full, the resulting equations in matrix form are,

$$\begin{bmatrix} M_{eff} & 0 \\ 0 & I_{eff} \end{bmatrix} \begin{Bmatrix} \ddot{z} \\ \ddot{\beta} \end{Bmatrix} +$$

$$\begin{bmatrix} c_z + 1/2\,\rho\,eU_0(\sin\Omega t + \lambda) & a\Omega + 1/2\rho fU_0\cos\Omega t \\ -a\Omega + 1/2\rho gU_0\cos\Omega t & c_\beta \end{bmatrix} \begin{Bmatrix} \dot{z} \\ \dot{\beta} \end{Bmatrix} +$$

$$\begin{bmatrix} K_z + b\Omega^2 + 1/2\rho\Omega eU_0\cos\Omega t & 1/2\rho f\Omega U_0(\sin\Omega t + \lambda) \\ 0 & K_\beta + d\Omega^2 + 1/2\rho h\Omega U_0\cos\Omega t \end{bmatrix} \begin{Bmatrix} z \\ \beta \end{Bmatrix} = \begin{Bmatrix} 0 \\ 0 \end{Bmatrix}$$

where e f g and h are system constants.

As can be seen the system equations are no longer linear but contain periodic coefficients. They can be solved either by numerical integration techniques for a given set of initial conditions or they can be analysed for system stability. The latter can be done for the system of equations containing all periodic terms using techniques such as Floquet or for a quasi-linearised system by using, for example, a freeze state approach.

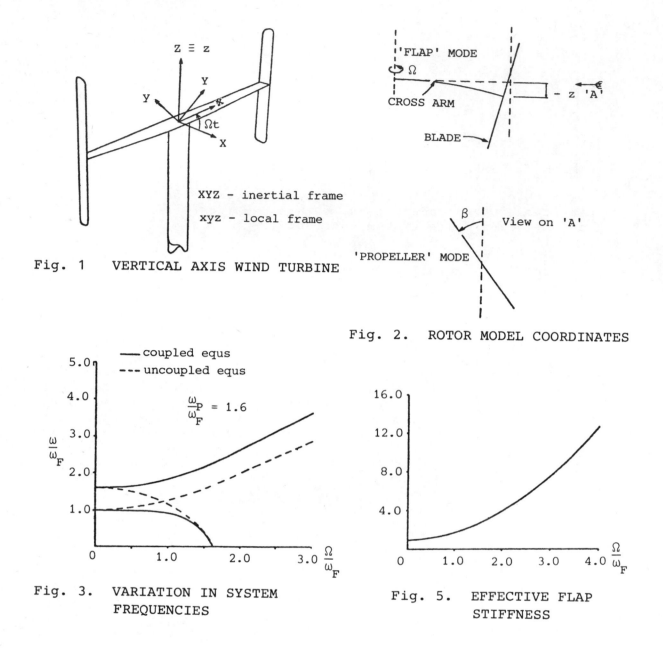

Fig. 1   VERTICAL AXIS WIND TURBINE

XYZ – inertial frame

xyz – local frame

Fig. 2.   ROTOR MODEL COORDINATES

Fig. 3.   VARIATION IN SYSTEM
FREQUENCIES

Fig. 5.   EFFECTIVE FLAP
STIFFNESS

Fig. 4.   VARIATION IN SYSTEM STABILITY

Fig. 6.   EFFECTIVE PROPELLER
STIFFNESS

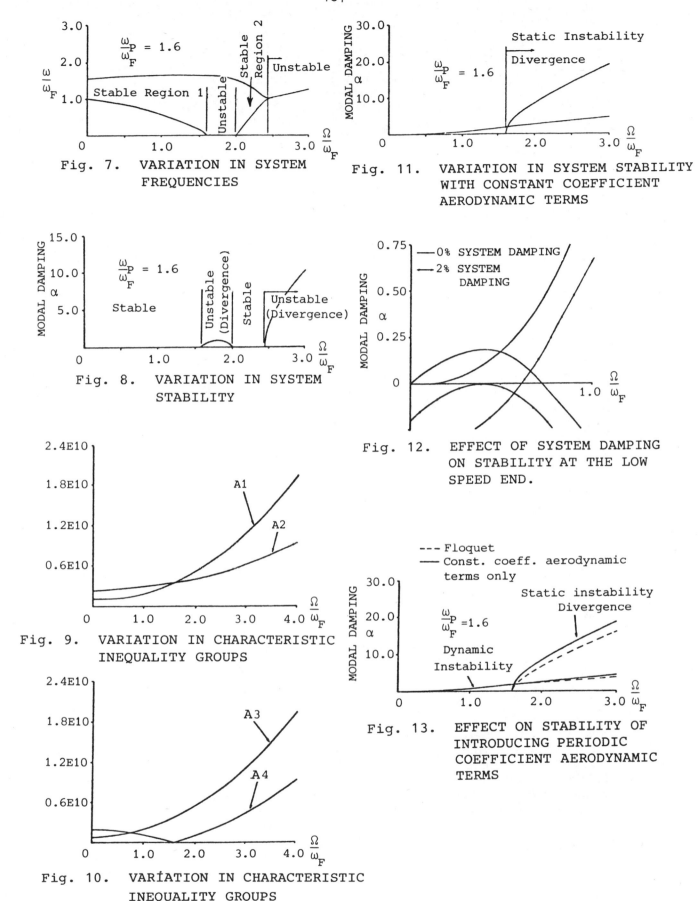

Fig. 7. VARIATION IN SYSTEM FREQUENCIES

Fig. 8. VARIATION IN SYSTEM STABILITY

Fig. 9. VARIATION IN CHARACTERISTIC INEQUALITY GROUPS

Fig. 10. VARÍATION IN CHARACTERISTIC INEQUALITY GROUPS

Fig. 11. VARIATION IN SYSTEM STABILITY WITH CONSTANT COEFFICIENT AERODYNAMIC TERMS

Fig. 12. EFFECT OF SYSTEM DAMPING ON STABILITY AT THE LOW SPEED END.

Fig. 13. EFFECT ON STABILITY OF INTRODUCING PERIODIC COEFFICIENT AERODYNAMIC TERMS

TURBULENCE INDUCED LOADS IN A WIND TURBINE ROTOR

A.D. Garrad and U. Hassan
Wind Energy Group
Greenford House
309 Ruislip Road East
Greenford
Middlesex
U.K.

Abstract

The response of a three degree of freedom model of a 2-bladed teetered horizontal axis wind turbine excited by wind turbulence is reported. It is demonstrated that turbulence induced cyclic loads may be comparable to deterministic loads in the flatwise direction. Shortcomings and difficulties in the analysis are discussed.

Introduction

What is the rise time for a wind gust? What is its maximum velocity? Are "down gusts" different in shape to "up gusts"? All these, and many more questions, arise when the stochastic behaviour of the wind is modelled as a series of discrete gusts. The answer to each question is that we don't know. A discrete gust is a useful concept for understanding the dynamic behaviour of the system that it excites, but it does not aid understanding of the wind itself.

For many years the frequency domain approach to wind loading on buildings has been recognised as the most desirable approach for analysing dynamically active structures, see for example Wyatt (1). For this type of structure methods exist for prediction of normal and extreme loads and these may, in principle, be applied to the analysis of wind turbine loads. There are, however, some important differences between stationary structures and wind turbine systems that tend to complicate this already relatively complex type of analysis:

- The blades that rotate through the turbulent wind

- The dynamic and aerodynamic coupling between the various modes of vibration

- The periodic terms in the equations of motion that describe a tower - rotor system

These three characteristics make the application of spectral techniques to the analysis of wind turbines considerably more complex than the equivalent task for stationary structures or indeed fixed wing aircraft, and they may well explain with good reason the fact that spectral analysis has been slow to find application in this technology, and also why it is still some way from being available as a standard design tool.

In fact Rosenbrock (2) laid the foundation for the analytical treatment of wind turbines in a turbulent velocity field some thirty years ago. The resurgence of wind energy has led to renewed interest in this work and various authors - Kristensen and Frandsen (3), Anderson (4) and Connell (5) have rediscovered the Rosenbrock approach and applied it together with the modern computing techniques to produce a reasonable description of the turbulence field encountered by a rotating wind turbine blade. It is not the intention of this paper to re-iterate these analyses but rather to demonstrate how the approach described by them may be used to assess the importance of turbulence on the loading of a wind turbine rotor, and to draw attention to the difficulties - some still unresolved - that are encountered in the process.

Dynamic Model

This paper addresses the problem of performing a dynamic analysis of a two-bladed, teetered horizontal axis wind turbine rotor excited by a turbulent wind. The results of such an analysis may take the form of displacement, velocity, acceleration or load spectra and their associated characteristics. Here they will be limited to teeter excursions and flatwise blade loads. The analysis of teeter excursions alone has already been treated in some detail by Anderson, Garrad and Hassan (6) and hence will only be of passing interest here. Estimates of fatigue damage and extreme loads can result from the analysis of dynamic response to turbulent wind. In the context of fatigue, edgewise blade loads are dominated by gravitational effects and hence turbulent excitation is not considered to be of significance. For this reason the 3 modal degrees of freedom used here are limited to the flatwise direction.

The dynamic model is derived from that given by Garrad (7) omitting the support structure. The resulting model may be described in terms of the first three rotor modes - teetering, symmetric bending and asymmetric bending - illustrated in Figure 1. In terms of these modes the displacement of a point on the blade is given by:

$$w(r, t) = \sum_{i=o}^{2} q_i(t) \phi_i(r) \tag{1}$$

A fairly standard linear, aerodynamic perturbation method is used here which is illustrated in Figure 2. The most important influence that the blade motion w, and the turbulence perturbations u have on the aerodynamic loads is via a change in the inflow angle:

$$\Delta\beta = (u + \sum \ddot{q}_i \phi_i) / |r| \, \Omega \tag{2}$$

The resulting perturbation in distributed thrust is:

$$\Delta T = \tfrac{1}{2}\rho \, W^2 \, c \, a \, \Delta\beta \, d \, r \tag{3}$$

where r is the radial distance from the hub, Ω the rotational speed of the rotor, W the magnitude of the apparent wind vector, c the chord, ρ the air density and a the lift curve slope.

Equation of Motion

The equation of motion of the rotor neglecting structural damping may now be formulated as:

$$
\begin{bmatrix} M_{21} & 0 & 0 \\ 0 & M_{22} & 0 \\ 0 & 0 & I \end{bmatrix} \ddot{q} + \begin{bmatrix} K_1 & 0 & 0 \\ 0 & K_2 & 0 \\ 0 & 0 & I\Omega^2 \end{bmatrix} q = \begin{bmatrix} -\int \Delta T\, \phi_1\, dr \\ -\int \Delta T\, \phi_2\, dr \\ -\int \Delta T\, \phi_0\, dr \end{bmatrix}
\tag{4}
$$

The subscript o denotes teetering and 1 and 2 denote symmetric and asymmetric bending respectively. I is the inertia and K_i are the generalised stiffnesses.

It was shown by Anderson, Garrad and Hassan (6) that some dimensionless parameters emerge naturally in the description of the rotor behaviour when excited by turbulence and these will be adopted here. Since vibrational modes have also been included here, other parameters are also required for a complete description. The analysis of an arbitrary blade has been undertaken, but since the object of this paper is to provide an illustration of this technique, rather than to perform a comprehensive analysis, the added complexity necessitated by the arbitary description will be omitted and the analysis limited to a uniform blade. Under these conditions the behaviour may be described by the following dimensionless parameters:

- Lock number $\qquad \gamma = 4\rho R^4 a\, c/I$

- Tip speed ratio $\qquad \lambda = r\,\Omega\,/\,U$

- Eddy size $\qquad \eta = 1/R$

- Mode shape $\qquad \Phi_i = \phi_i/R$

- Space $\qquad x = r/R$

- Blade frequency $\qquad P_i = \omega_i/\Omega$

plus numerous modal parameters such as:

$$D_i = \int x^2 \, dx \int \Phi_i^2 \, |x| \, dx \left/ \int \Phi_i^2 dx \right.$$

which will not be listed in detail. R is the radius of the rotor, l the length scale of longitudinal turbulence, U the mean wind speed and w_i the frequency of mode i.

Formulation in the Frequency Domain

Equations (2) and (3) are substituted into equation (4) and the Fourier transform of the resulting equation is taken to yield a system of equations that describe the behaviour in the frequency domain. For an arbitary blade these equations demonstrate that the symmetric bending mode is uncoupled from the other two, but that the teetering and asymmetric modes are coupled together. It is a fortunate characteristic of the uniform rotor that permits these modes to become uncoupled and results in three single degree of freedom oscillators. The last step is to transform the frequency domain equations into spectral form to obtain the result:

$$S_{jj}(p) = \left(\frac{\gamma}{8\lambda} \right)^2 c_j^2 \frac{S_{jj}^F(p)}{\overline{|A_{jj}|}^2} \tag{5}$$

where

$$A_{jj} = (P_j^2 - p^2 + i\, D_j\, p)$$

$S_{jj}(p)$ is the power spectrum of q_j;

$$S_{jj}^F(p) = \int_{-1}^{1}\!\!\int_{-1}^{1} |x_k| \, |x_1| \, \Phi_j(x_k) \, \Phi_j(x_1) \, S^u(x_k, x_1, p) \, dx_k \, dx_1$$

and $S^u(x_k, x_1, p)$ is the spectrum of longitudinal turbulence suitably transformed to the rotating frame of reference.

Results

Figures 3, 4 and 5 show the input spectra, the corresponding transfer functions and the resulting output spectra for the generalised co-ordinates for a rotor with the following characteristics:

$$\gamma = 3 \quad \lambda = 6 \ , \ \eta = 3, \ p_1 = 3.5, \ p_2 = 6.5$$

Reference to Figure 5 shows a large peak in teeter motion at $p = 1$ (the rotational speed) indicating that the resonant nature of the teeter motion concentrates the available energy at its resonant frequency. This is due to both the rotor response and the redistribution of turbulence energy at the rotor speed and its harmonics. Above $p = 1$ there is a rapid decay, although small peaks are clearly visible at $p = 3, 5, 7$, etc.

The rotor has been designed so that its resonant frequencies do not coincide with any harmonics of the rotor speed and hence the major peaks are at the resonant frequencies themselves. Again minor peaks are also visible at odd harmonics for the asymmetric mode and even harmonics for the symmetric mode. It should be noted that since the asymmetric and teetering modes only respond to differential loading they contain very little low frequency energy. This is because the scale of turbulence at low frequency is large with respect to the rotor and hence tends to engulf both blades. The symmetric mode on the other hand responds to symmetric loading and hence contains the large scale low frequency element.

Interpretation of Spectra

Many authors have concluded their analyses at this point with the derivation of output spectra. These are of little use to the design engineer. Some effort was made in reference (6) to attempt an interpretation of the spectra by considering "zero crossing" analysis that allows frequency of exceedances of specific values (in this case teeter excursions) to be calculated. Similar estimates may be made for frequencies of peaks and their values. For buildings subject to wind loading these calculations are relatively straightforward since the time varying component of the load is purely stochastic. Unfortunately for a wind turbine the excitation is a mixture of periodic (deterministic) loads, arising from yaw misalignment, shaft tilt, tower shadow and shear, and stochastic turbulent loads. A mixture of these types precludes a straightforward analysis of zero crossing and although this problem is by no means peculiar to wind turbine technology it does require some further work to attempt to provide a useful working tool for design engineers.

Given these qualifications it is still possible to obtain some results about the importance of turbulent loads on wind turbine rotors. The statistical parameters that provide this insight are the moments of the spectra, or rather combinations of them. Their definition and interpretation is given in many standard texts and will not be reproduced here. The most important of these parameters are the variance, m_0; the apparent up-crossing frequency, N^+; and the irregularity factor ε .

The simplest of these is the variance, which is shown in Figure 6 plotted for the three degrees of freedom in the present model for a variety of rotor characteristics for unit turbulence intensity. Various trends emerge from the Figure. The change of teeter variance with Lock Number is relatively small, whereas the corresponding change in the bending modes is quite large - sometimes up to 3 orders of magnitude as γ changes from 1 to 10. The dependence on the other parameters, tip speed ratio and turbulence scale, is very mild - the four graphs in Figure 6 all have very similar shapes. The variance reduces with increasing λ for both bending modes; the variation in the symmetric bending mode with length scale is scarcely discernible, but the asymmetric variance clearly decreases with increasing length scale, as the loading becomes more uniform across the disc.

Figure 7 shows a similar series of graphs for the up-crossing frequency. This parameter gives a guide to the dominant frequency of the spectrum. Reference to Figure 5 shows clear peaks at distinct frequencies which can be seen plottedin the graphs of Figure 7. The coincidence of the apparent up-crossing frequency and the resonant frequency indicates that the response is narrow band - the teetering response is an excellent example of this. The frequencies plotted for both bending modes, but particularly the symmetric mode, display a considerable variation with Lock number, tip speed ratio and turbulence length scale indicating that the response to turbulence is quite sensitive to rotor and site characterisitcs.

The final parameter of interest is the irregularity factor ε from which the likely distribution of peaks may be deduced. The value of ε ranges from 0 to 1, 0 indicating that the peaks from a Rayleigh and 1 indicating a Gaussian distribution. For the cases considered here ε varies from 0.08 to 0.3 for the teeter and asymmetric bending and from 0.6 to 0.8 for the symmetric bending, demonstrating the effect of the low frequency content on the symmetric bending that is not present for the other two modes.

Numerical Example

To put these calculations in context it is illustrative to assess the importance of the turbulence induced loads with respect to the deterministic loads. Consider a machine characterised by the following parameters:

$$\eta = 3, \quad \gamma = 4, \quad \lambda = 6, \quad I_u = 12\%, \quad \Omega = 34 \text{ rpm}$$

Taking typical mass values, an rms value of the stochastic flatwise root bending moment of 0.17 MNm is obtained compared with an rms value of 0.14 MNm for the deterministic part due to a typical shear profile, tower shadow and yaw angle. The precise value of these should be treated with some caution since the approximations in modelling the blade may have a considerable effect on their size. The figures do, however, demonstrate that turbulent and deterministic loads may be of comparable magnitude.

Conclusions

This paper has outlined a model that is capable of predicting the dynamic loads induced in a teetered rotor by turbulence. The analysis is approximate but it has demonstrated that these loads are of importance in the flatwise direction. However, the major fatigue loads are due to gravity and the gravitational edgewise loads are usually the dominant factor in blade fatigue design.

Further work is required to assess the influence of turbulence on the support structure and consideration is required to develop an adequate means of using this type of analysis for basic blade design.

Acknowledgements

The authors wish to thank the UK Department of Energy and Taylor Woodrow Construction Limited for their support and permission to publish this work.

References

(1)   Wyatt, T.A.

"The Dynamic Behaviour of Structures Subject to Gust Action"

Proc. CIRIA Conference "Wind Engineering in the Eighties" November 1980

(2)   Rosenbrock, H.H.

"Vibration and Stability Problems in Large Wind Turbines Having Hinged Blades"

ERA Report C/T 113 (1955)
ERA, Cleeve Road, Leatherhead, Surrey.

(3)   Kristensen, L.
         and
      Frandsen, S.

"Model for Power Spectra of the Blade of a Wind Turbine Measured from the Moving Frame of Reference"

Journal of Wind Engineering and Industrial Aerodynamics, 10, No.2, (1982)

(4)   Anderson, M.B.

"The Interaction of Turbulence with a Horizontal Axis Wind Turbine"

Proc. 4th BWEA Wind Energy Conference, Cranfield, UK (1982) pp 104-118.

(5)   Connell, J.R.

"The Spectrum of Wind Speed Fluctuations Encountered by a Rotating Blade of a Wind Energy Conversion System: Observation and Theory"

Battelle Pacific Northwest Laboratory Report PNL 4983 (1981)

(6)   Anderson, M.B.,
      Garrad, A.D.
         and
      Hassan, U.

"Teeter Excursions of a 2-Bladed Horizontal Axis Wind Turbine Rotor in a Turbulent Velocity Field"

To be published

(7)   Garrad, A.

"An Approximate Method for the Dynamic Analysis of a 2-Bladed Horizontal Axis Wind Turbine Systems"

Proc. 4th International Symposium on Wind Energy Systems, Stockholm, September 1982.

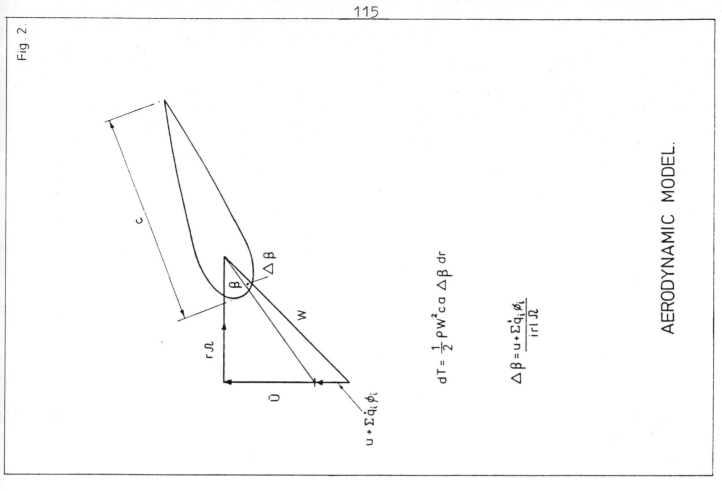

Fig. 2.

$$dT = \frac{1}{2}\rho W^2 ca \, \Delta\beta \, dr$$

$$\Delta\beta = \frac{u + \Sigma \dot{q}_i \phi_i}{irl \, \Omega}$$

AERODYNAMIC MODEL.

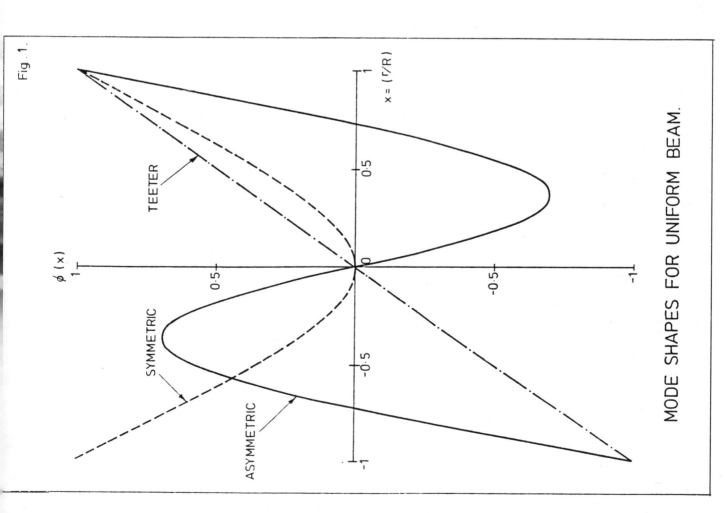

Fig. 1.

$x = (r/R)$

TEETER

SYMMETRIC

ASYMMETRIC

MODE SHAPES FOR UNIFORM BEAM.

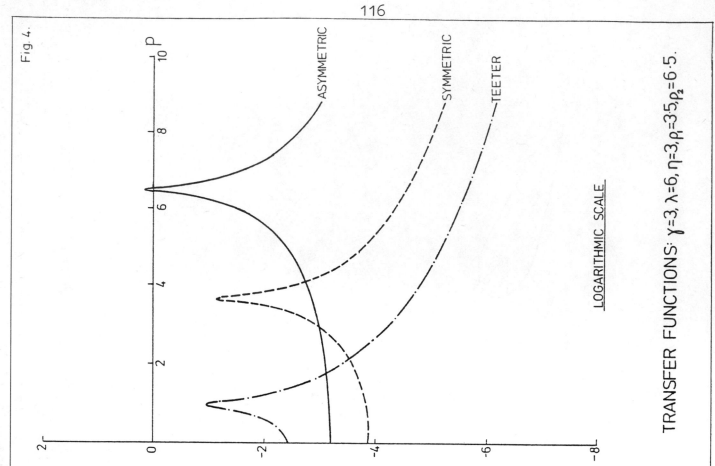

Fig. 4.

TRANSFER FUNCTIONS: $\gamma=3$, $\lambda=6$, $\eta=3$, $\rho_1=3\cdot5$, $\rho_2=6\cdot5$.

LOGARITHMIC SCALE

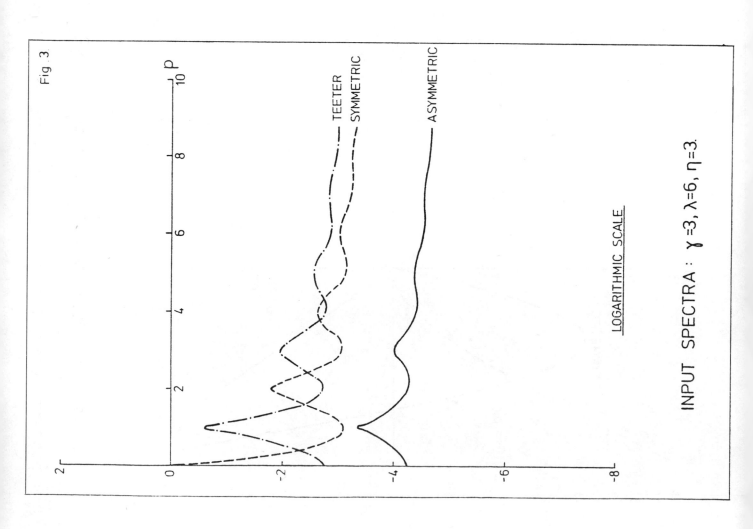

Fig. 3

INPUT SPECTRA: $\gamma=3$, $\lambda=6$, $\eta=3$.

LOGARITHMIC SCALE

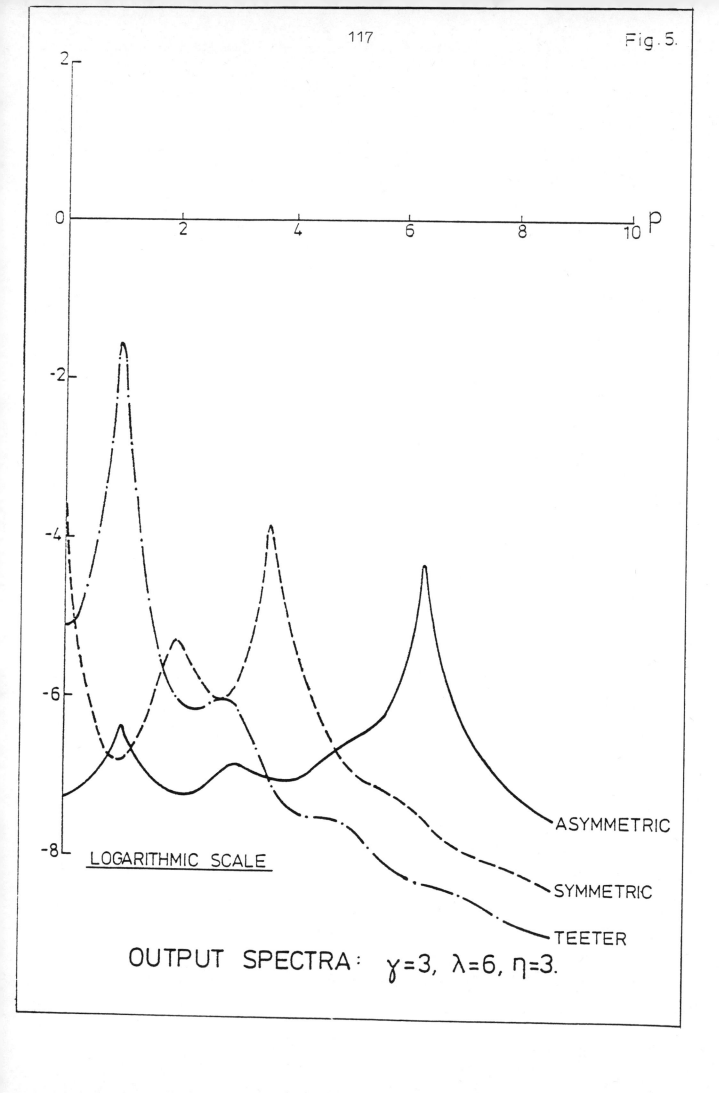

OUTPUT SPECTRA: $\gamma=3,\ \lambda=6,\ \eta=3.$

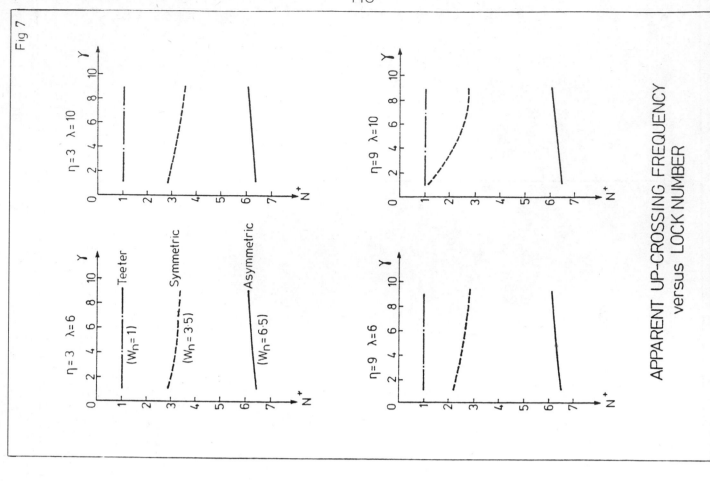

Fig 7

APPARENT UP-CROSSING FREQUENCY versus LOCK NUMBER

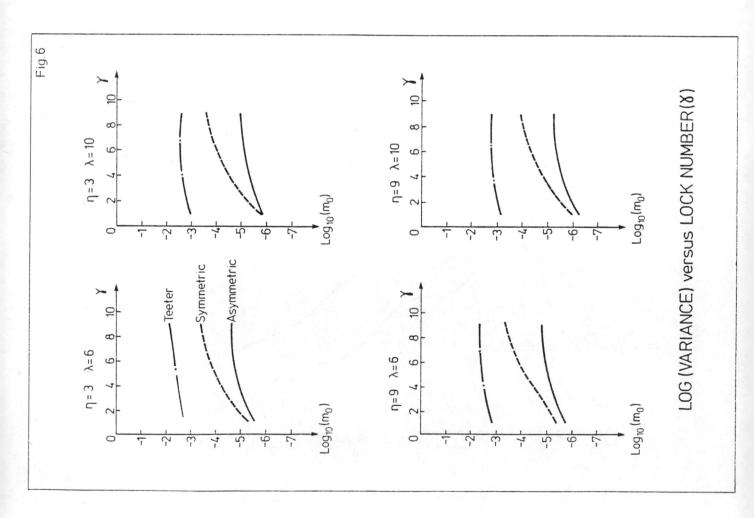

Fig. 6

LOG (VARIANCE) versus LOCK NUMBER(ɣ)

A NUMERICAL SIMULATION OF THE RESPONSE OF A LARGE HORIZONTAL
AXIS WIND TURBINE TO REAL WIND DATA.

S J R POWLES, E J FORDHAM, Cavendish Laboratory, Cambridge

M B ANDERSON, Sir Robert McAlpine and Sons Ltd.

Abstract

To design large wind turbines it is important to understand the rotor response
to fluctuating forces and moments due to the turbulence in the natural wind.
This problem is approached here by using a combination of blade element theory,
developed by Anderson and others, to predict the steady state loads, and a blade
dynamics analysis, developed by Powles et al (1), to calculate the rotor dynamic
response to half-hour stretch of real wind data supplied by Pacific Northwest
Laboratories, and previously analysed by Fordham and Anderson (2). The area
over which the wind data were recorded makes it appropriate to model a rotor of
approximately 60m diameter. The statistics of the dynamic response may be compared with the input data, for a variety of rotor configurations, and useful
design feedback obtained.

Introduction

A suite of computer programs has been developed to predict the aerodynamic
performance of the blades of a Horizontal-Axis wind turbine, and to use this to
determine the dynamic response of the blades to measured real wind data. Data
input is of a very general nature, allowing a wide range of areofoil section,
blade geometry, and root attachment (ie fixed, hinged or teetered hub) to be
used for the rotor, and any available measured wind data to be used to calculate
the fluctuation of forces on the rotating blades with time. The programs are
written in standard FORTRAN IV, and in all comprise of over 2500 lines of FORTRAN
coding, requiring approximately 250K bytes of store, when compiled on an IBM 3081.

The steady state aerodynamic loads, and particularly the axial and tangential
interference factors a and a' are calculated first, for the required blade
geometry, over a wide range of operating tip speed ratio, using tabulated values
of the variation of lift and drag coefficients with angle of attack: several
types of tip loss model can be included, and solution is by an iterative
procedure for a and a'.

Wind speed data need interpolating and transforming to polar coordinates, to
enable the instantaneous wind speed at every point on the blade to be known, as
the blade steps round in azimuth angle with time. We have used data recorded
over a ring 49m in diameter (Fig 1), and interpolated this using a form of least
squares fitting to a limited expansion in eigenfunctions of the operator ∇^2
described below. The original data may also be processed in a variety of ways:
eg by subtraction of the mean wind variation with height across the rotor disc,
the wind shear can be removed, or different wind shear models substituted;
a tower shadow model could also be superimposed on the wind data, to simulate
the extra blade dynamics problems of the rotor passing behind a tubular of
lattice tower (Ref 3).

Using the aerodynamic performance, and the quickly varying point wind speed data, the rotor dynamic response to this real wind regime may be predicted. Normal mode shapes and eigenfrequencies for the rotating blade are first calculated using a "Lumped Parameter" method, dividing the blade span into discrete parts after Holzer(4) and Myklestad(5), again for any desired blade input geometry. Alternatively, experimental mode shapes and frequencies may be used as input. These modes are used, in conjunction with the calculated aerodynamic forces, to build up a picture of the variation of blade bending deflection out of the plane of rotation with time. A recurrence relation may be formed for the blade shape over several steps in azimuth angle (see Ref 1), and a time history of required blade response (eg teeter excursions, root bending moment, blade tip vibration, cone angle variation) may be saved for statistical analysis. For example, a quick analysis could be done to find the number of crossings of a particular level of teeter angle (this being the sort of information interesting from a fatigue point of view), or a least squares calculation to find the mean and variance of teeter excursion.

Initial Input Data

The size of the ring used in the Pacific Northwest Laboratories experiment (49m diameter - see Fig 1) suggests a useful blade diameter of 60m, which is a realistic useful size for a large turbine (eg the W E G 60m Orkney machine). The outer anenometer circle is at 82% of the blade span, making wind data extrapolation to the blade tip unlikely to be difficult. A "middle range" tip speed ratio (ie 5-10) was required for generality, and the frame rate of the wind data sampling (0.2 seconds) meant that a fixed rotor speed of 25 RPM was very convenient, giving a tip speed ratio of 7, in the 11 ms^{-1} wind speed average. The wind data sampling rate is therfore once every 30^0 in azimuth angle, giving 12 frames per revolution (2.4 seconds); this was interpolated by six, to give wind data at every 5^0 in blade azimuth angle - this had been found, from previous experience, to be a sufficiently small step length, over which to work out the blade bending to ensure continuity in the blade shape recurrence equations. For all calculations, the blade is divided into 20 radial sections, though this number could be reduced for less accurate, and less time-consuming calculations.

The variation, with angle of attack, of the lift and drag coefficients were tabulated for a NACA 4415 aerofoil, at a Reynolds number of 3.10^6, after Jacobs and Sherman(6), and the axial and tangential interference factors were calculated for a range of TSR from 3 to 12 in steps of 0.25, for a Tapered Chord Zero Twist blade(7) of TSR 7, of 30m radius, at 20 discrete sections along the blade span. For the blade dynamics calculation, only the first few eigenfrequencies and modes are required for accurate calculation, and the vibration frequencies used are: teeter - 1p (0.42 Hz), 1st symetric - 3.5p (1.46 Hz), 1st asymetric - 6.5p (2.71 Hz).

Aerodynamic Performance

The aerodynamic performance prediction program (HAWT) was written by M Anderson, and uses blade element theory to calculate steady state aerodynamic blade loading, for a wide range of input conditions, over a range of tip speed ratio. Blade-element theory calculates the forces on the blade due to its motion through the air (combination of the wind velocity, induced velocity and the rotational velocity of the blade), to determine the performance of the entire rotor. Basically, blade-element theory is lifting- line theory applied to a rotating blade. It is assumed that each blade-element behaves like a two-dimensional aerofoil to produce aerodynamic forces (lift and drag), with the influence of the wake and the rest of the rotor contained entirely in an induced velocity at the element. The aerodynamic forces at each element are equated to those derived from momentum considerations enabling the induces velocities to be determined. This is based on the underlying assumption that each element is radially independent and the pressure in the far wake is equal to the free stream pressure, which is only strictly true for non-expanding wake, and a rotor consisting of an infinite number of blades. For a finite number of blades the application of lifting-line theory is not strictly valid near the blade tip. When the chord is finite at the blade tip, blade-element theory gives a non-zero lift all the way out to the end of the blade. In fact, however, the circulation drops to zero at the tip over a finite distance due to the cross-flow around the tip, which reduces the pressure between the upper and lower surfaces of the aerofoil. The flow, relative to the blade at a radius r, will be composed of an axial velocity $V_\infty(1 - a)$ and a rotational velocity $\Omega r(1 + a')$.

The standard method of applying tip loss correction(8) is to assume that the maximum change of the axial ($2aV_\infty$) and tangential ($2a'r\Omega$) velocities in the slipstream occurs at the vortex sheets, and that the average decrease is only a fraction F of this velocity. Applying the equations of axial and angular momentum to an elemental annular streamtube of radius r and thickness dr we obtain the elemental components of thrust and torque. Combining these with the corresponding blade-element equations gives a solution for a and a'. From geometical considerations, and the equations for a and a', it can be shown that the total induced velocity is perpendicular to the relative velocity. This is what would be expected from momentum considerations, ie opposite to the applied force (lift).

Other methods of applying tip loss can be used in the program: eg the "linear" method suggested by Wilson and Lissaman(9), based on the assumption that the axial and tangential induced velocities are localised at the blade, and only a fraction F of these occur in the plane of the rotor. The resulting equations from momentum considerations give a different (a), but the same (a') as before. However, the induced velocities in the plane of the rotor are likely to be periodic, and a more accurate method would be to take into account the angular variation of the induced velocities, and integrate the elemental thrust and torque accordingly. This "cyclical" method may be chosen when using the program.

Assuming that the blade geometry, aerofoil properties and tip speed ratio are specified an iterative procedure is followed to determine the axial and tangential interference factors at each radial station and hence the blade loading. Once these have been determined, the coefficients of power, thrust and torque can be evaluated.

Interpolation of the available wind data from the Buttelle experiment

The available wind data consisted of measurements of u,v,w- components of the wind at 12 points, 8 in an outer ring, 4 in an inner ring (Fig 1). The data from this experiment have been comprehensively investigated by Fordham and Anderson(2). The data were sufficient to obtain a fairly clear picture of the two-point correlation structure of the wind in this experiment; however, in the present application we aim to use the data to compute the turbine blade motion from one time-step to the next; furthermore, information is required at several radii along the blade span. Some rational form of interpolation of the available data is clearly necessary to use the data as input to the blade dynamic calculations. The procedure under trial is a form of least-squares fitting to a limited expansion in eigenfunctions of the operator ∇^2, described in more detail below.

We aim to represent the wind (the u-component in particular) by a surface of the form:

$$u(r,\alpha) = \sum_{m=1}^{N} a_m \Psi_m(r,\alpha) \tag{1}$$

where r,α are polar co-ordinates centred on the hub, the Ψ_m are N suitable elementry functions, and the a_m are coefficients to be determined.

The coefficients a_m are determined by a least-squares criterion using the available data. From equation(1) data value estimates at the ith data point D_i are:

$$D_i = \sum_m a_m \Psi_m(r_i,\alpha_i) = \underline{a} \cdot \underline{\Psi}_i \text{ say} \tag{2}$$

Forming the χ^2 statistic $\chi^2 = \sum_{i=1}^{12} (D_i' - D_i)^2/\sigma_i^2$ \hfill (3)

Where the D_i' are the actual data values and σ_i^2 the variance of the expected noise in each measurement (here assumed uniform over the anemometer array) and minimising with respect to each of the a_m, we obtain:

$$\frac{\partial \chi^2}{\partial a_m} = \frac{\partial}{\partial a_m} \left\{ \sum_i \left(D_i' - \sum_n a_n \Psi_{n,i} \right)^2 \right\}$$

$$= 2. \sum_i \left(D_i' - \sum_n a_n \Psi_{n,i} \right) \Psi_{m,i} = 0 \tag{4}$$

Hence under the least-squares criterion

$$\sum_i D_i' \Psi_{m,i} = \sum_i D_i \Psi_{m,i} = \sum_{i=1}^{12} \Psi_{m,i} \sum_{n=1}^{N} a_n \Psi_{n,i} = \sum_{n=1}^{N} \left[\sum_{i=1}^{12} \Psi_{m,i} \Psi_{n,i} \right] a_n$$

or, in vector/matrix rotation:

$$\underline{\beta} = \underline{\underline{A}} \cdot \underline{a} \tag{5}$$

where $\underline{\beta} = (\sum\limits_{i=}^{12} D_i' \Psi_{m,i}; \; m = 1, \ldots\ldots N)$

$\underline{a} = (a_m; \; m = 1, \ldots\ldots N)$

$\underline{\underline{A}} = \{\sum\limits^{12} \Psi_{m,i} \Psi_{n,i}; \; m = 1, \ldots\ldots N; \; n = 1, \ldots\ldots N\}$

Hence the required coefficients a required can be found from:

$$\underline{a} = \underline{\underline{A}}^{-1}\underline{\beta} \tag{6}$$

Whatever the choice of the functions $\Psi_m(r,\alpha)$, $\underline{\underline{A}}$ and therefore $\underline{\underline{A}}^{-1}$ (provided $\underline{\underline{A}}$ is non-singular) are simply constant, and can be worked out in advance. The computation of the $\underline{a}$ coefficients thus requires the computation of $\underline{\beta}$ using the real data for each frame, and in principle a matrix multiplication; however, by a suitable choice of the Ψ_m, $\underline{\underline{A}}$ can be made a very sparse matrix reducing computation still further.

In the present application, an ideal choice for the Ψ_m would be the low-order members of some complete set of eigenfunctions of an appropriate self-adjoint operator; this rationale is just an extension of the well-known approximation of a function of a single-variable on a finite interval by the first members of its Fourier series; indeed, the azimuthal variation assumed here will indeed be a Fourier series.

Eigenfunctions of ∇^2 seperable in polar co-ordinates ie. solution of $\nabla^2\theta = -k^2\theta$ say are:

$J_\ell(kr) \begin{cases} \cos\ell\theta \\ \sin\ell\theta \end{cases}$

where ℓ is an integer and J_ℓ is a Bessel function of order ℓ. (The companions $Y_\ell(kr)$ are excluded owing to their singularity at the origin.) Here k (which determines the position of the zeroes of J_ℓ would be chosen to satisfy any necessary radial nodes.

However, the availabe data here have extremely limited radial information; it was thus decided to use simple power laws for the radial variation, choosing the first terms in the series expansions of the Bessel functions:

$J_0(x) = 1 - \frac{1}{4}x^2 + 1/64x^4 - \ldots\ldots$

$J_1(x) = x/2 - 1/16x^3 + \ldots\ldots$

$$J_2(x) = x^2/8 - 1/96x^4 + \ldots\ldots \tag{7}$$

The Ψ_m functions have thus been chosen from:

$\text{const.}(1 + \alpha_0 r^2 + \ldots\ldots)$ \qquad (like J_0)

$$\begin{matrix} \cos\theta \\ \sin\theta \end{matrix} \quad (r + \alpha_1 r^3 + \ldots\ldots) \qquad\qquad (\text{like } J_1)$$

$$\begin{matrix} \cos 2\theta \\ \sin 2\theta \end{matrix} \quad (r^2 + \alpha_2 r^4 + \ldots\ldots) \qquad\qquad (\text{like } J_2) \qquad\qquad (8)$$

The representation equation(1) was thus chosen to be the first few terms of:

$$a_1$$
$$+ a_2 r \cos\theta + a_3 r \sin\theta$$
$$+ a_4 r^2 + a_5 r^2 \cos 2\theta + a_6 r^2 \sin 2\theta$$
$$+ a_7 r^3 \cos\theta + a_8 r^3 \sin\theta + a_9 r^3 \cos 3\theta + a_{10} r^3 \sin 3\theta \qquad\qquad (9)$$

$N = 6$ has been chosen in the first instance. This choice of the Ψ_m means that with the present disposition of the data points (Fig(1)) the matrix $\underline{A}$ has only two off-diagonal components. The composition of the $\underline{a}$ from (9), and thus the required data interpolation from equation (1), is thus fast and efficient, an important consideration in any attempt to build a dynamic simulation model requiring step-by-step interpolation of real data.

The closeness of the fitted surface equation (1), to the real data can be measured by calculating χ^2 for each data frame; the effect of the above procedure on the statistics of the wind can in principle be calculated, knowing the experimental statistics; however, a simple empirical calculation of auto- and cross-covariances of the interpolated wind probably gives the most direct comparison with the statistics of the raw data.

A danger with the simple power law radial variation is that extrapolating outside the circle of array data (ie. towards the blade tips, assuming the outer ring in Fig(1) to represent values at ∿0.8 blade radius) runs the risk of unrealistic large values being generated. A check is made on the occurrence of large excursions in the blade tip region; apart from this precaution, little can be done about the problem, other than refraining from extrapolating too far, which has been done in the present exercise by choosing the outer ring in Fig(1) to be at a large fractional blade radius.

Wind Shear

The analysis of Fordham and Anderson(2) concerned only variations from the local expected mean wind, which varies with height. In the present analysis, wind shear has been accommodated by firstly subtracting from each data point the (slowly varying) mean wind for that height before performing the surface fit outlined above. Any chosen mean wind shear can then be added on to the fitted turbulence, enabling the effect of wind shear, and different wind shear models, to be distinguished from that of turbulence. Using the whole original data with its own wind shear has been retained as a further option.

Blade Dynamics

Normal modes and frequencies of out-of-plane vibration are required as input to the blade dynamics program, DYNAMO, writted by S.Powles. These may be empirical or calculated; one method of calculation is to use a 'Lumped

Parameter' method, as described in Ref(1), with blade section masses and stiffnesses grouped at equally spaced points along the blade: the deflection, shearing force, and bending moment at each element may be found, when the appropriate boundary conditions are satisfied.

Solution for the blade shape, at each step in azimuth angle, is built up in the form of sums of these normal mode shapes, with the weighting depending on the aerodynamic force on the blade element, and the vibration frequency. To achieve an accuracy of a few percent in blade dynamic response, it is only necessary to use the first three or four blade eigenfrequencies and mode shapes - the contribution of the higher order modes to blade bending decreases rapidly with vibration frequency. For example, for a teetered hub, just the following could be used: the 1p straight bladed asymetric rotation about the teeter hinge, the 1st symetric vibration mode (same as for a fixed hub), and the 1st asymetric vibration mode (same as for a hinged hub). The vibration frequencies for these modes are approximately in the ratio 1: 3.5: 6.5.

By expressing the blade shape as a sum of normal modes:

$$Z = R \sum_{n=1}^{\infty} S_n(x) . \phi_n(\alpha) \tag{10}$$

a recurrence relation may be built up for the deflection of each blade element at one azimuth step, in terms of the deflections over the previous few steps. This is explained in greater detail in Ref(1). However, the interference factors a and a', must be calculated from the _mean_ wind speed over a short time up to the present blade rotational position (eg. use the moving average wind speed over one previous full blade rotation), because the induced velocities will remain nearly constant over the very short time scale of one blade azimuth step ($5^0 = 0.033$ seconds), despite the rapid variation of the individual point wind speed data. The angle of attack and apparent wind speed at each blade element will therefore depend upon the quickly varying point wind speed, the slowly varying induced velocities, and any previous blade movement out of plane (at constant RPM):

$$\tan \theta = (V_{point} - V_{average} . a - \frac{dz}{dt}) / (\Omega r(1 + a')) \tag{11}$$

and

$$V^2 = (V_{point} - V_{average} . a - \frac{dz}{dt})^2 + (\Omega r(1 + a'))^2 \tag{12}$$

and the axial thrust on each blade element can be found, using the aerofoil characteristics at this angle of attack:

$$\frac{dF}{dx} = \tfrac{1}{2} \rho V^2 c R (C_L . \cos \theta + C_D . \sin \theta) . \cos \beta \tag{13}$$

This element thrust is used to solve for the ϕ_n. The program therefore generates the exact shape of the deflected blade (at 20 radial elements) at each small step in azimuth angle, α (5^0). Any, or all, of this large amount of blade bending information may be saved for further examination, eg. level crossing analysis, or mean and variance calculations, using the shape of the blade to calculate bending moments and shearing forces at each element from the original normal modes used.

Obviously, with a large program of this type, it is wise to carry out a few initial tests, before running with full data input, to check the program is functioning as expected. Fig 2 shows a simulation of a Cambridge Research Turbines 5m freely hinged blade, being released from zero cone angle at high RPM, and smoothly taking up its design (TSR 10) cone angle of $\sim18^0$, in about one full revolution - the blade is just over critically damped in flapping motion. Another test was to simulate a wind speed steadily increasing over several revolutions, and observe the blade cone angle variation at constant RPM - this showed the cone angle increasing nearly linearly with time over a TSR from 6 to 12, as expected.

Results and analysis from running the 60m diameter teetered blade in a half hour stretch of real wind data from the Pacific Northwest array will soon be available.

Program Structure

Initial set up:

Calculate induced velocities, given C_L and C_D for blade aerofoil and geometry, over a range of TSR. Use this as a 'look-up table' for the Blade Dynamics program.
Calculate blade normal modes and frequencies.

Once per 10 revolutions:

Read 10 revolutions worth of wind data.
Work out fitting coefficients for all frames (120 x 6).

Once per revolution:

Calculate moving mean wind speed.
Generate point wind data (20 radial positions x 72 azimuth positions).

Blade dynamics caclucation, using:
      C_L and C_D tabulated ;
      TSR from moving average wind speed;
      hence a and a' at each blade radius;
      point wind speed data for angle of attack and apparent wind speed.
      Hence thrust on each blade element.

      Recurrence relation for blade shape each azimuth step (5^0)

Save required information (eg. teeter angle, root bending moment, tip deflection)
till all wind data used up.

Perform required statistical analysis on saved data.

Nomenclature

a = axial interference factor

a' = rotational interference factor

a_m = coefficients of elemental functions

c = blade chord

C_L = coefficient of lift

C_D = coefficient of drag

D_i = ith data point

dF = axial thrust

J_ℓ = Bessel function of order ℓ

N = no. of elementary functions

r = blade element radius

S_n = nth normal mode shape

$u(r,\alpha)$ = u component of wind

V = undisturbed wind velocity

V = apparent wind speed

V_{point} = point wind speed data

$V_{average}$ = moving average wind speed

dz/dt = blade element velocity

α = blade azimuth angle

β = blade cone angle

θ = angle of attack

σ_i^2 = variance of expected noise

ϕ_n = multiplier of normal modes

Ψ_m = elementary function

Ω = angular velovity of rotor

References

1) Powles SJR, Anderson MB, Wilson DMA, Platts MJ. Articulated wooden blades for a horizontal-axis wind turbine: design, construction, and testing. Proc. 4th BWEA Conference, Cranfield UK. 1982.

2) Fordham EJ, Anderson MB. An analysis of results from an atmospheric experiment to examine the structure of the turbulent wind as seen by a rotating observer. Proc. 4th BWEA Conference, Cranfield, UK. 1982.

3) Powles SJR. The effects of tower shadow on the dynamics of a horizontal-axis wind turbine. In the press.

4) Holzer H. Die Berechnung der Drehschwingungen. Berlin 1921.

5) Myklestad NO. Vibration analysis. McGraw Hill 1944.

6) Jacobs EN, Sherman A. Aerofoil characteristics as affected by variation of th Reynolds number. NACA Rep. no. 586. 1939.

7) Anderson MB. Blade shapes for horizontal-axis wind turbines. Proc. 2nd BWEA Workshop, Cranfield UK. 1980.

8) Glauert H. Aerodynamic theory. Julius Springer, Berlin 1935.

9) Wilson RE, Lissaman PBS. Applied aerodynamics of wind power machines. Oregon State University, National Science Foundation Rep. no. NSF-RA-N-74-113. 1974.

Acknowledgements

The financial support of an SERC CASE award with the Central Electricity Generating Board for S Powles and E Fordham, and from the University of Cambridge for E Fordham, is acknowledged.
We are indebted to Dr S F Gull for suggesting the wind data interpolation procedure, and to Dr J C Doran at Battelle Pacific Northwest Laboratories for supplying the raw data.

Wind components:
w
u
v
× functioning anemometers
(×) other anemometers not used here
1
2
3
4
5
6
7
8
9
10
11
12
(×)
34·5 m
49m
61·1 m
53·9 m
36·6 m
19·4 m
12·1 m
6·1 m
3 m
7·25 m
17·25 m
24·5 m
10 metres

Fig. 1 Geometry of the Battelle vertical plane array.

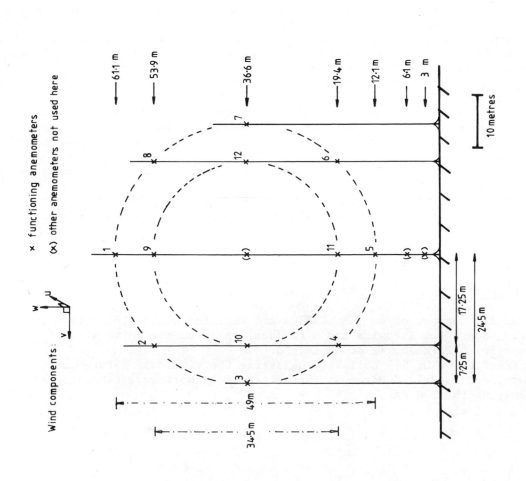

Blade Cone Angle (degrees)
20
15
10
5
0
0
1
2
3
4
Nubmer of Full Revolutions

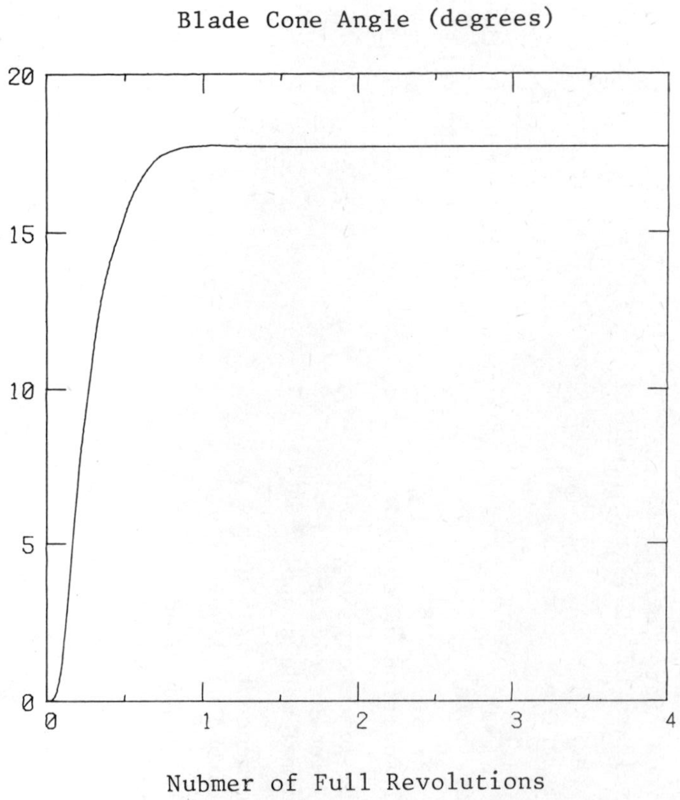

Fig. 2 A freely hinged blade 'released' from zero cone angle takes up its design cone angle in about one full revolution.

THE INFLUENCE OF TURBULENCE ON THE DYNAMIC
BEHAVIOUR OF LARGE WTGs

R H Swansborough
Energy Systems Department
ERA Technology Ltd., Leatherhead, Surrey.

Abstract

The wind field in which large megawatt size wind turbine generators (WTGs) operate is complex and requires extensive anemometry to obtain sufficient wind data to enable designers to predict the behaviour of WTGs with confidence.

Insufficient wind data are presently available to understand the nature of turbulent wind over the turbine swept area, especially in the offshore environment.

Synthetic wind models have been developed in both the time and frequency domain to study the influence of turbulence on WTG behaviour. These are being applied to dynamic models of horizontal axis WTGs which include the generator and electrical supply system.

The main features of the models are described and samples of the study results so far obtained are presented in graphical form. A sensitivity analysis leading to better definition of wind data requirements and guidance to WTG designers is planned; this includes variations in WTG design parameters and wind representation.

Introduction

Initial studies have demonstrated that a very substantial wind energy resource exists in British offshore waters. However, it is generally accepted that the wind data presently available for these offshore areas are inadequate if the possibilities are to be pursued in more detail. This is especially true in view of the strong dependence of energy production upon annual mean wind speed, and because of the detailed knowledge of wind structure which is necessary for the successful design and operation of large (megawatt-scale) wind turbine generators.

To meet this need for improved data and understanding, National Maritime Institute are undertaking for the Department of Energy a suite of projects with the general objective of providing wind data for the assessment of wind energy conversion systems in British offshore waters.

One of the projects, which is being undertaken for NMI by ERA, is a theoretical investigation of the dependence of wind turbine generator (WTG) design upon meteorological parameters. This investigation has two principal objectives:

- to establish the sensitivity of design to assumptions of meteorological parameters (specially the turbulent characteristics of wind) and to define the accuracy of data required for design purposes;

- to provide general guidance on the application of meteorological data for the design of WTGs.

The general approach to this project is to undertake a simplified theoretical analysis of dynamic WTG behaviour and to input to this analysis a range of wind conditions in order to check the range and sensitivity of the predicted response. The analysis is being confined to rotational dynamics of the turbine, mechanical transmission and generator and is being undertaken in both time and frequency domains.

The modelling is being based upon generalised parameterisations of a large horizontal axis WTG. Detailed data for large overseas WTG designs (especially the Boeing MOD2 in USA) have been sought with a view to analysing and comparing the response of different styles of machine, each with their own distinctive and detailed set of parameters. However, insufficient data have become available to permit this approach and the features of the 60m, 3.7MW phase 2 machine designed by the Wind Energy Group (this being the only definitive large UK design available at the time of the study) will be used as the baseline in the generalised parameterisation. This analysis will include the variation of parameters such as turbine diameter and speed, whilst retaining major features such as two blades and fixed pitch.

Time Domain Analysis

One of the most important features of a model for studying the dynamic behaviour of WTGs is an adequate representation of the wind field in which the WTG will operate. An ideal wind model would consist of wind data from a dense two dimensional array of fast response anemometry continuously recording all three wind components. Although some wind measurements are available from vertical arrays of anemometers (Ref.1), these are generally insufficient to understand the nature of turbulent wind over the entire turbine swept area of large WTGs without making unrealistic assumptions such as lateral uniformity across the turbine disc. The alternative which has been adopted in this work, is to revert to classical theory and produce a purely synthetic wind model. Although in a synthetic wind model there is no clear link between real measurements and the parameters used in the model it can be arranged to have the expected spectral and correlation properties (Ref.2). The particular attraction of the synthetic model is that its describing parameters can readily be varied to represent different conditions.

The incident wind model adopted is a time-stepping model based on the power spectrum of wind fluctuations as seen by a rotating observer in homogeneous isotropic turbulence. Random time series are synthesised from independent gaussian random variables so that the expected time correlations and cross correlations along the blades are satisfied. For WTG behaviour analysis the simulation period for which the time series are generated is considerably greater than the response time of the entire wind turbine system, and so that the time series can be tested for correct spectral properties a period of several minutes has been adopted. Time steps of about 0.1s are used to ensure that rotational eigen frequencies in the WTG are included.

The design of WTGs needs realistically to take into account the fluctuating loads imposed by turbulence which could give rise to fatigue damage. In order to study these effects a suitably detailed and self consistent model of the WTG is necessarily complex.

A number of possible designs for offshore use exist, each with its own distinctive features. Of the horizontal axis type the Boeing MOD2 typifies a compliant design with sophisticated control features while the UK 60m, 3.7MW phase 2 machine is representative of a much simpler and more rigid design. In the absence of adequate MOD2 data the model has been based on the latter machine. The main feature of the design is a two bladed fixed pitch turbine driving an asynchronous generator through a helical spur gearbox. Consideration was given to the inclusion of vertical axis WTG analysis in the work but the required wind model and turbine representation is completely different, and coverage will be limited to a brief commentary in the final report on the significance of the horizontal axis WTG results for vertical axis designs. A similar proof analysis to that described here is in fact being undertaken within the McAlpine consortium (Ref.3).

The computer program devised for the analysis comprises separate complementary modules for the incident wind, turbine, drive shaft and gearbox, generator and electrical network (Fig.1).

The turbine module includes a segmented representation of the blades in which the aerodynamic forces at each segment are calculated from local lift and drag coefficients with allowance for Reynolds number effects and local blade stall. For structural dynamics analysis, the blades are modelled as simple offset spring-hinged cantilevers which enable first mode response in the flap and lead-lag directions to be represented. Damping in these two directions is included. The resulting blade bending moments and shaft torque are calculated by aggregation. The mechanical power train is represented by conventional equations of rotational motion which include shaft stiffness, inertia and damping coefficients. Separate models of the turbine shaft, gearbox and generator shaft are included each with their equivalent lumped stiffness and inertia values. The forcing functions are the turbine torque at one end and the electrical air gap torque developed in the generator at the other.

The induction generator model includes electrical parameters of rotor, air gap and stator and is based on Park's classical two-axis representation. Generator impedances are determined by changes in rotational speed and changes in electrical system frequency.

An impedance matrix representation is used to model the electrical transmission and distribution system together with some local load. Impedance values are chosen to be typical of those expected with an offshore WTG installation connected to the UK National Grid.

Output from the model includes plots of time histories of wind speed, blade loading, blade displacement, shaft torque, shaft speeds, generated power and voltage at selected network points. A sample output is shown in Figs.2-9. In this case the initial condition is that the WTG is rotating at nominal speed and is decoupled from the wind and electrical network models. At zero time the wind is applied and the connection to the network is made simultaneously. The incident wind shown in Fig.2 is a test case taken from a fast response anemometer record over an arbitrary 10 second period and applied uniformly over the turbine's swept area.

A number of the synthetic wind models described above are under development. The models will cover a range of turbulence intensities and shear profiles at different mean wind speeds up to 25 m/s.

A comprehensive sequence of computer runs in the time domain is planned in order to test the sensitivity of WTGs to wind parameters. The different wind models will initially be applied to the model representing the baseline 60m design. Further runs will be carried out to evaluate the influence on the program output parameters of variations in the turbine design. Design parameters including turbine tip speed ratio, hub height, turbine diameter and turbine mass will be varied over a wide range. In addition, a limited number of runs will be made to investigate the influence on dynamic behaviour of different dynamic stalling characteristics and blade damping.

To investigate the wind data required to analyse WTG behaviour and hence arrive at guidelines for suitable anemometry the wind model will be gradually 'coarsened' to represent decreasing intensity and response of anemometry until the predicted WTG behaviour is significantly impaired.

Frequency Domain Analysis

Of necessity, a time domain analysis examines the time response of the WTG to a fairly short time record of wind speed. To include long term considerations a complementary frequency domain model has been developed. This introduces a statistical approach at the expense of detail in non-linear effects such as aeroelastic effects, blade stall and variations in rotational speed.

The wind model is again based on the assumption that turbulence is homogenous and isotropic but this time uses turbulence statistics. The turbine representation again includes segmented blades. For every possible pair of blade segments the cross correlation of turbulence is evaluated following the method described by Anderson (Ref.4). The statistics of fluctuating wind speed are converted first to statistics of fluctuating angle of attack, then to statistics of fluctuating aerodynamic force and finally to statistics of fluctuating torque.

An estimate of the effect of wind shear is included in the model by an evaluation of torque variations about the blade passing frequency. The dynamic response of the turbine blades has also been included to represent the first mode of blade flapping vibration. Tower shadow effects cannot be readily included in the model.

The equations of motion for the mechanical power train and generator are combined to give a second order transfer function which links the fluctuations in turbine torque and electrical network frequency with fluctuations in the power output of the generator. The method is based on that described in Ref.5 which has been applied to the 200kW Gedser wind turbine and compared with measured data.

Spectral transfer functions for the mechanical drive train including the gearbox could not be obtained for the baseline 60m design. In the absence of suitable transfer functions a simple equivalent lumped model has been developed for the drive train from a knowledge of the individual inertia

and stiffnesses of the turbine shafts, gearbox and generator.
The resulting simplification is not expected to significantly influence the
results as the turbine exhibits high inertia and the asynchronous generator
introduces a high degree of damping in comparison with the other elements
in the drive train.

The asynchronous generator is modelled by a linearised torque - speed curve
over the normal generating range.   Damping in the generator is inherent
due to the change in operating speed with power output.   The electrical
network is modelled by the spectral density of the variations in electrical
frequency, which for the UK National Grid is that given in Ref.6.

Output from the frequency domain model includes power spectrum plots of
torque developed by the turbine and power produced by the generator.

Sample results from two of the preliminary runs at different mean wind speeds
are shown in Figs.10-15.   Common to each run is the power spectrum of
frequency variations for the UK National Grid (Fig.10) and the transfer
function for the mechanical drive train (Fig.11).   In both cases the blades
are divided into nine segments.   Sensitivity of model predictions to
number of blade segments has been checked, which indicates very little
improvement is obtained beyond 9 or 10 segments.

The sensitivity analysis in the frequency domain will involve a series of
program runs with various wind inputs covering a range of mean wind speeds,
turbulence intensities and shear profiles.   Variations in turbine design
will be covered by a range of non-dimensional parameters including turbine
tip speed ratio and the ratio of turbulent length scale to turbine diameter.

Discussion

The results of the limited number of program runs carried out so far provide
the following main indicators:

- cyclic blade vibrations arise from gravity, sudden
  changes in incident wind speed, wind shear profile
  and tower shadow.   The precise level of blade loading
  fluctuations and resulting torque fluctuations is therefore
  expected to be significantly influenced by the particular
  design concept adopted for the turbine and tower;

- a high degree of damping is introduced by the asynchronous
  generator so that the electrical power output does not
  exhibit rapid fluctuations;

- partial dynamic blade stalling occurs towards the blade
  tips and is likely to be more pronounced with a compliant
  blade design;

- increasing the number of blade segments in the turbine
  model much beyond 10 does not significantly influence the
  turbulence-induced torque fluctuations.

At the conclusion of the sensitivity analysis the results will be presented in two separate reports.   One will be a full report covering the detail of the work carried out and including a definition of the accuracy of wind data (specifically the turbulent characteristics) required for design purposes. The other report will be in the form of a guide to WTG designers, indicating the relationship of WTG loads and power output fluctuations to the major characteristics of the incident wind, and the extent and quality of wind data which is necessary for adequate treatment of rotational dynamics.

Acknowledgments

The work is being undertaken under contract to NMI for ETSU and the Department of Energy. NMI have engaged the help of Wind Engineering Services in an advisory capacity and this contribution has assisted considerably in determining the direction of the work.

Permission from the Directors of ERA Technology Ltd to publish the paper is gratefully acknowledged.

References

1.   Wills, J A B : 'The measurements of wind structure at West Sole'. NMI Report R79, June 1980.

2.   ESDU data sheet 75001 : 'Characteristics of atmospheric turbulence near the ground - Pt III'.   Engineering Sciences Data Unit, July 1975.

3.   Vaahedi, E, and Barnes, R : 'Dynamics behaviour of a 25m variable-geometry vertical-axis WTG'.   IEE Proc., Vol. 129, Pt.C, No.6, November 1982.

4.   Anderson, M B : 'The interaction of turbulence with a horizontal axis wind turbine', Cavendish Laboratory, Cambridge.

5.   Lundsager, P., Frandsen, S and Christensen, C J : 'Analysis of data from the Gedser wind turbine 1977-1979'.   Report No. M2242 Risø National Laboratory, Denmark, August 1980.

6.   Mogridge, L and Farmer, E D : 'Modelling of grid frequency histories for generation studies'.   CEGB Report PL-ST/1/79, November 1979.

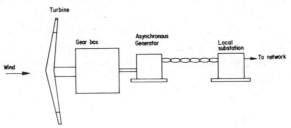

Fig.1: Schematic diagram of model

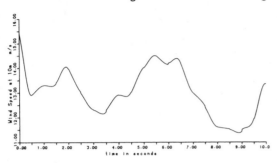

Fig.2: Wind speed

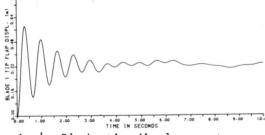

Fig.3: Turbine hub torque

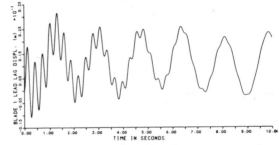

Fig.4: Blade tip displacement
(flap)

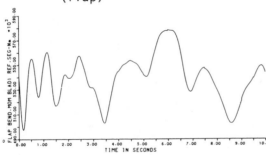

Fig.5: Blade tip displacement
(lead-lag)

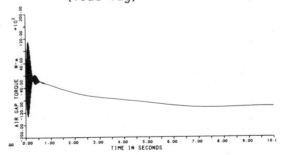

Fig.6: Blade bending moment at
mid-span (flap)

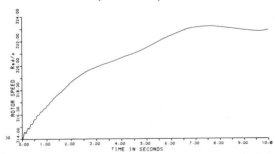

Fig.7: Generator air gap torque

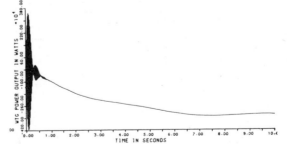

Fig.8: Generator rotational speed

Fig.9: Generator power output

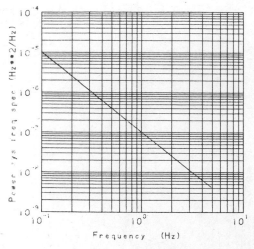

Fig.10:Spectral density of network frequency variations

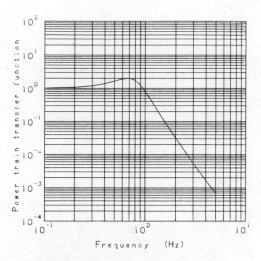

Fig.11: Power-train transfer function

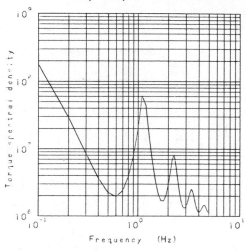

Fig.12: Spectral density of turbine torque variations(15m/s)

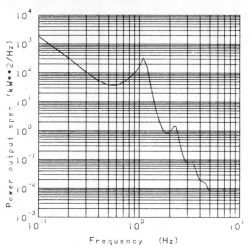

Fig.13: Spectral density of output power variations(15m/s)

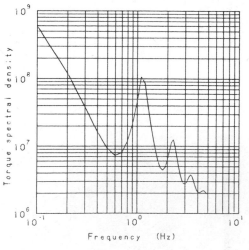

Fig.14: Spectral density of turbine torque variations(20m/s)

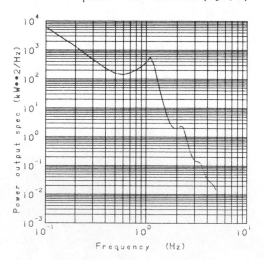

Fig.15: Spectral density of output power variations(20m/s)

THREE YEARS PERFORMANCE RECORD OF A SMALL (10 KW) WECS IN THE EAST OF IRELAND

T. O'Flaherty
Kinsealy Research Centre
Dublin

Abstract

A multi-blade 10 kW WECS has been in operation at Kinsealy Research Centre just north of Dublin since March 1980. It produces electricity which is used to provide root zone warming for a tomato crop in a glasshouse. The rotor diameter was originally 5.7m, but in December 1980 larger blades were fitted, increasing the diameter to 6.4m. A number of breakdowns occurred between then and August 1982, but since the latter date trouble-free operation has been experienced. Details are given of the energy yield of the machine since its installation, while the causes and remedies of the more serious breakdowns are fully discussed. Finally, the various practical difficulties involved in the operation of small-scale WECS are considered in the light of the experience gained.

Introduction

In the late 1970's the firm SJ Windpower ApS, of Frederikshavn, Denmark, achieved a good level of sales on their home market of their 10 kW electricity-generating WECS. In 1979 they agreed to make one of their units available to Kinsealy Research Centre near Dublin for evaluation of its performance on a site at the Research Centre, and as a demonstration unit with a view to hoped-for sales in Ireland. Accordingly the unit was installed at Kinsealy in January 1980, and connected to load on the following March 16.

Subsequently the monitoring of the installation was incorporated within the national wind energy programme which developed in 1980 and following years, and improved instrumentation for data collection has been financed under that programme. The SJ machine is the longest-running monitored wind energy installation in Ireland.

The machine has a 16-blade wind turbine which drives through a 30:1 gearbox - a 4-pole self-excited alternator rated to deliver 10 kW at 380 volts , 3-phase, at a running speed of 1500 rpm. Turbine, gearbox and alternator are mounted on a 14 m lattice steel tower.

Overspeed control is provided by means of the tail-vane which turns the turbine out of the wind under the action of a hydraulic system. This system is actuated in the first instance by an electronic control unit, but in case of failure of this unit is actuated mechanically as rotational speed increases.

The output from the alternator is of variable frequency and voltage, and therefore is suitable only for connection to heating elements. It is used to supply root zone heating for a glasshouse tomato crop grown in a nutrient solution (hydroponic culture). The solution is circulated to the plant rows from a tank in which it can be conveniently heated by immersion elements.

This application has the advantage that all the energy supplied by the windmill can be put to effective use without the need for any special means of energy storage, since the root heating system in the 30m x 9m glasshouse provides a large enough demand to absorb the full output of the 10 kW alternator. When wind energy is insufficient to maintain the required temperature a mains back-up system is activated thermostatically.

Machine Performance: Occurrence of Faults

A vital factor in the economic success of any WECS is that the availability of the installation should be high i.e. that it should not be out of service for significant periods for any cause. Therefore it is important at the outset to consider the major interruptions to continuous operation which have affected the SJ machine since it was first put on load (Table 1).

The first of these breakdowns was of somewhat catastrophic proportions. Its root cause stemmed from a desire on the part of the manufacturers to improve the energy output of the machine, given that the site at Kinsealy is one of only moderate mean windspeeds. As initially supplied, the machine's rotor diameter was 5.7m, and its energy yield was somewhat disappointing, as discussed below. The manufacturers therefore supplied and fitted a set of larger blades, increasing the diameter by 0.7m, equivalent to an increase in swept area of some 26%.

This change was made on December 13, 1980, and the resulting improvement in energy output was indeed very marked. But the corresponding increase in all of the stresses in the turbine framework was of course equally marked, and nothing had been done to strengthen the framework to withstand the increased stresses. It survived for four months, but on April 27, in a strong but not exceptional gale, it failed in spectacular manner. The radial members supporting the blades failed under the torque they were transmitting to the hub, and tried to wind themselves round the hub. The plane of the blades was then folded over the hub like an umbrella, under the wind thrust on the up-wind rotor. The effect of this was that each blade as it rotated made violent impact with the tower, smashing several blades. As quickly as possible the mechanical brake was applied and the frame secured to the tower to prevent rotation, but it was immediately clear that the turbine was a "write-off".

On reporting this event to the manufacturers it was learnt that a similar failure had occurred in Denmark on a machine which had also been fitted with longer blades. A strengthened frame had been designed and this was dispatched, together with new blades. The main change in frame design was the use of webbing pieces between the radial members near the hub to give increased torsional strength to the frame. A new generator and gearbox assembly was also supplied as the original unit was found to be not in correct alignment. Vibration due to this non-alignment was a likely cause of several bolts in the tower having worked themselves loose shortly before the occurrence of the major failure. The new installation, incorporating only the tower and tail-vane of the original unit, was re-commissioned on September 25, 1981.

hydraulic system, which caused the rotor to remain partially out of the wind.

The third phase of operation, from August 10, 1982 until the present, gives some grounds for hope that the installation's worst troubles are in the past and that continuous operation at full output levels may now persist for a considerable time.

Machine Performance : Energy Production

The energy yield of the installation has to be considered separately for the periods before and after the increase in rotor diameter on December 13, 1980 (Table 2). The change in diameter led to a marked increase in mean output when the machine was operational, but this increase was not fully reflected in total units generated because of the relatively low availability of the machine during much of this period. It is noteworthy that there was also a change in output per unit of rotor swept area between the two periods, which must be mainly accounted for by a difference in wind conditions.

The relationship of mean output to prevailing wind conditions is indicated on a monthly basis in Table 3, for the period 10/8/'82 to 1/3/'83. Comparison of mean annual wind-speeds at Kinsealy with those in other parts of Ireland (Table 4) shows clearly that higher outputs could be expected from the SJ machine at other sites, especially near the west or north coasts.

At a mean output of 1.05kW (Table 2) total annual energy production would be 9,200kWh. Valued at 6p/kWh, the current Irish domestic electricity tariff, this yields £552 as return on a capital investment of some £4500 (all values in sterling). A better return might be expected on a more favourable site. On the other hand, the above estimation assumes 100% machine availability, and also assumes that all the energy produced throughout the year can be utilised for an application which would otherwise have to be supplied by mains electricity. Both these assumptions may be difficult to realise in practice.

Wind Energy for Glasshouses

The heat requirement of a glasshouse is very substantial. A typical 1-acre (4047 m^2) glasshouse of traditional design has a peak heat requirement of nearly 1 MW. Even using the best modern techniques of energy conservation the peak demand is unlikely to be lower than 500 kW, with a total energy consumption over the growing season of some 5000 GJ. To obtain a contribution of 25% of this from wind energy would require a 200 kW windmill working at a load factor of 20%. A machine of such a size does not seem realistic for this application.

A recent development in glasshouse crop production which offers the possibility of a significant overall energy saving is the technique of heating the root zone of plants, while maintaining minimum night air temperature at a considerably lower level than would be necessary in the absence of root zone warming (RZW). The heating load for RZW is approximately 10% of that for the heating of the whole glasshouse, and is also a less fluctuating load. Therefore a WECS with a rating in the range 20 – 50 kW might supply a useful proportion of this requirement over the heating season.

At Kinsealy the 10 kW WECS supplies a RZW system in a 270m^2 glasshouse. In 1983, for the period February 8 - 28, 39.5% of the heat requirement of the RZW system was supplied from wind energy. While this is encouraging, clearly it is not until a full season's results are available that a general conclusion can be reached as to the contribution that the WECS can make to RZW heat requirement.

Conclusion

In spite of - or perhaps to some extent because of - the various difficulties which have been encountered, a great deal of valuable practical experience has been gained in the three years since the SJ WECS was installed at Kinsealy. Perhaps more than anything else it has demonstrated how seriously the economic attractiveness of a WECS can be diminished by loss of energy production due to breakdowns.

When a breakdown occurs the length of time needed to restore an installation to service, and the cost involved, can be increased by a number of factors:
* in the worst cases damage can be substantial;
* replacement parts may not be locally or speedily available, particularly at the present stage of development of the technology;
* difficulty of access may hinder the diagnosis and/or repair of a fault, particularly during a period of unfavourable weather conditions;
* personnel with the necessary skills and the ability to work at considerable heights may not be readily available.

With accumulating experience of wind energy installations the ability to cope with problems such as these is bound to improve, leading to a much improved pattern of availability of equipment. The key benefit from work on installations such as that at Kinsealy is that it permits progress on the learning curve which is associated with any new technology, so that users of small-scale WECS in Ireland will more quickly be able to derive real economic benefit from our undoubtedly large wind energy resource.

Acknowledgements

Thanks are due to SJ Windpower ApS for making available their wind energy conversion system, and to the Irish Department of Industry and Energy, and Electricity Supply Board, who supported the purchase of instrumentation. The expert technical assistance of Mr. P. McCormack and Mr. J. Grant is also gratefully acknowledged.

Table 1:   Major breakdowns

Event	Period of outage	
	from	to
Rotor frame failure	27/4/81	25/9/81
Loosened hub bolts	23/12/81	12/2/82
Electronic controller fault	12/3/82	10/8/82

The next serious fault arose on December 23, 1981, when it was noticed that the main flange bolts holding the turbine hub to the gearbox/generator assembly had loosened to an alarming extent. While attention was partly drawn to this by a slightly altered running noise from the machine, it was largely by good fortune that it was discovered before the bolts worked completely loose, with what would have been disastrous results – the detachment of the whole turbine from the top of the tower. Under extremely difficult working conditions the bolts were re-tightened as much as possible, and the machine was left partially braked to minimise the danger of them working loose again until bolts of a better type could be fitted. Due to continuing bad weather it was not until February 12 that new bolts with self-locking nuts could be fitted, and the machine allowed to run at full output again.

Unfortunately the machine operated normally for only a month from this date before another serious fault occurred. In this case no danger of damage to the installation was involved, but delivery of energy was interrupted. On March 12, the output voltage from the generator dropped suddenly to only a fraction of its normal level. At full rated speed an output of only about 30 volts was being generated instead of the normal 380 volts, and power output was correspondingly diminished.

Fault-finding guidelines and advice from Danish sources indicated the probable cause of this behaviour as failure of rectifier diodes in the alternator field circuit. However, replacement of the rectifier assembly on the alternator did not correct the fault. Further tests were carried out in search of a presumed fault in the alternator, before it was discovered that disconnection of the electronic controller eliminated the problem, the cause of the trouble being evidently a fault in the controller.

The installation has operated without the electronic controller since August 10 last, speed control of the machine relying on simple mechanical actuation of the hydraulic controller. A pump driven from the rotor shaft increases hydraulic pressure sufficiently to turn the rotor out of the wind as rotor speed increases.

Experience to date indicates that this method of control, while less sensitive, is quite satisfactory. Indeed it seems likely that energy production is marginally improved, because the electronic controller may have turned the rotor out of the wind more than was necessary. Recent information from Denmark has confirmed that installations there have been found to work well without the electronic controller in use. Therefore it has been decided to continue running the machine in this way.

In summary, it is clear that the operating record of the machine since its first installation has been a very interrupted one. Three distinct phases of its history can be identified. In the first, up to April 1981, quite a high level of availability was achieved. Output was lost occasionally due to tripping out, and some data was lost in early months through instrumentation teething troubles. The most significant problem was a fault in the connector through which power is taken off from the alternator. Due to difficulty of access this caused loss of output from March 24 to April 7, 1981.

The second phase, from April 1981 to August 1982, presents a particularly gloomy picture. It covers the period of all the major outages. During it the machine operated at full efficiency for only some 103 days in all, and the longest spell of trouble-free operation was only 41 days. On top of the major faults, output was reduced for about 5 days during October and November due to an incorrectly-installed valve in the

Table 2: Energy production

	16/3/80-13/12/80	13/12/80-1/3/83
Total energy produced (kWh)	2218	11212
Machine availability (%)	89	52
Mean output when fully operational (kW)	0.38	1.05
Mean output per unit of rotor swept area (W/m^2)	15.2	32.8

Table 3: Monthly mean output and windspeed

Month	Mean output (kW)	Mean windspeed (m/s)
1982		
August (10-31)	0.91	4.7
September	0.59	3.5
October	0.99	4.1
November	1.40	5.2
December	1.35	5.4
1983		
January	2.32	6.6
February	1.29	4.9

Table 4: Mean annual windspeeds at Irish meteorological stations

Station	Mean windspeed (m/s)
Malin Head (N. coast)	8.1
Belmullet (W. coast)	6.8
Rosslare (S.E. coast)	6.0
Valentia (S.W. coast)	5.6
Kinsealy (E. coast)*	4.3
Kilkenny (inland)	3.3

* Value for Kinsealy based on 2-metre readings corrected (factor 1.31) for 10 metres; other stations 10-metre readings.

DYNAMICS AND CONTROL OF AUTONOMOUS WIND-DIESEL GENERATING SYSTEMS

A. Tsitsovits and L. L. Freris
Department of Electrical Engineering
Imperial College, London

Abstract

The investigation of the dynamic behaviour and stability of WECS has so far been restricted to one or several units in parallel connected to an infinite bus [1,2,3]. If the WECS penetration is large, as would be the case in a small diesel supplied network, the frequency and voltage of the system depend heavily on wind fluctuations. As a consequence control policies have to be developed for the system as a whole taking into account the dynamic properties of the diesel set(s) and the network configuration.

To assist the development of control policies in autonomous wind-diesel systems a detailed simulation program has been developed at Imperial College.

In this paper the stability of the system is investigated through an eigenvalue analysis and the results examined in the light of the participation matrix of the system's state variables.

Finally, a simplified linear model is developed and general conclusions on the control system are drawn.

Description of the system

As shown in Fig.1 the system consists of:

- (a) Diesel machine driving a
- (b) Synchronous generator
- (c) Dump resistive load connected to the bus by a power transformer and controlled by thyristor rectifier
- (d) Variable pitch wind turbine driving through a
- (e) Gear box an
- (f) Induction generator
- (g) Static two stage switchable capacitor connected to the induction generator
- (h) Transmission line between WECS and bus
- (i) Electromagnetic clutch between diesel machine and synchronous generator

Eigenvalue analysis

To perform the eigenvalue analysis the system was linearized at characteristic steady state operating points and modes of operation given in Table 1. Three wind speeds chosen were (a) near the cut-in speed, (b) nominal and (c) near the furling speed. Following the eigenvalue analysis, the participation matrix for the state variables [4] was calculated to enable an assessment of the contribution to stability by each state variable. Some numerical results are given in Table 2.

The system is extremely stiff with, ratio of the largest to the smallest non-zero eigenvalue being of the order of 10^4. The general structure of the eigenvalue distribution depends on the mode of operation but does not change significantly for different operating points. The zero eigenvalues are attributed by the participation analysis to shaft angles. This is to be expected as the shaft angles are pure integrals of the angular velocities and their transfer function is a pole on the origin.

There appear in all cases complex eigenvalue pairs with high imaginary parts and relatively weak damping attributed to the torsional characteristics of the shafts. However, for high torsional frequencies, a large inherent natural dumping due to the shaft material is expected. Very weakly damped are the torsional oscillations of the gear-box which is characterised by the highest eigenfrequency. This suggests possible fatigue problems and requires measures to enhance the damping (e.g. flexible gear-box mounting, etc.), or gear-box over dimensioning.

The eigenvalues attributed to the excitation system are all real and negative in a range of -0.12 to -100 which suggests satisfactory voltage stability in all cases.

The comparison of the eigenvalues for the cases with and without the diesel machine indicates that, as expected, the system with the two generators operating in parallel shows better inherent stability properties due to the stabilizing effect of the diesel machine speed governor.

In the mode in which the WECS is supplying all the load and the synchronous generator operates as synchronous capacitor, we observe a complex pair $(-0.07\pm j\ -0.05)$ attributed mainly to the induction generator voltage behind the transient reactance. As the eigenfrequency of this mode is low, it could be excited by low frequency wind turbulence. Therefore the enhancement of the damping of this mode should be considered a control policy objective.

It should be noted that the eigenvalue analysis performed above, only predicts the onset of small signal oscillations. As our system is non-linear, the results of this analysis predict only possible sources of instability. Confirmation of these results can only be obtained through fully fledged simulation studies.

The simplified linearized system

The linear models of Figs. 2 and 3 are derived on the basis of the following assumptions:

- (a) Infinite shaft rigidities
- (b) Negligible friction
- (c) Constant bus voltage maintained by the synchronous generator excitation system
- (d) Linear slip-power characteristic for induction generator
- (e) Diesel governor without delay

1. System with WECS alone

Letting $K_{C_p} = \dfrac{\partial C_p}{\partial \omega T}$ and after some algebra we obtain the following equations for the Laplace transforms of the angular speed

$$\Delta\omega_S = \frac{K_{IG}}{H_S H_T} \cdot \frac{\Delta P_{WO} - (1 - \frac{K_{cp}}{K_{IG}} + \frac{H_T}{K_{IG}} S)\Delta P_L}{S^2 + \left|(\frac{1}{H_T} + \frac{1}{H_S})K_{IG} - \frac{K_{cp}}{H_T}\right|S - \frac{K_{IG}K_{cp}}{H_T H_S}} \tag{1}$$

$$\Delta\omega_T = \frac{K_{IG}}{H_S H_T} \cdot \frac{(1 + \frac{H_S}{K_{IG}} S)\Delta P_{WO} - \Delta P_L}{S^2 + \left|(\frac{1}{H_T} + \frac{1}{H_S})K_{IG} - \frac{K_{cp}}{H_T}\right|S - \frac{K_{IG}K_{cp}}{H_T H_S}} \tag{2}$$

$$\Delta P_{IG} = \frac{K_{IG}}{H_S H_T} \cdot \frac{H_S\, S\Delta P_{WO} + (-K_{cp} + H_T S)\Delta P_L}{S^2 + \left|(\frac{1}{H_T} + \frac{1}{H_S})K_{IG} - K_{cp}/H_T\right|S - \frac{K_{IG}K_{cp}}{H_T H_S}} \tag{3}$$

Where H_T, H_S are the turbine and synchronous generator inertial time constants respectively and $K_{IG} = 1/\text{slip}_{nom}$ (p.u.)

We can observe that the system behaviour depends on the slope of the C_p, λ characteristic K_{cp}

(a)   if $K_{cp} > 0$ (i.e. we are operating on the ascending part of the C_p, λ curve) there is a positive real system eigenvalue which indicates instability.

(b)   if $K_{cp} = 0$ (i.e. we are operating on the flat top part of the C_p, λ curve) there is a zero eigenvalue and the system has an integral behaviour.

(c)   if $K_{cp} < 0$ (i.e. we are operating in the descending part of the C_p, λ curve) the eigenvalues are both real and negative and the system is stable.

The negative real zeros in Eqns.(1) and (2) indicate a derivative behaviour in the response of $\Delta\omega_S/\Delta P_{WO}$ respectively, with the behaviour of ΔW_S more accentuated as the zero is closer to the origin.

2. System with parallel operation of WECS and Diesel generator

Noting $K_D = \dfrac{P_{Drated}}{Droop}$ (p.u./p.u.), we obtain the following system transfer equations

$$\Delta\omega_S = \frac{K_{IG}}{H_T H_D} \cdot \frac{\Delta P_{WO} - (1 - \frac{K_{cp}}{K_{IG}} + \frac{H_T}{K_{IG}} S)\Delta P_L}{S^2 + \left|(\frac{1}{H_T} + \frac{1}{H_D})K_{IG} - \frac{K_{cp}}{H_T} + \frac{K_p}{H_D}\right|S + \frac{K_{IG}K_D - K_{cp}(K_{IG}+K_D)}{H_T H_D}} \tag{4}$$

$$\Delta\omega_T = \frac{K_{IG}}{H_T H_D} \frac{(1 + \frac{K_D}{K_{IG}} + \frac{H_D}{K_{IG}} s)\ \Delta P_{WO} - \Delta P_L}{s^2 + |(\frac{1}{H_T} + \frac{1}{H_D})\ K_{IG} - \frac{K_{cp}}{H_T} + \frac{K_D}{H_D}|s + \frac{K_{IG}K_D - K_{cp}(K_{IG}+K_D)}{H_T H_D}}$$

(5)

$$\Delta P_{IG} = \frac{K_{IG}}{H_T H_D} \frac{(K_D + H_D s)\ \Delta P_{WO} + (-K_{cp} + H_T S)\ \Delta P_L}{D(s)}$$

(6)

Similar observation as previously apply but with an important distinction:

The system is stable for $K_{cp} > 0$ but only if $K_{cp} < \frac{K_{IG}K_D}{K_{IG}+K_D}$

as a numerical example with $K_{IG} \tilde{=} 33$ (nominal slip 0.03) and $K_D = 25$ (Droop 0.04) the margin of stability is

$$K_{cp} = 14.2$$

Considering that $K_{cp} = \frac{R_T}{W} \cdot \frac{\partial C_p}{\partial \lambda}$

for a wind machine with rotor radius $R_T = 19$ m and a cut-in speed of $W = 4.5$ m/s the stability margin for the slope of the C_p, λ curve is

$$\frac{\partial C_p}{\partial \lambda} = 3.36$$

This is obviously for larger than any realistic value. The maximum K_{cp} expected for the model machine is of the order of 0.5. For both modes of operation we observe that the dynamic behaviour of the frequency (W_S) is influenced more rapidly by load changes and the turbine speed by wind power changes.

The induction generator power shows an attenuated derivative behaviour for small t when the diesel machine is decoupled and a similar but attenuated behaviour when the diesel machine is coupled.

Control policies considerations

As the mode of operation of the WECS alone shows far worse stability properties than the mode with diesel parallel operation, we can restrict the discussion to the first mode and expect the control policies to be applicable to both cases. As previously outlined the frequency is rapidly influenced by load changes and the turbine speed by wind power changes.

A simple but proven effective control policy is therefore to construct the frequency feed-back loop through the rectifier firing angle controller (strong influence on the load) and the turbine speed/power feedback loop through the blade pitch controller (strong influence on the converted wind power).

As the rate of change of the blade pitch must be limited because of aerodynamic considerations and the rate of change of the thryistor firing angle
is influencing the voltage (change in reactive power absorbed by the bridge)
only PI controllers are considered.

The controller parameters should be ideally adapted to the operating point
because of the inherent nonlinearities in the relations between control
and controlled variables. Fixed controller parameters have nevertheless
shown acceptable results in dynamic studies. Ideally, the set-points for
the controllers have to be adjustable according to an operating table of
wind speed and load mean values stored in the control memory and calibrated
according to recent measurements.

References

1. Pantalone, D. K., and Fouad, A. A., "Modes of Oscillations of Wind
    Generators in Large Power Systems", IEEE Paper A 78579-5, July 1978.

2. Gilbert, L. J., "Response to Three Phase Faults on a Wind Turbine
    Generator", Ph.D. Thesis, Toledo University, January 1978,

3. Javid, S. H., Murdoch, A. and Winkelman, J. R., "Control Design for a
    Wind-generator using Output Feedback", 20th IEEE Control and Decision
    Conference, San Diego, California, December 1981.

4. Perez-Arriaga, I. J., et al, "Determination of Relevant State Variables
    for Selective Modal Analysis," IEEE PAS-101, N.9, September 1982.

Table 1

Operating points for eigenvalue analysis

Wind speed (m/s)	5.0	8.0	16.0
Load (kW)	40	80	
Load power factor	0.9lag	1.0	0.9lead

Table 2

WECS supplies alone all real load
Diesel generator operated as synchronous capacitor

Parameter Values

Real Load = .644 PU
Reactive Load = -.322 PU
Voltage = 1.003 PU
Frequency = 1.0000 PU
Diesel Real Power= -.030 PU
Diesel Reactive Power = .020 PU
Excitation Voltage = 1.023 PU
Power Angle = -1. Degrees
Real Dump Load = .240 PU
Reactive Dump Load = .204 PU
SCR Firing Angle = 40. Degrees
Induction Generator Real Power = .914 P.U.
Induction Generator Reactive Power = -.561 P.U.
Capacitor Switch Factor = 1.5
Static Capacitive Power = .422 PU
Wind Speed = 16.0 M/S
Wind Turbine Power = 1.058 PU
Blade Pitch Angle = 14. Degrees
Number of Iterations for Last Case = 79

MATRIX EIGENVALUES ARE:

	REAL PART	IMAG. PART	MAGNITUDE
1.	0.00000	0.00000	0.00000
2.	-.00000	0.00000	.00000
3.	-.06836	.05193	.08585
4.	-.06836	-.05193	.08585
5.	-.12262	0.00000	.12262
6.	-.46376	0.00000	.46376
7.	-.73766	0.00000	.73766
8.	-7.03033	0.00000	7.03033
9.	-44.61628	0.00000	44.61628
10.	-9.77494	48.21538	49.19627
11.	-9.77494	-48.21538	49.19627
12.	-62.03045	0.00000	62.03045
13.	-100.00000	0.00000	100.00000
14.	-.94703	707.02981	707.03044
15.	-.94703	-707.02981	707.03044

MAGNITUDE OF PARTICIPATION MATRIX

	1	2	3	5	6	7	8	9	10	12	13	14
1	0.00	0.00	.02	0.00	.14	1.07	0.00	0.00	.04	0.00	0.00	0.00
2	0.00	.28	.05	0.00	.05	.63	0.00	0.00	.04	0.00	0.00	0.00
3	0.00	0.00	0.00	0.00	0.00	0.00	0.00	0.00	.01	0.00	0.00	.48
4	0.00	0.00	0.00	0.00	0.00	0.00	0.00	0.00	.01	0.00	0.00	.48
5	0.00	0.00	0.00	0.00	0.00	.05	0.00	0.00	.40	.20	0.00	.02
6	0.00	.71	.05	0.00	.05	.63	0.00	0.00	.63	.16	0.00	.02
7	0.00	0.00	.61	0.00	.04	.01	0.00	0.00	0.00	0.00	0.00	0.00
8	0.00	0.00	.63	0.00	0.00	.02	0.00	0.00	0.00	0.00	0.00	0.00
9	0.00	0.00	0.00	0.00	0.00	.03	0.00	0.00	.13	.94	0.00	0.00
10	1.00	0.00	0.00	0.00	0.00	0.00	0.00	0.00	0.00	0.00	0.00	0.00
11	0.00	0.00	.02	0.00	1.11	.14	0.00	0.00	0.00	0.00	0.00	0.00
12	0.00	0.00	0.00	0.00	0.00	0.00	0.00	0.00	0.00	0.00	1.00	0.00
13	0.00	0.00	0.00	0.00	0.00	0.00	.15	1.14	0.00	0.00	0.00	0.00
14	0.00	0.00	0.00	.89	0.00	0.00	.11	0.00	0.00	0.00	0.00	0.00
15	0.00	0.00	0.00	.10	0.00	0.00	1.04	.14	0.00	0.00	0.00	0.00

State variables index in the participation matrix

1. Turbine speed
2. Turbine shaft angle
3. Gear box speed
4. Gear box angle
5. Induction generator speed
6. Induction generator shaft angle
7. Induction generator transient voltage (real part)
8. Induction generator transient voltage (imag. part)
9. Synchronous speed
10. Synchronous generator shaft angle
11. Transient excitation voltage
12. Voltage regulator output
13. Excitation amplifier output
14. Field voltage
15. Exciter stabilizer output
16. Diesel machine speed
17. Diesel machine shaft angle
18. Diesel machine power
19. Diesel machine speed governor output

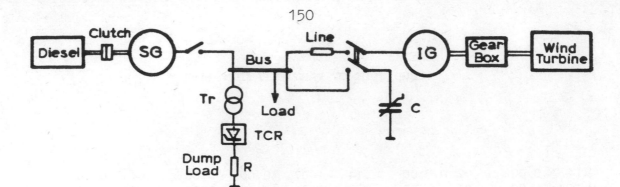

Fig.1 System lay-out

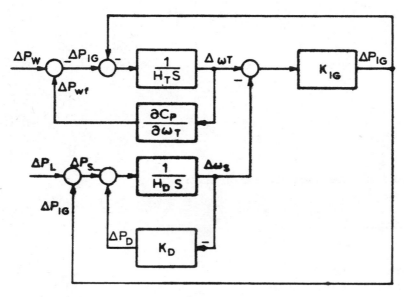

Fig. 2

Simplified model for parallel WECS/Diesel operation

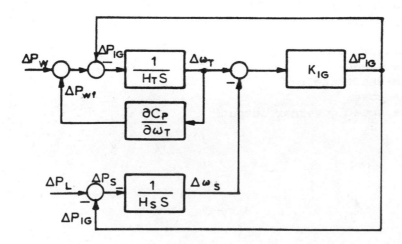

Fig. 3

Simplified model for WECS alone operation

SMALL SCALE WIND/DIESEL SYSTEMS FOR ELECTRICITY GENERATION IN ISOLATED COMMUNITIES

D G Infield
Energy Research Support Unit
Rutherford Appleton Laboratory, Chilton, Didcot, UK

Abstract

A large potential demand exists for electricity in remote areas isolated from grid supply. The high cost of diesel generation, especially in the developing countries, has prompted an examination of integrated wind/diesel systems, with the wind turbine viewed primarily as a fuel saver.

Integrated systems encounter severe matching problems due to the inherent variability of the wind turbine output and the community load. In this context the role of energy storage is emphasised both to deal with unacceptable operating conditions for the diesel back up, and to improve wind energy utilisation. A simple methodology for assessing the economics of competing energy storage systems is presented, together with some preliminary conclusions.

Lastly it is demonstrated that a multiple diesel back up, with carefully chosen control strategy, can obviate the need for storage, but at the expense of some increased fuel consumption.

Introduction

Wind energy, in common with most renewable energy forms, is intermittent and thus cannot alone be relied upon to provide a continuous electricity supply. Given that an uninterrupted supply is desirable in many applications, some means must be adopted to overcome this limitation. In addition the electricity requirement of an isolated community will in general be continually changing and it is clearly desirable to match the supply and the demand. Otherwise either the demand requirement will not be met or surplus energy will be wasted. (For a general discussion of these problems see Ref 1).

The solution to these problems for an isolated system is the introduction of energy storage. Most commonly it takes the readily available and high density form of fossil fuel, ie the diesel oil used to fuel the back up generating set. The first and most pressing of the two problems is dealt with in this way (since surplus wind power cannot be converted to diesel fuel the second but not crucial problem remains). The attraction of a diesel back up immediately suggests the retrofitting of wind turbines to the many cases of grid isolated diesel generation to be found in rural communities.

Hence the interest in wind/diesel systems shown by a number of research teams including those at Reading/RAL, ECN (Holland), DAF Indal (Canada) and Imperial College. Research, however, has indicated a number of obstacles preventing the immediate realisation of a successful wind/diesel package. These can be summarised as follows:

(i) The back up diesel generator set is forced by the variability of the wind turbine output and the consumer load, to undergo an unacceptably high number of stop/start cycles (Ref 2).

(ii)   The diesel is often forced to run under low load conditions which is expected to lead to increased wear and maintenance. Particular problems are carbonisation and bore-glazing (Ref 3).

(iii) Low wind energy utilisation (for the reasons mentioned above).

(iv)   It is difficult to identify the optimum relative size of system components (eg the wind turbine size in relation to the installed diesel capacity), particularly when the turbine is being integrated into an existing electricity system.

(v)   Short term (less than a few seconds) variation of the wind turbine output, in response to wind turbulence, can cause unacceptable variation in the voltage and frequency of the supply (Ref 4).

This paper will principally address the first three problems. Two approaches to these problems have been developed by the Reading/RAL team, the first involving energy storage, and the second a multiple diesel back up.

The Application of Energy Storage

A possible way to deal with the adverse operational characteristics covered by (i), (ii) and (iii) is the introduction of additional storage, over and above that associated with the diesel fuel. The storage medium should preferably be convertible both **to** and **from** electricity. Although situations can be envisaged where a role exists for one-way storage media such as thermal storage. The diesel operating/control problems will then be transferred to the storage system (storage plus power flow control) and such a change will only be desirable if the particular storage device is capable of tolerating frequent and rapid changes in the direction and magnitude of energy flows without undue wear and maintenance. Of course such a storage system must also be available at a reasonable price. Bearing in mind these specifications a number of storage forms have been considered; their properties are summarised simply in Table 1.

Table 1

Sumary of energy storage media for wind/diesel systems

	Energy Density kWh/litre	Characteristic Time T - hours	Costs £/kWh
Sensible heat storage	10^{-1}	1 to 15+	9
Phase change storage	3×10^{-1}	1 to 15+	20
Flywheel storage systems	10^{-1}	0.1+	90
Lead/acid batteries	5×10^{-2}	5+	100
Nickel/cadmium batteries	10^{-1}	3+	300
Silver/zinc batteries	10^{-1}	0.1+	2000
Hydrogen (gas at 200 Bar)	1	–	55
Hydrogen hydride storage	5	?	150+
Hydraulic/pneumatic systems	10^{-1}	0.01+	5100
Diesel fuel	10	–	0.1

NB: The characteristic time T referred to in the above table is defined as:

$$\frac{\text{Storage capacity (kWh)}}{\text{Rated output of store (kW)}}$$

Unfortunately storage characteristics are often unsuited to the wind/diesel
system requirements which demand high charge/discharge rates as well as a
specified storage duration. For example it has been established that the
high number of start/stop cycles can be reduced to an acceptable level with
the introduction of only a few minutes of storage. Batteries are the most
easily available form of electricity storage but the five hour discharge
time of a lead acid cell is far too long to smooth the short term
variability, and high cost precludes their use for simply increasing the
level of wind energy utilisation. Moving to high rate batteries such as
nickel cadmium does not provide a viable solution due to the higher costs
of these batteries. Moreover recent work suggests that nickel cadmium
cells exhibit lower efficiencies (Ref 5). Silver/zinc batteries are
capable of even higher rates of discharge (as low as 6 minutes) but costs
of approximately £2000/kWh coupled to a much reduced cycle life at high
discharge rates, eg lifetimes of 15 - 20 cycles at 1 hour rate and 5 - 10
cycles at the 10 minute rate, preclude their application to wind/diesel
systems. (Data from Yardney Tungstone Special Batteries Limited).

Energy Storage - A Methodology for Objective Assessment

In order to objectively assess competing storage technologies the following
simple methodology has been developed following Ref 6. Use is made of the
reasonable assumption that a storage system/device can be adequately
described by two characteristics: storage capacity (kWh), and
charge/discharge rate (kW).

To facilitate clarity, the simple pay-back period is used rather than a
discounted cash flow analysis. For this initial study it is further
assumed that wind energy used to charge the store is free from cost, ie the
energy would otherwise be dumped. In this instance the efficiency of the
store does not appear in the analysis.

The economic value of the store is determined in this assessment by the
diesel fuel saving which it facilitates. No attempt is made at this time
to include the other important contributions of the store, namely security
of supply and an improved operating regime for the diesel back-up. The
following notation is used:

r = rated output of store (kW)
c = storage capacity (kWh)
T = c/r is time to discharge store at rated output (h)
d = fraction of time that store is available
$\bar{p}$ = average discharge rate of store as a fraction of rated output
t = time to discharge store at rated output as a fraction of the storage
   cycle time
C_d = fuel cost of electricity generation displaced (£/kWh)
λ = $d.\bar{p}.t$ is a measure of the energy contributed by the store as a
   fraction of the energy that would be supplied if the store was
   discharging continuously at rated output

Let the total cost of the storage system be given by:

$$C_t = cC_s + rC_p$$

where C_s is cost per unit storage capacity (£/kWh), and C_p is cost per unit
of rated output (£/kW).

The energy displaced (saved) is, $r\lambda \cdot (365 \times 24)$ kWh/annum, and the annual
saving is thus,

$$C_d \; r\lambda \; (265 \times 24) \; \text{£/annum}$$

The simple pay-back is given by:

$$SPP \;=\; \frac{cC_s + rC_p}{rC_d\lambda} \; \frac{1}{(365 \times 24)} \;=\; \frac{(TC_s + C_p)}{8760 \; C_d\lambda} \; \text{years}$$

When λ and C_d, C_s and C_p have been fixed the pay-back period is seen to be a linear function of T, the time to discharge the store at rated output. In practice considerable effort will be required to evaluate λ for different storage applications and different degrees of wind turbine over-size. For the time being, values of λ have been assumed in order to allow a preliminary comparison of different storage technologies.

Efficiencies found for diesel sets vary considerably with size, condition and operating regime. Large diesel sets (MWs) used for example on the Shetlands can be up to 37% efficient on full load, on the other hand small sets (< 5 kVA) operating in isolated rural locations can have average efficiencies as low as 10% (Ref 7). For the purposes of this analysis an average efficiency of 25% has been assumed as representative of a reasonably maintained set of about 10kVA. Taking diesel fuel costs at £1 per gallon we calculate the cost of displaced electricity as:-

$$C_d = \frac{\text{diesel fuel cost} \times (\text{calorific value})^{-1}}{\text{diesel efficiency}}$$

$$= \frac{\text{£1/gal} \times 0.021 \; \text{gak/kWh}}{0.25} = \text{£0.084/kWh}$$

More difficult to estimate are the values of the parameters which contribute to λ. In the absence of any research aimed at calculating these values the following set of assumptions has been adopted.

$$d = 0.9, \; \overline{p} = 0.5 \text{ and } t = 0.44$$

which gives a value of $\lambda = 0.2$. To be exact λ is not in general constant, but a weak function of T.

Values for C_s and C_p have been taken from the references listed separately in Appendix 1 (corrected where necessary in order that all figures refer to 1982 costs). Their values are shown in Table 2 overleaf.

Fig 1 displays the results of this analysis. The dashed sections indicate that physical limitations of the storage device prevent it from being operated at such a high charge/discharge level. This, of course, does not preclude a low rate system being used to provide short term storage but the consequence, as with a lead acid battery providing 5 minutes storage, is that storage capacity will be wasted. In the case of the battery, less than 10% of its available capacity will be used and this would reflect poorly on the economics of such an application.

The figure indicates hydraulic storage (see Ref 8 for details) to be the most attractive for low values of T (< 6 minutes); and, at the other end of the scale, thermal storage provides the only obvious form of long term storage. (Phase change systems have no advantage over sensible heat storage when there is no premium associated with space saving, as in rural communities. Since they are far more expensive and, moreover, not tried and tested in the field, they have not been included in the analysis).

Table 2

	C_s(£/kWh)	C_p(£/kW)
Sensible heat storage	9	50
Flywheel storage systems	90	200
Lead acid batteries	100	210
Nickle cadmium batteries	300	210
Silver zinc batteries	2000	210
Hydrogen electrolysis and storage		
(i) presently available	55	2000
(ii) projected costs	20	250
Hydraulic/pneumatic storage system	5100	60

If the optimistic projections for hydrogen electrolysis and storage are to be believed then a role should exist for such technology in the future. For the time being it is out of the question.

Flywheels also may have some role to play especially if the advantage of compatible power transfer (ie rotating torque in the wind turbine, the diesel generator and the flywheel storage) can be exploited. Coupling of the flywheel (especially if it is of the 'advanced' type that rotates at up to 30,000 rpm) to the rest of the system is far from trivial and much work is needed on this topic.

To summarize we must wait for results of experiments on the hydraulic system by G Slack of the RAL/Reading team before judging this technology. And we are left with thermal storage as the most attractive approach to longer term energy storage. An interesting related approach is load control. Here the community demand is controlled to match the supply available at any particular time (see Ref 9, for further details). In practice such an approach makes use, to a great extent, of the intrinsic energy storage associated with the building fabric, hot water tanks, fridges, freezers, etc. The wind installations on both Fair Isle and Lundy have adopted this form of system control.

Twin Diesel Back Up as an Alternative to Short Term Energy Storage

The role of short term energy storage (eg hydraulic accumulator) in wind/diesel systems is primarily to ease operational problems, in particular to reduce the potentially high number of diesel starts. It has been suggested that the operational difficulties could also be dealt with by using a twin, rather than single, diesel back up. As such the twin diesel system would provide an alternative to short term energy storage. In order to investigate this possibility a suitable computer simulation has been developed.

The model was based upon G Slack's wind/diesel simulation programme (for details see Ref 10). A number of operational modes are now incorporated.

Three options for mode of electricity supply are available:

(A1) Diesel only operation (no wind contribution).
(A2) Either wind or diesel operation (wind and diesel cannot simultaneously supply the load).
(A3) Simultaneous wind and diesel generation possible (ie synchronised operation).

When the system includes two diesels, three control options are available.

(B1) Fixed sequence diesel operation in which the second diesel only starts if the first diesel is already running.
(B2) Free option diesel operation in which the diesel (or diesel combination) satisfies the load with minimum spare capacity.
(B3) Continuous operation (between maximum and minimum loading) of the first diesel.

When both diesels are operating three options are available:

(C1) First diesel fully loaded
(C2) Second diesel fully loaded
(C3) Power sharing in proportion to diesel ratings

Additional options relating to diesel operation are:

(D) A minimum permissible loading can be specified for the diesels. (Should the load drop below the specified level the diesel will continue to operate at the minimum set loading, with surplus energy being dumped.)
(E) A minimum diesel run time can be specified, ie once started the diesel will run for a minimum of X minutes.

The model was run for a period of one week in which the mean wind speed was 6.95 m/s. A repeated daily electrical load profile with peak to mean ratio of 2.5 and average value of 1.0 kW was used (for exact specification see Ref 10). For all runs it was assumed that the total installed diesel capacity was equal to the peak load of 2.5 kW. These sizes are intended to be viewed as scaled down versions of larger systems - all results being available through appropriate scaling.

For all runs the following fixed wind turbine characteristics have been used:

$$V_R/V_M = 1.35 \text{ (ie } V_R = 9.38 \text{ m/s)}$$

$$\text{Swept Area} = 15 \text{ m}^2 \text{ (ie } 15 \text{ m}^2/\text{kW average load)}$$

The profile used, and calculation of transmission and generation losses can be found in Ref 10.

Discussion of results

Fig 2 shows the application of twin diesels operating in the free option mode (B2) to meeting the load in the absence of a wind contribution (the other relevant option is C1). Only the rating of the first diesel is indicated, the rating of the second being dictated by the 2.5 kW total capacity. The case of a single diesel is represented by taking the first diesel rating as zero. The number of starts in this instance is one, and follows from the fact that the load profile never drops to zero. It can be seen that significant fuel savings (up to 28% for the optimal combination) can be made by using two diesels in the free option mode rather than one. The optimum occurs with one diesel of 0.75 kW and the other 1.75 kW. At 21 and 15 the respective numbers of starts during the week are quite acceptable. The greater fuel efficiency of the twin diesel system derives from the higher average load factors experienced by the diesels. These in turn should ease maintenance and prolong the useful lifetimes of the diesel sets concerned. Putting to one side for a moment consideration of wind energy, it appears that the twin diesel combination provides an attractive alternative to the single diesel.

Changing to the fixed sequence mode of diesel operation (option B1) still provides a fuel saving as shown in Fig 3. In this instance the optimum choice has both diesels sized at 1.25 kW. The fuel saving of 19% is less than with the free option mode of operation. This follows from a less precise matching of total diesel capacity to instantaneous load. It can be seen that the first diesel runs continuously (one start) for all size combinations, again as a result of the non-vanishing load.

The introduction of wind energy in parallel with the diesel (option A3) provides the greatest utilisation of wind energy (neglecting storage), but presents the most difficult operational problems for the diesel back up. This is because the resultant load to be met fluctuates with variation in the wind. The free option mode of diesel back up encounters problems as shown by the very high numbers of starts indicated in Fig 4 (curve e), over 330 per week for the first diesel when sized for maximum fuel saving. These starts arise, to a great extent, because the diesels are alternating in an attempt to match the load. The fuel savings are however significant. A 32% saving is provided by the optimal combination (0.75 and 1.75 kW) when compared with the single diesel back up for this situation. Comparison with the diesel only supply (ie no wind) indicates savings of 65% and 51% in relation to the single diesel and optimal twin diesel combination, respectively.

Adopting the wind or diesel strategy (option A2) eases the starts problem with some inevitable increase in fuel consumption. As shown in Fig 4 (curve a) the maximum fuel saving is reduced to 24% of the corresponding single diesel value. This is 20% greater fuel consumption than the equivalent wind plus diesel system but a 41% saving when compared to the optimal diesel only combination. For reference, the wind plus diesel system uses 7% less fuel than the wind or diesel system when only one diesel is considered. It appears that greater advantage is taken of the additional wind, from the wind plus diesel mode, if a twin diesel back up is provided. Although reduced significantly, the number of diesel starts still exceeds 110/week for one or other of the diesels; for maximum fuel saving the starts are roughly equal.

The results of running wind plus diesel, together with fixed sequence diesel starts, are shown in Fig 5. The limitation here is that the number of starts for the first diesel remains constant at 180 although the starts for the second diesel drop to an acceptable level (ie below 50/week) as the capacity of the first diesel rises above 1.5 kW. The maximum fuel saving is predictably reduced with this control strategy. With a first diesel rating of between 1.25 and 1.5 kW the saving is 21% in comparison with the single diesel back up.

One approach to reducing the remaining high number of first diesel starts is continuous running. A number of different control strategies have been investigated, including power sharing, and minimum run times for the second diesel. The results, presented in Figs 6 and 7, suggest the most attractive strategy to be one utilising power sharing (as in C3), a minimum run time of 30 minutes and a minimum diesel loading of either 30% or 40%. Very limited data on part and variable load diesel performance exists and, furthermore, little is known about the increased maintenance resulting. Thus a definitive selection of operating strategy cannot be made at this time. Daf Indal (Ref 11) have carried out preliminary studies and research recently initiated at RAL should provide valuable information on this topic for future research. On the basis of present modelling assumptions, the two strategies above give satisfactory diesel cycle frequency (ie less than 50 per week) although some fuel penalty is incurred in comparison with single diesel back up.

To conclude, the studies of twin diesel systems indicate that, if sufficient care is taken in defining the appropriate control strategy, they could provide a feasible alternative to expensive storage systems.

Acknowledgement

I would like to acknowledge my obvious debt to the Reading Wind Energy team, in particular to G Slack on whose work the twin diesel model is founded. I am also grateful to J A Halliday, Professor N H Lipman and Dr F M Russell of Rutherford Appleton Laboratory for helpful criticism, and to Joan Mackie for the typing.

References

1. Infield D G, Lipman N H. Energy supply for rural applications - the UK experience. To be presented at Rural Power Sources (UK-ISES), Newcastle Polytechnic, 29,30 March 1983.

2. Slack G W. Small diesel generation systems: the cost and fuel savings possible by the addition of a wind turbine, with and without short term storage. Reading Energy Group Report 83/2.

3. Marshall E L (BP Research Centre, Sunbury). Private communication, 1982.

4. Tsitsovits A J, Freris L L. Dynamics of diesel-wind systems. Presented at Autonomous Wind Power Systems (BWEA), London, 15 October 1982.

5. Sexon B A (Reading University). Private communication, 1983.

6. Russell F M. Priority R and D areas in energy storage. Draft RAL Report, RL-82-032.

7. Barbour D, Twidell J. Energy use on the Island of North Ronaldsay, Orkney, pp 39 - 51, Energy for Rural and Island Communities, ed Twidell J, Pergamon, 1981.

8. Lipman N H, Musgrove P J, Dunn P D, Sexon B, Slack G. Wind generated electricity for isolated communities - a study of integration strategies. Soon to be published as a Department of Energy report.

9. Wyper H. Draft MSc thesis, Department of Applied Physics, Strathclyde University, 1982.

10. Slack G, Lipman N H, Musgrove P J. The integration of small wind turbines with diesel engines and battery storage. 4th BWEA Conference, Cranfield, 1982.

11. Development, installation and testing of a wind turbine diesel hybrid. DAF Indal Limited, report for Ontario Ministry of Energy and National Research Council, Canada.

Appendix 1: References on Energy Storage

Energy Storage for Wind Energy Conversion Systems, Zlotnick M. Proc of Second Workshop on WECs (1975)

An Assessment of SWECS/Mechanical Heating Systems, Schroeder M P and Tu P K C, US DoE Report RFP-3261 (1981)

Wind Power Recent Developments, ed De Renzo D J, Noyes Data Corporation (1979)

Institute of Gas Technology, Report PB259 318 (1979)

John Wilkinson (CJB Developments Limited, private communication)

Hydrogen Storage in Metal Hydrides, Reilly J J and Sandrock G D, Scientific American (February 1980)

The Economic Value of Hydrogen Produced by Wind Power, Stodhart A H, ERA 75-35 (1975)

Wind Generated Electricity for Isolated Communities, Lipman N H et al, Report to DoE (December 1981)

Demonstration of a Low Cost Flywheel in an Energy Storage System, Rabenhorst D W, p485, Energy Storage ed Silverman J, Pergamon (1980)

Metal Hydrides for Energy Storage, Ivey D G et al, J Materials for Energy Systems 3, p3 (December 1981)

Energy Storage Using Metallic Hydrides, Angus H C, Future Energy Concepts, IEE (January 1979)

Priority R and D Areas in Energy Storage, Russell F M, Draft RAL Report, RL-82-032

A Design Procedure for Wind Powered Heating Systems, Manwell J F and McGowan J G, Solar Energy, 26, p437 (1981)

Modelling of a Domestic Wind Power System Including Storage, Bogle A W et al. Energy Research 3, p113 (1979)

The University of Massachusetts Wind Furnace Project: A Summary Statement, Heronemus W E, Proc of Second Workshop on WECs (1975)

Storing Wind Energy as a Liquid Fuel, Miller D, p143, Proc of Second BWEA Wind Energy Workshop (April 1980)

Development Status of the General Electric Solid Polymer Electrolyte Water Electrolysis Technology, Nuttall L J, Proc 15th Intersociety Energy Conversion Engineering Conference, Seattle (1980)

Figure 1: Energy Storage Economics

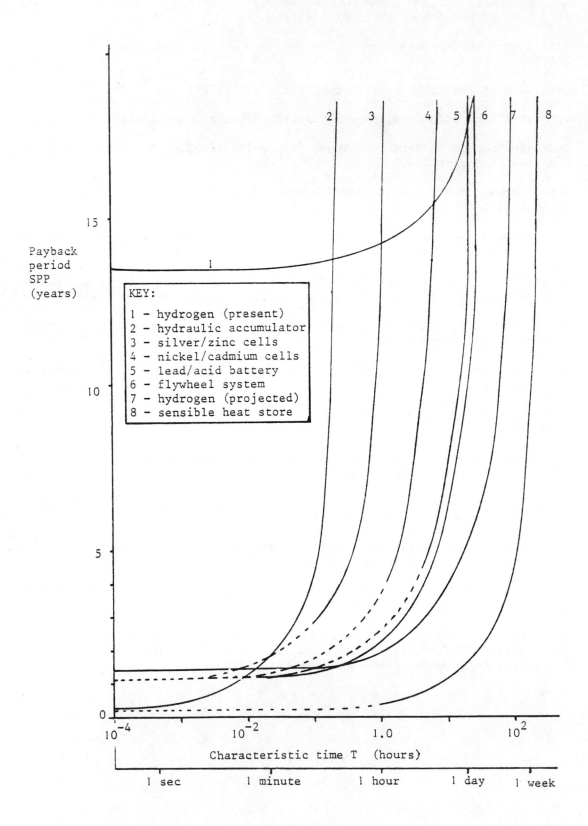

KEY:

1 – hydrogen (present)
2 – hydraulic accumulator
3 – silver/zinc cells
4 – nickel/cadmium cells
5 – lead/acid battery
6 – flywheel system
7 – hydrogen (projected)
8 – sensible heat store

Fig. 3. Diesel Only - Sequential starts

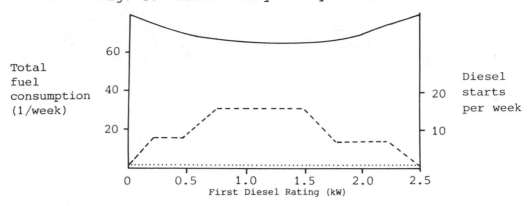

Key to graph: a) —— diesel only fuel consumption (no wind)
b) ······ first diesel starts per week with no wind contribution
c) --- second diesel starts per week with no wind contribution

Fig. 2. Diesel Only (B2)   Fig. 4. Wind/Diesel (B2)

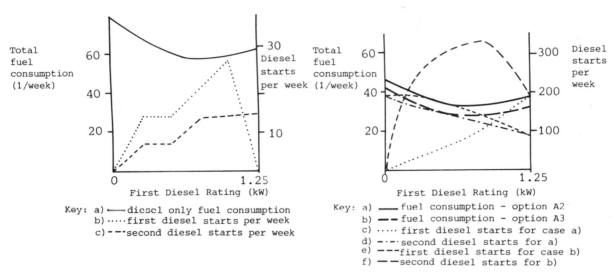

Key: a) —— diesel only fuel consumption
b) ······ first diesel starts per week
c) --- second diesel starts per week

Key: a) —— fuel consumption - option A2
b) -- fuel consumption - option A3
c) ······ first diesel starts for case a)
d) -·-· second diesel starts for a)
e) --- first diesel starts for case b)
f) —— second diesel starts for b)

Fig. 5. Wind/Diesel - Sequential starts

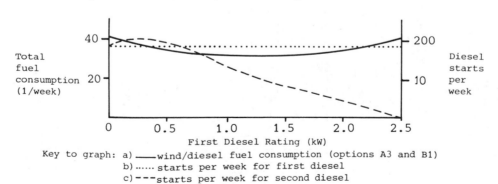

Key to graph: a) —— wind/diesel fuel consumption (options A3 and B1)
b) ······ starts per week for first diesel
c) --- starts per week for second diesel

Fig 6:   Continuous First Diesel

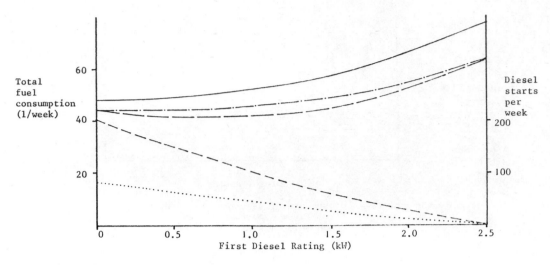

Key to graph: a)———— wind/diesel fuel consumption (min. load 50%)
                b)——·—— as above but with min. load 40%, min. run time 30 minutes
                c)—— —— as for b) but with power sharing option C3
                d)——— starts per week for second diesel in case a)
                e)········· starts per week for second diesel in cases b) and c)

Fig 7: Diesel Load Histories – 1st diesel 1.0kW (cont. running)
                                    2nd diesel 1.5kW (min. run 30 mins)

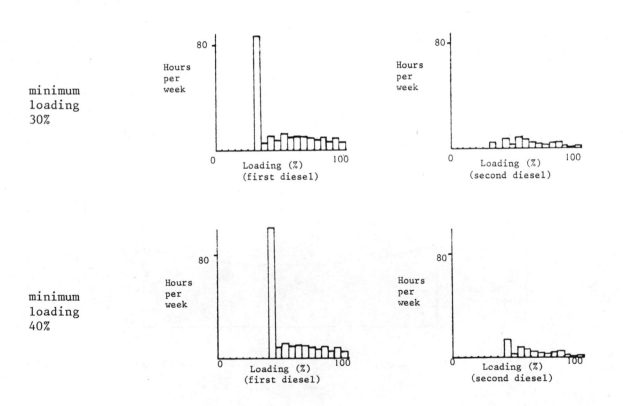

A TRANSPORTABLE WIND-BATTERY SYSTEM
FOR MONGOLIA

G.R. Watson
Northumbrian Energy Workshop Limited
Tanners Yard, Gilesgate, Hexham

In 1981 Northumbrian Energy Workshop were approached to develop a wind-battery system suitable for supplying Nomadic tribesmen on the plains of Mongolia. Late that year a visit to the U.K. by delegates from the Mongolian Peoples Republic finalised details on the mark I systems and 4 systems were delivered to Mongolia early in 1982. An engineer from N.E.W. visited Mongolia late in 1982 and from site assessment and discussions a Mark II system has been developed which it is hoped will be delivered to Mongolia in the middle of 1983.

Mongolia is situated some 11,000 km from London sandwiched between Russia and China. The country covers an area of 1,500,000 square km and is 2,400 km long from East to West. It is divided into 3 separate regions; The Gobi Desert in the South, an area of flat steppe land in the centre and a mountainous region in the North. Being 2,500 km from the sea the climate is very severe and one of the main design problems was coping with the extremely low temperatures. Luckily these are not matched by very high temperatures in the Summer, for example Ulan Bator, the capital city, has an annual average temperature of -3 deg. C. Figure 1 shows the wind information for the country and the area of primary interest, the Steppe region, is shown shaded. Winds are not high, although Spring storms can produce locally high velocities for short periods, but they are very consistent with few calm periods. This is to be expected with the cyclonic conditions that Mongolia experiences almost all year.

Since the revolution in 1921 the country has been organised in a de-centralised manner with a system of appointing represent-atives from localised communities to local area government and then appointing a representative from each area government to a central government in Ulan Bator. The way that the Mongolians organise their economy means that the local co-operative is primarily responsible for providing living accommodation and other equipment (including windmills) to their members. Thus although the working people do not earn a high salary they do have the benefit of a developing infrastructure, including schools and doctors, but also including schemes like district heating for the living accommodation in the capital city.

In the vast central plains area of Mongolia the majority of agriculture is carried out and in total this employs some 40% of the workforce and earns Mongolia some 63% of its foreign currency. There is no wood in the area and thus it is not possible to provide fencing, so the people of the plains have to live a nomadic existance, moving their living accommodation some 8-12 times per year. For the Winter period (November-March) they retire to a Winter quarters location where the animals are housed for the extreme cold period. With such a lifestyle it is very difficult to provide conventional power sources for the Nomads and the use of renewable energy obviously has considerable attractions to them. The requirement was primarily to provide a small amount of power that could be used for a range of activities from high power spot lamps to illuminate wolves at night through to portable television, as Mongolia is served by a three channel satellite. Without providing at least basic electrical requirements then the Mongolian infrastructure would suffer severe strains due to the movement of younger people from the agricultural areas and into the city.

Defining system requirements for the Mark I system proved to be extremely difficult and several assumptions had to be made. No detailed climatic information was available to us but the temperature range was obviously a problem, as was the likelihood of not having skilled personnel or tools and equipment. It was therefore decided to make the system capabilities as flexible as possible within the constraints of a probable un-skilled workforce and an unknown carrying capability between sites.

System 1 is as shown in figures 2 & 3. It is a 24 volt system using a modified Winco Wincharger powering sealed lead-calcium cells. The control module provides 24 volt power outlets and also 6, 9 or 12 volts for powering low current electronic equipment.

The Winco is a machine that has been around for many years and has generated enough kWh over that time to show up some of the basic design faults. Although not all minor faults could be rectified the blade and governor assembly were replaced with up-graded units and, together with modifications to the field winding and the yaw bearing assembly, it was felt that this was the best system for Mongolian conditions as known at that time. The tower is an aluminium tube construction with nylon plugs providing a male/female connection at tower split points. The whole assembly can be put together without the use of spanners and then it is feasible to erect, using the falling derrick arrangement, with only two people. Finding out climatic information was difficult enough so obtaining reliable data on soil conditions proved totally impossible. Each tower was therefore supplied with both stake and screw-in ground anchors for use in appropriate ground conditions.

The transportable energy module housed the controller, a sealed lead-calcium battery pack, housed in its own separate wooden containment, and the ballast load heater. The controller sensed battery condition and compensated for temperature on all the set points by means of a sensor housed in the battery module. The main control system was a switching regulator using a resistance heater as ballast but with additional linear regulators to provide a variable voltage output depending on pin connections. Wherever possible overload and circuit reverse-connection protection was built-in by means of manually re-set cut-outs.

The systems were duly despatched for their long journey to Ulan Bator and in the Autumn of 1982 a visit was arranged to Mongolia to inspect conditions and advise on installations. Some 1,500 km were travelled around the central plains area of Mongolia and discussions were held with Nomadic end-users, as well as officials of the M.P.R. As well as finding out relevant meteorological information from the academic institutes in Ulan Bator it was possible to ascertain actual ground conditions and to be involved in the dismantling and construction of the nomadic accommodation. Some time was also spent inspecting and evaluating potential local manufacturing capabilities which included a trip to the North of the country to obtain potential wood supply and manufacturing information. The wind turbines arrived during the visit and, due to some logistical problems, it was not possible to be present right from the 'opening crate' stage. When the test site was eventually visited the local staff had, without any instructions, successfully erected and commissioned the first system. They were, however, in the process of incorrectly connecting loads to the transportable energy storage module and this strengthened the design philosophy that all electrical connections must be capable of self-protection. As is usual in discussions with potential wind turbine users, the load requirement was far in excess of a potential wind system's capabilities. However, it was possible to ascertain the critical loads which would provide 80% of the requirements and there was general acceptance that certain equipment (like electric cookers) would never be feasible when utilising a wind energy system capable of being transported by camel or simple cart. The Mongolians were very impressed with the capabilities of the Mark I system but it was obvious that some major modifications to the design could take place, to the benefit of end users and to increase potential local manufacturing content.

The Mark II system comprises a modified Rutland WG910 wind turbine mounted on a predominantly wooden tower with metal attachment plates for guys and fittings. By modifying the wind turbine it is possible to do away with the ballast heater and the entire energy management module is housed in wooden containment all of which can be locally replaced. The 50W wind turbines have been re-engineered for potential local manufacture and increased durability during transport, but the same concept of a fairly high solidity rotor coupled with an iron-free permanent magnet alternator, which provides a 'start charging' wind speed of less than 2 m/s, has been retained.

Figure 1 shows a comparison between the Mark II and Mark I wind systems operating in the wind regime experienced in the Steppe region of Mongolia. It underlines the importance of not only selecting wind turbines on nominal rated output but also considering their cut-in and rated wind speeds. It can be seen that for a large proportion of the year the Mark II system, all be it equipped with a wind turbine of 25% the rated output, is producing much the same energy output in terms of kWh. As with the Mark I systems the batteries are sealed lead-calcium plate and all the interconnecting cabling is flexible down to $-60^{\circ}C$. By utilising a blade shape which is effectively a constant chord curved plate then the Mongolians have the option of replicating these blades either in wood or metal, both of which they have the skills and equipment to do. Hopefully, the turbine could eventually be totally locally produced probably excluding some low-cost electrical and electronic components. If, after gaining some experience with the systems, the local users were able to manage with automotive type lead-acid batteries (and keep them warm!) then much of the electronic control gear could also be dispensed with.

Although working on the Mongolian systems is a rather different activity to the majority of work carried out by N.E.W., many of the conclusions from the experience with the project to date, apply generally to all wind systems. It is imperative when considering a major installation that has specific requirements, particularly if these involve local manufacture in a Third World country, to have the resource and user requirements closely examinded. Small details like ground earthing conditions and the operating voltage of widely available portable radios are often overlooked but are of essential concern to those seriously designing a successful wind system.

Our co-operative does not manufacture any particular wind turbine and we always try to develop specialist units in collaboration with appropriate producers. This is not always possible as often the specialist requirement will be for one or perhaps two units, whereas the turbine manufacturer has to consider the added cost of interruption to batch production processes. However, whenever practicable, standard components are used, both to ease spares support and to minimise the number of potentially unreliable 'unknowns'. The first solution is thus not always the one of lowest technology nor one that necessarily maximises local build potential.

To provide a useful source of energy the systems must be reliable even in the face of that worst of hazards - those with a little knowledge and a screwdriver. Only once experience has been gained is it possible to prune back on any type of wind system, as to do so from the beginning inevitably involves possible system failures and gives the technology a stigma of unreliability which is totally undeserved.

Figure 1 - Annual average windspeed bands over M.P.R.

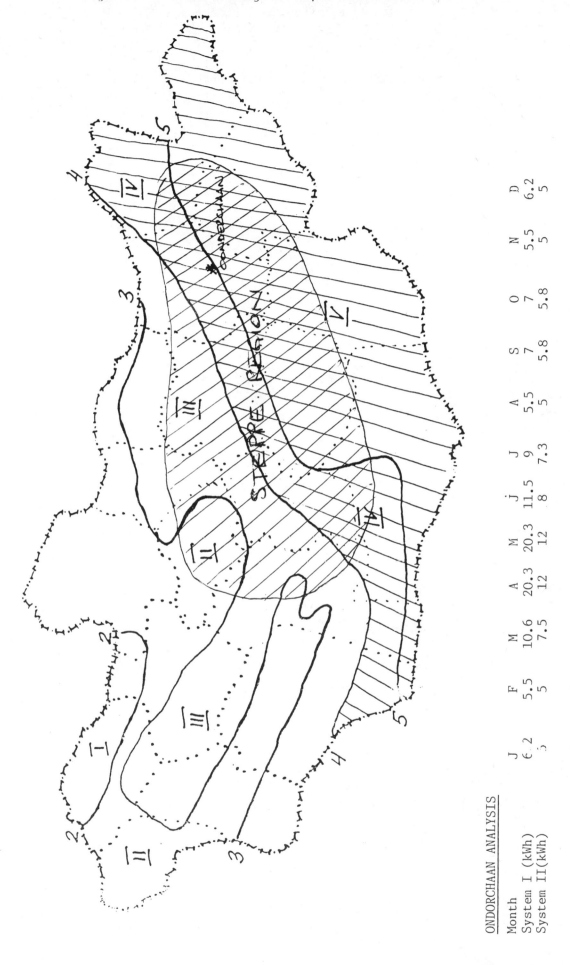

ONDORCHAAN ANALYSIS

Month	J	F	M	A	M	j	J	A	S	O	N	D
System I (kWh)	6.2	5.5	10.6	20.3	20.3	11.5	9	5.5	7	7	5.5	6.2
System II(kWh)	5	5	7.5	12	12	8	7.3	5	5.8	5.8	5	5

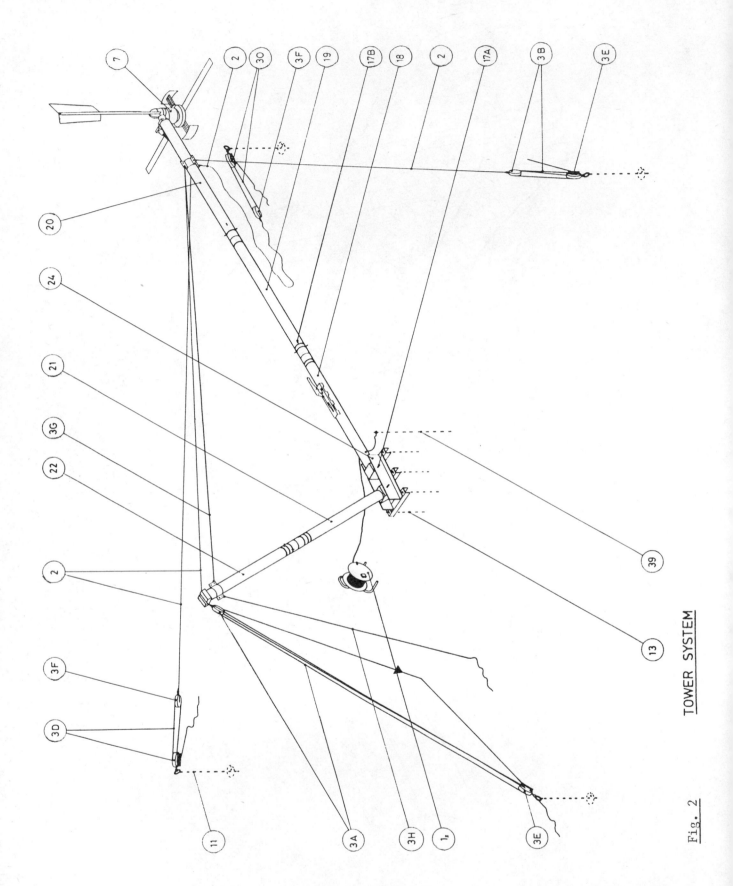

TOWER SYSTEM

Fig. 2

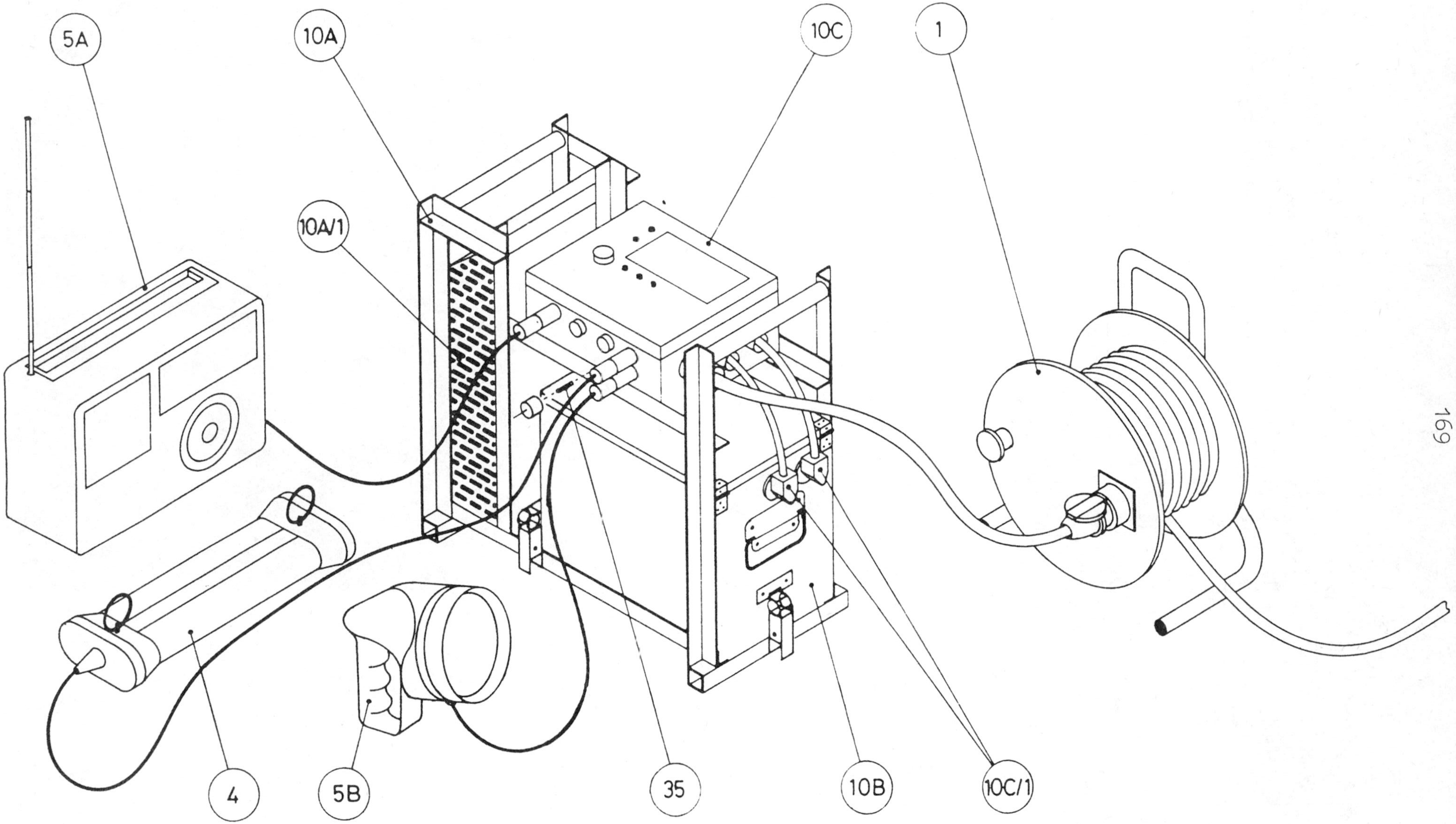

169

TRANSMISSION, STORAGE AND CONTROL EQUIPMENT

Fig. 3

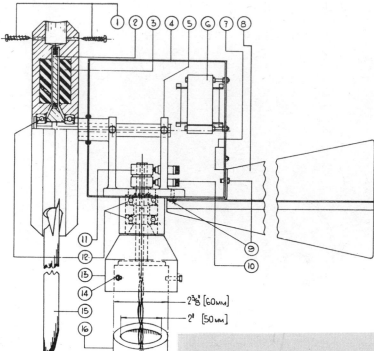

Fig. 4

Cut-away of
Mk II system
Wind Turbine

$2\frac{3}{8}$" [60mm]

2' [50mm]

Fig. 5

MPR Delegation
inspecting Mk.I
System at N.E.W.
Test Site.

THE FAIR ISLE WIND POWER SYSTEM

W G STEVENSON, NORTH OF SCOTLAND HYDRO-ELECTRIC BOARD

W M SOMERVILLE, INTERNATIONAL RESEARCH DEVELOPMENT LTD
(FORMERLY NEI CLARKE CHAPMAN LTD)

ABSTRACT

This paper describes the installation of and operational experience with a 55 kW wind turbine generator and associated equipment commissioned on Fair Isle in June 1982. Although not designed to operate in parallel with existing diesel generators the control system is such as to optimise, as far as practicable, the use of both means of generation. The wind turbine generator is the first unit to be operated on a commercial basis in the UK.

INTRODUCTION

Fair Isle is situated approximately halfway between the two island groups of Orkney and Shetland off the North of Scotland. It is approximately rectangular in shape and some 3 kilometres by 3.5 kilometres in size. The population is about 70 and research carried out on the island by the resident freelance meteorologist shows that the island is probably one of the windiest inhabited places in the British Isles.

The National Trust for Scotland became responsible for Fair Isle in 1954 and at that time there was no electricity on the island. In 1962 a limited supply was installed by the Trust to all the crofts by means of 5 small diesel generators. With the technical advice of the NSHEB this scheme was replaced in 1975 by 2 diesel generators centrally located in a power house. Distribution was by underground cable. The cost of generation dictated that only a limited supply (in terms of hours of generation per day) be maintained. The private electricity supply is run by an elected Fair Isle Electricity Committee.

In recent years, due to the large increases in the price of fuel and freight, consideration was given to alternative means of generating electricity on the island. In collaboration with the NSHEB, who subsequently acted as Engineers to the Project, NEI Clarke Chapman Ltd evolved a scheme to link the existing diesel system with a new wind turbine generator. Funding was provided by the Shetland Islands Council, the Highlands and Islands Development Board and the European Regional Development Fund for the supply and installation of a wind turbine generator. The contract was placed with NEI Clarke Chapman Ltd by the National Trust for Scotland in November 1981. The Trust provided the project management.

THE WIND TURBINE GENERATOR

The wind turbine generator is a 14 m diameter 3 bladed type with a nominal rated output of 55 kW and a maximum output of 65 kW in wind above the rated wind speed of 13 m per second (see Figure 1). It uses the step-up gear transmission and braking system developed and used by Windmatic A/S of Denmark. However, since the standard transmission is designed to drive an induction generation at 1000 rpm a further speed increase to 1500 rpm is provided to drive a standard 4 pole brushless alternator complete with inbuilt automatic voltage regulator. The control for the speed of the machine and thus the frequency of output is accomplished by automatically varying the electrical load applied to the generator in response to small changes in speed, the voltage being independently controlled by the voltage regulator.

The support tower is of lattice construction and consists of fabricated sections bolted together. The tower height is 15 metres. The two upper sections are each 6 metres tall and fabricated as complete units. The base section is 3 metres tall and fabricated as four corner units. Each corner unit is bolted to a concrete plinth secured by an 8 metre rock bolt.

Figure 2 shows a diagrammatic arrangement of the wind turbine machinery which is largely self explanatory. The design of the machine lays strong emphasis on safety. There are dual braking systems both arranged to fail safe. Built in safety devices continuously monitor for:

1	Excessive vibration	2	Low air pressure
3	Low gear case oil level	4	Belt damage
5	Worn brakes	6	Excess twist on the power cable
7	Excitation failure	8	Overspeed
9	Alternator Overheat		

For item 8 dual tachometer systems are provided.

SYSTEM MONITORING

The National Trust for Scotland with financial assistance from NEI Clarke Chapman Ltd contracted the NSHEB to design, manufacture, install and test a wind generator data logger which records the following parameters:

a)	Wind speed	b)	Wind direction
c)	Generator output (kW)	d)	Output to dump load (kW)

Wind speed and direction are obtained from a Vector Instruments Type A100 anemometer and type W200 windvane respectively. These are mounted at hub height (15.5 metres) on a pole located approximately two diameters (30 metres) from the wind turbine generator. Generator output and output to dump load are obtained from NEI kW transducers.

The four data inputs are fed to a multiplexer unit with analogue to digital converter and controlled by a microprocessor which also drives the output cassette recorder.

Power for the logger is obtained from the 24 V diesel start battery but a reserve battery within the logger keeps it operating for up to 3 hours when the diesel battery is removed for maintenance or replacement. The programme is stored in non-volatile PROM (programmable read-only memory).

The four parameters are sampled every 5 seconds and their average, maximum and minimum values computed every 10 minutes (120 samples) and recorded on cassette. The flexibility provided by the microprocessor on site facilitates amendments to the logging procedure when required.

The facility to display the parameters on demand from the keyboard is provided together with a printed record with a date and time heading.

The cassettes which require to be changed every 5-6 weeks are returned to the NSHEB Research Laboratory where the data is processed to give, for example, a graph of generator output versus wind speed arranged over 10 minutes or multiples thereof, thus defining machine performance (see Figure 3). This, or any other analysis can be performed, as and when required.

DISTRIBUTION SCHEME

Two diesel generators are available to supply electricity. The larger of these is a 52 kW unit which was utilised in winter and the smaller 20 kW unit in summer. Cost of generation in 1981/82 was estimated at 10.6 p/kWhr. The cost to the consumers was reduced by a Shetland Islands Council grant. Supply was made available only during the "guaranteed periods" of two hours in the morning and from dusk to 11 pm in the evenings. The supply provided lighting, entertainment and freezer power after dark and allowed washing machine and vacuum operation in the mornings. The maximum demand varied from 10-11 kW in summer to 16 kW in winter. The total annual consumption was of the order of 40,000 kWhrs.

With the advent of the wind turbine generator it was decided that, at least until a reasonable period of operating experience had been achieved, the "guaranteed periods" would be maintained on the existing network. The original cable network to supply essential services was closely rated on volt drop and unable to carry any additional power to the 30 consumers on the system. To obtain the maximum utilisation of the wind turbine generator, new cables were laid to service heating appliances at each location. Each consumer has the facility to install an immersion heater and up to three heating appliances of the storage radiator type, the rating of each being 750 watts.

Standard storage radiators supplied at cost by the NSHEB were used with new 750 watt elements. The cable and ancillary equipment for the new distribution network was supplied at cost by the NSHEB who also assisted with the cable jointing. The new underground cable system is provided with Protective Multiple Earthing and although the existing essential services system remains separate and is separately metered it was converted to PME to standardise the earthing.

Figure 4 shows the wind energy distribution scheme linked to the existing diesel generator system. It can be seen that the system makes provision for four groups of load. Group 1 is the "old" network supplying the essential services load. Group 2 is the new distribution system supplying the heating load. Group 3 is designated a local distribution with wind priority. This facility is not utilised at present. Group 4 is the dump load purely for wind turbine generator safety. This consists of a series of 5 kW resistors totalling 75 kW.

The Group 4 load is permanently connected to the wind turbine generator so that it is always possible to govern the turbine speed. The spot control frequencies for the Group 4 load are set so that the energy is only dumped if the system cannot absorb it all.

Group 1, the "old" system, was modified by the addition of a main contactor switch D which was added to isolate the diesel engines when it is possible to feed wind power through the other section switches E and F to the Group 1 load. Switch D can only be closed when the "guaranteed" hours are in force and then only if there is not sufficient wind to feed Group 1.

When switch D is closed, either switch E (20 kW diesel running) or switch F (50 kW diesel running) must be open.

When the 50 kW set is being run, switches D and E, are closed to ensure a good load factor.

If the Group 1 load is high when the 50 kW diesel is running, the Group 2 load will automatically be shed as the diesel speed is reduced. Switches D, E and F are so arranged that it is not possible for more than two to be closed at any time.

It is thereby not possible to directly couple the wind turbine generator to either diesel generator.

The excess power produced by the wind turbine generator over and above the power demand is controlled and automatically distributed by special automatic frequency sensitive relays. Figure 5 shows how this may be accomplished by dividing the distributed heating load into twelve discrete and approximately equal loads, each allocated a specific cut in frequency. Group 1 load is approximately 10 kW and the remainder of the output is divided into 12 approximately equal steps spread across all three phases to maintain a balanced load on the generator.

Each consumer has a frequency relay with three discrete steps as shown typically in band A. Three other groups of frequencies, band B, band C and band D are allocated to load Group 2 so interlaced to ensure that all consumers have at least one section energised before the band A group is energised on the second section and so on. Load Group 3 when used would be interlaced with these set frequencies to achieve a distribution and balanced load.

If due to disconnected consumer loads, the distribution networks are unable to accept all the power produced by the wind turbine generator then the dump load comes into effect for safety and to prevent loss of wind energy due to an overspeed shut down. The dump load set frequencies are such that energy is only dumped when the remainder of the system is fully energised.

Figure 6 shows the Group 2 distribution network and the frequencies allocated to each of the three heating appliances at each of the premises supplied. Figure 7 indicates diagrammatically the benefits which accrue from the wind turbine generator project. The "old" network (Group 1) remains as before with the "guaranteed periods" ensured by the use of the diesel generators when there is no wind. When there is sufficient wind Group 1 is supplied outside the "guaranteed periods". Experience may show that savings in diesel fuel due to wind energy supply during the "guaranteed periods" justify extending the "guaranteed periods".

The new Group 2 load network distributes low cost energy which is well suited for heating purposes, either to provide hot water for both washing and heating or for space heating where the dry heat by this electrical means is much appreciated.

The total loading shown in the typical household on the Group 2 feeder is 2.25 kW with the first option automatically switched from a 750 watt immersion heater to a 750 watt storage heater under the control of the water heater thermostat.

The load control relay units fitted at each of the premises are crystal controlled for reliability and accuracy. They respond only to frequency and not to voltage. They have indicator lights indicated when each channel is energised. This means that each consumer has an indication of the power available and also the possibility that increased load on the Group 1 system will cause a drop out on the Group 1 system in non guarantee periods.

THE CONTROL SYSTEM

Figure 8 shows the control scheme for the wind turbine generator and load management in a basic form. On the command to start, the wind turbine generator systems check that the machine is safe and then releases the brakes. The machine is self starting and as the turbine runs up to speed, the output voltage is monitored when there is sufficient speed to secure excitation. At this point the nacelle power is on and all services are self maintaining. Sections of dump load are energised as the frequency rises until a stable state is reached. At this stage contactor F is closed and connects the wind turbine generator output to load Group 2. The system then checks to see if a diesel generator is running and if it is not, it will then lock out contactor D and close contactor E to provide power through to load Group 1. This action also prevents the starting up of the dieel generator at the commencement of the guarantee period and will continue to do so until the load on the Group 1 network is greater than the available wind power causing the machine to slow down below 90% of the normal running speed. Should this occur, the control system automatically disconnects Group 1 load from the wind turbine generator and permits the diesel to start in the normal way.

A time delay not shown on the diagram prevents another attempt to supply the Group 1 load for 10 to 15 minutes.

If a diesel generator is already running the control system compares the wind turbine output with the load on the diesel generator and if it is at least 10 kW greater and maintained for 15 minutes, an automatic sequence takes place disconnecting the diesel generator and applying the wind turbine energy to the Group 1 network. A successful takeover is monitored for 3 minutes before the signal is given to the engine to shut down. If the wind power becomes insufficient after the diesel has been shut down, there is an enforced delay of 15 to 30 minutes before the diesel automatically restarts.

If the connected consumer load or dump load or combination of both, because of a particularly strong wind reaches 65 kW the spoilers on the blades of the wind turbine generator are activated thereby limiting the power output from the wind turbine generator to a lower level thereby protecting the alternator from excessive output.

ERECTION PROCEDURES

Because there is no heavy lifting equipment on the island and the local transport is limited to tractor trailers and two light lorries, special consideration was given to the erection procedure from the outset.

The tower was provided with a hinge at the base of the West face and with a reinforced lifting point at the top. A concrete foundation with a ground anchor eye and bolts to secure a hand winch was built 14 m East of the tower foundation. The necessary tackle including a 7m high timber A frame was prepared and shipped with the main equipment. All heavy lifts were carefully oriented before despatch as there was no means to turn them later.

The exact placing of all heavy items on selected vehicles was worked out before hand and rehearsed with the island labour so that each item could be correctly positioned using the swinging derrick on the carrying vessel MV Bussant on arrival.

The tower was erected from the hinge point in a horizontal attitude using the transport vehicles for support as required and the lifting tackle was then rigged so that the upper end of the tower could be raised and lowered to facilitate the mounting of the turning plate and the nacelle. When the nacelle assembly was complete, with the rotor spider in place, it was then rotated 90° to a nose down attitude using the lifting tackle to obtain ground clearance and the wings were then lifted into place in turn by hand and the turbine rotor rigged and set as required.

When assembly was complete, a light tirfor rope winch was attached to control the movement as the tower swung over the hinge pin into the upright position and the complete machine was ready for the main lift. This was accomplished without incident using a hand winch with volunteer island labour in relays in a time of 55 minutes.

OPERATING EXPERIENCE

The whole erection procedure and the ensuing commissioning period were characterised by a period of unusual calm on Fair Isle which necessitated extending the commissioning period by some four days. The wind came back gradually allowing a progressive build-up to full power operation. As with any new system, a number of minor teething problems were encountered and attended to during this period.

The success of this type of system depends to a degree on the collaboration of the consumers as it is necessary to reduce the essential services demand in periods when there is little wind energy in order to maintain minimum services such as lighting during non guarantee periods. During the first days of operation, there was a natural tendency among the consumers to try and see how much power they could get and, in these light wind conditions, there were a number of unpredicted outages. It was most gratifying to find that the basic requirements were appreciated and the need to unload essential services understood by all in only two days of operation.

During the initial running period, from June through to September, things settled down well and the machine demonstrated its ability to produce a considerable amount of energy in varying wind conditions. The provision of hot water and space heating was appreciated almost as much as the extended period of electricity supply provided by the machine. It was also shown that the diesel running hours could be reduced to as little as 20% of that required during the previous regime of interrupted supply.

During this period two electrical storms occurred, virtually unknown on the island, causing minor damage to the data logger equipment and to a section of the windmill control system.

During the latter part of the year, a number of severe gales were experienced which led to some further minor problems mainly associated with the power limiting system using the spoilers when operating in severely gusting winds. The spoilers operated by compressed air are subject to a minimum time when extended to avoid depletion of compressed air reserve. The nominal time delay is set between 15 and 30 seconds. A gust triggering the spoilers at 65 kW so reducing the output to about 25 kW can pass in 4 seconds and be followed by a lull which reduces the power output further causing drop out of the supply. These phenomena are associated with narrow sectors of wind approach from the Western cliffs: one some 10° wide due West and other 25° wide North-West, the latter being more troublesome since strong North-Westerly winds have been uncommonly predominant during the testing period causing an exceptionally high use of the spoiler device which has highlighted some mechanical weaknesses in this system which have not been observed elsewhere.

Fortunately, all the weaknesses highlighted are readily solved by good engineering practice and the experience gained has been valuable in appraising the design.

After 2175 hours of operation, a gearbox fault developed and the machine was taken out of service on the 20 October with some 61,000 kWhrs generated. A new gearbox was provided and replaced under warranty with the machine being returned to service in mid-December. Of the total 7 weeks of generation lost, more than half is attributable to problems associated with the severe weather conditions affecting delivery of parts and delaying the progress of the exchange operation.

At the time of writing, the precise reason for the gearbox failure has not been determined and all possible causes are being investigated thoroughly in order to avoid a recurrence. In the meantime, the replacement unit is being monitored closely.

The operational experience with the load distribution system and the frequency controlled load relays in the consumers' houses has been most gratifying as the benefits of the hot water and dry electric space heating provided by this means is much appreciated by the householders in the island. The considerably extended supply hours giving a lighting service during the dark winter days and extended TV is also much appreciated.

CONCLUSIONS

This project, the first of its kind, has shown that considerable benefit in improved service and amenity can be provided to remote communities by the use of an aerogenerator and the ability to distribute all the collected energy to the community is a major factor in achieving this.

Teething troubles are to be expected on any new equipment and show how improved standards and reliability can be achieved.

The costs of providing for service and warranty in such a remote location are high but thre is no substitute for field experience in the proper evaluation of the design of any type of machine. Operation in severe wind conditions (10.5 metres/sec mean winter wind speed) with extreme winds frequently in excess of 30 metres/sec and occasionally over 40 metres/sec will rapidly expose even minor deficiencies in design and long term reliability of the system can be attained more quickly than by testing in less onerous conditions. Problems encountered so far can all be solved by well known methods and there is no reason to doubt the viability of this concept of wind generation.

ACKNOWLEDGEMENTS

The authors wish to thank the following organisations for information used in the production of this report:

NEI Clarke Chapman Ltd
International Research and Development Co Ltd
The National Trust for Scotland
The Fair Isle Electricity Committee
The North of Scotland Hydro-Electric Board

2502DQEWGS01

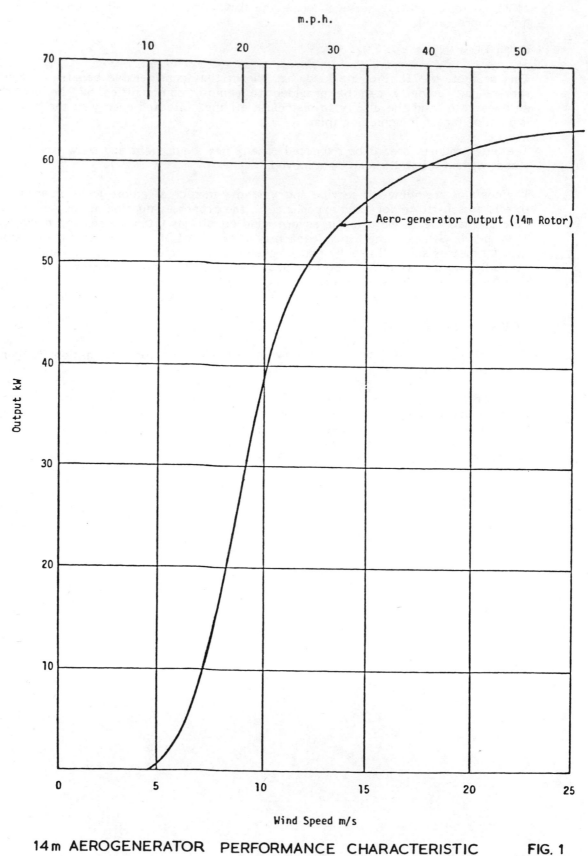

14m AEROGENERATOR PERFORMANCE CHARACTERISTIC   FIG. 1

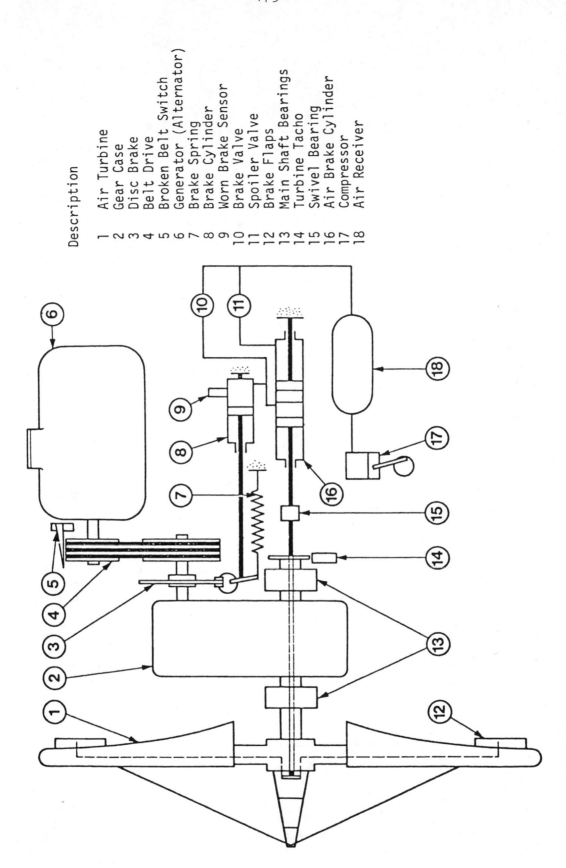

Description

1 Air Turbine
2 Gear Case
3 Disc Brake
4 Belt Drive
5 Broken Belt Switch
6 Generator (Alternator)
7 Brake Spring
8 Brake Cylinder
9 Worn Brake Sensor
10 Brake Valve
11 Spoiler Valve
12 Brake Flaps
13 Main Shaft Bearings
14 Turbine Tacho
15 Swivel Bearing
16 Air Brake Cylinder
17 Compressor
18 Air Receiver

DIAGRAMMATIC ARRANGEMENT OF NACELLE EQUIPMENT   FIG. 2

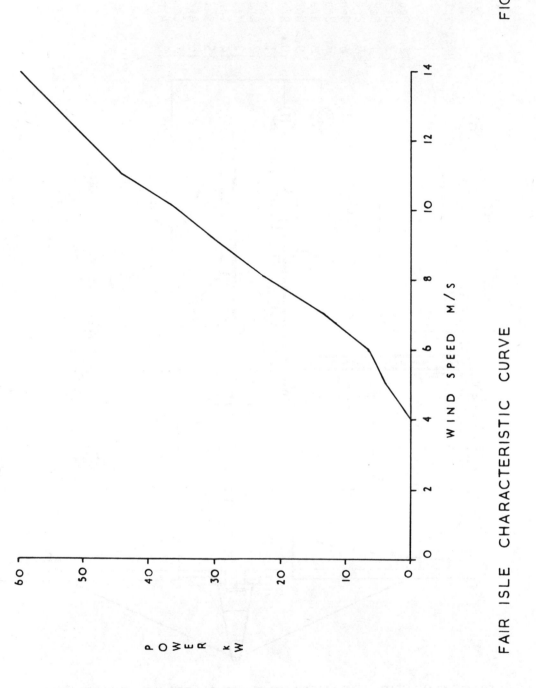

FIG. 3

FAIR ISLE CHARACTERISTIC CURVE

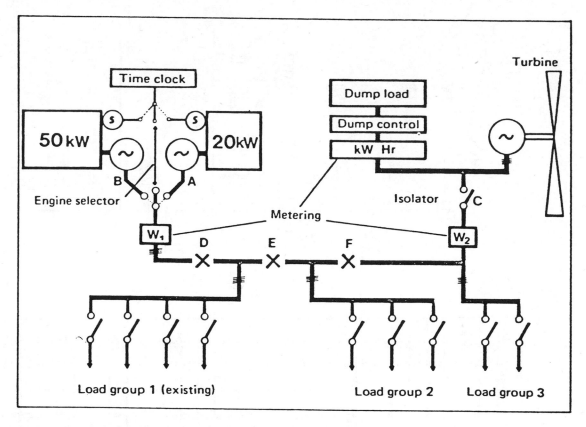

WIND AND DIESEL DISTRIBUTION ARRANGEMENT   FIG. 4

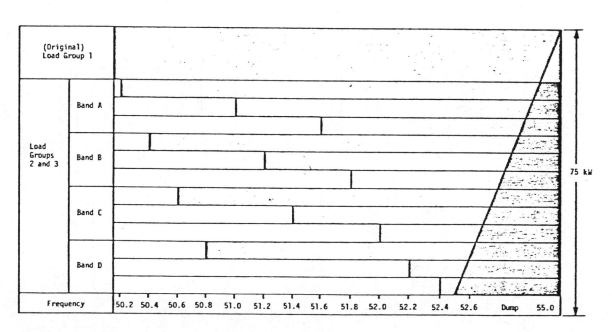

WIND ENERGY DISTRIBUTION SEQUENCE   FIG. 5

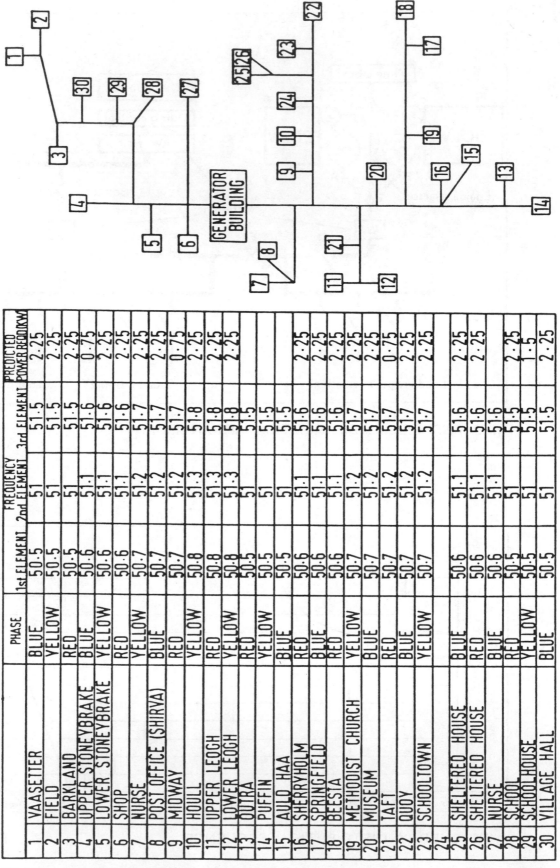

No.		PHASE	FREQUENCY 1st ELEMENT	2nd ELEMENT	3rd ELEMENT	PREDICTED POWER REQD KW
1	VAASETTER	BLUE	50.5	51	51.5	2.25
2	FIELD	YELLOW	50.5	51	51.5	2.25
3	BARKLAND	RED	50.5	51	51.5	2.25
4	UPPER STONEYBRAKE	BLUE	50.6	51.1	51.6	0.75
5	LOWER STONEYBRAKE	YELLOW	50.6	51.1	51.6	2.25
6	SHOP	RED	50.6	51.1	51.6	2.25
7	NURSE	YELLOW	50.7	51.2	51.7	2.25
8	POST OFFICE (SHIRVA)	BLUE	50.7	51.2	51.7	2.25
9	MIDWAY	RED	50.7	51.2	51.7	0.75
10	HOULL	YELLOW	50.8	51.3	51.7	2.25
11	UPPER LEOGH	RED	50.8	51.3	51.8	2.25
12	LOWER LEOGH	YELLOW	50.8	51.3	51.8	2.25
13	OUTRA	RED	50.5	51	51.5	
14	PUFFIN	YELLOW	50.5	51	51.5	
15	AULD HAA	BLUE	50.5	51	51.5	
16	SHERRYHOLM	RED	50.6	51.1	51.6	2.25
17	SPRINGFIELD	BLUE	50.6	51.1	51.6	2.25
18	BEESTA	RED	50.6	51.1	51.6	2.25
19	METHODIST CHURCH	YELLOW	50.7	51.2	51.7	2.25
20	MUSEUM	BLUE	50.7	51.2	51.7	2.25
21	TAFT	RED	50.7	51.2	51.7	0.75
22	QUOY	BLUE	50.7	51.2	51.7	2.25
23	SCHOOLTOWN	YELLOW	50.7	51.2	51.7	2.25
24						
25	SHELTERED HOUSE	BLUE	50.6	51.1	51.6	2.25
26	SHELTERED HOUSE	RED	50.6	51.1	51.6	2.25
27	NURSE	BLUE	50.6	51.1	51.6	
28	SCHOOL	RED	50.5	51	51.5	2.25
29	SCHOOLHOUSE	YELLOW	50.5	51	51.5	1.5
30	VILLAGE HALL	BLUE	50.5	51	51.5	2.25

TABLE OF DISTRIBUTION SEQUENCE ALLOCATION   FIG. 6

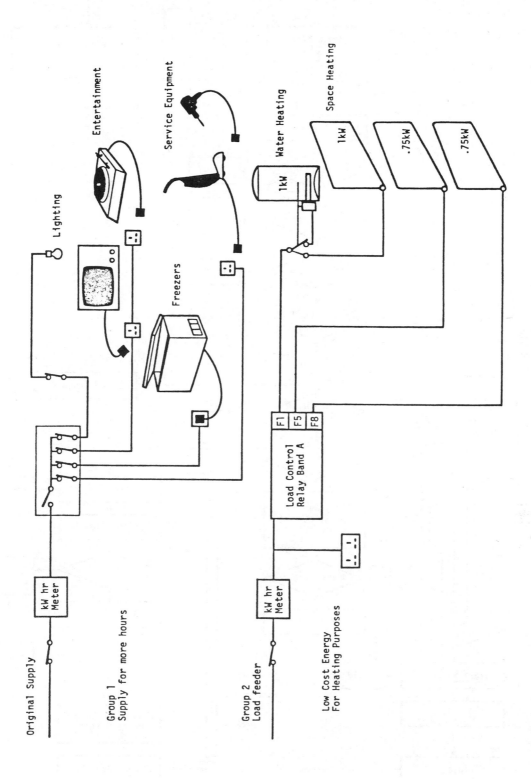

HOUSEHOLD WIND ENERGY UTILISATION   FIG. 7

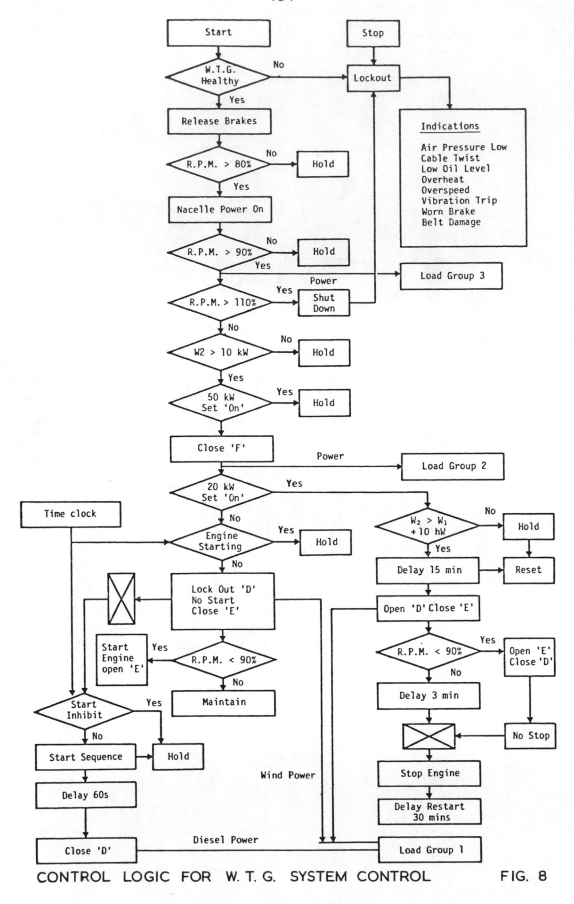

CONTROL LOGIC FOR W. T. G. SYSTEM CONTROL   FIG. 8

WIND POWER ON LUNDY ISLAND

W.M. Somerville,
Manager,
New Product Development,
International Research & Development Co. Ltd.
Newcastle upon Tyne,

and

J. Puddy,
Staff Engineer,
Lundy Island.

Abstract.

The placing of any new costly high technology plant in a remote, frequently isolated situation, regardless of the state of the art, must be an act of faith at the outset. The history of wind power is long and beset by failures and, even today, with greater knowledge of materials and stressing, the design lifetime of such machines has still to be proved in the aggressive environment in which they work. The benefits that a modern aerogenerator can bring to an isolated community can be quantified to a degree in financial terms by off-setting the costs of financing the installation against the operating and supply costs of conventional energy and power generation. There are, however, other less tangible benefits in the improvement of the amenity, such as longer periods of generation, improvement in working conditions, additional power for space heating to reduce building maintenance costs, and an improved standard of living for those in the community which will help to reduce the drift of the population towards mainland and larger conurbations. Lundy Island now have an aerogenerator and are making diligent efforts to extract the maximum benefit from the system installed.

The Island.

Lundy, in the approaches to the Bristol Channel, Figure 1, is just over 5½ kilometres long and a little over 1 kilometre wide. The island is predominantly granite, standing steeply out of the sea some 130 metres. The top is bare and gently undulating, with granite cliffs to the West and very steep sloping sides to the East.

To quote from the Handbook, "Lundy's climate is milder than that of the mainland, and the rainfall is lower ... but the wind is the real ruler of Lundy; the East wind can make it impossible to use the landing beach, upsetting everybody's plans and the West wind, rushing with unbroken force from the Atlantic, can, on occasion, blow roofs off buildings and sometimes cattle over the cliff edges."

When the Island came up for auction in 1969, it was purchased by the National Trust underwritten by the Landmark Trust. Further, the landmark Trust has undertaken to administer the island for 60 years and maintains a staff on the island. Income is derived mainly from the holiday trade through the letting of cottages and hostels, and from a small hotel and limited camping. Self-catering and social requirements are serviced by a shop and a tavern on the island. Some additional income is gained by farming the plateau. The nearest mainland is Hartland Point, some 18 kilometres South, but supplies are brought from Ilfracombe, some 38 kilometres East, on a small coaster, the "Polar Bear", owned by the Trust. There are, weather permitting, two sailings a week in the winter and three a week in the summer. There is no harbour on Lundy and, indeed, at high tide there is no beach.

Supplies and equipment are landed on the island with the aid of a specially built wheeled landing craft which is kept on the island on a slip, well above high water mark, on the East side of the island at the South end. This craft can accommodate an agricultural trailer and supplies are loaded directly into the trailer using the 6 Ton derrick on the coaster at anchor some 200 metres offshore. The craft then makes the run to the beach and the trailer can be towed away by tractor. The beach is normally accessible one hour after high tide and consists largely of shingle and rock, subject to considerable variation and movement when seas are running from East and North East. The beach is frequently unusable when easterly winds are blowing and when large swells are coming in from the Atlantic. Access to the top of the island is by a single track road cut into the cliff with gradients of 1 in 4 and a shear drop to one side complicated by some sharp bends, including two hair- pins, and the presence at one point of low overhanging branches from some of the few trees which survive on the lee side of the island and which the Trust are anxious to preserve.

Site Selection.

All the habitable buildings on the island are concentrated at the South end and mainly on the eastern side, Figure 1, taking some advantage of a slight declivity to obtain a little shelter from the prevailing westerly and south- westerly winds. The existing electrical services are provided from a generator house located in the centre of the community. This fact, combined with the desirability of avoiding high voltage transmission, limited the distance at which the aerogenerator could be sited because of the high costs which would incurred in providing a large section cable to operate at an acceptable voltage drop over a greater distance. A second factor was the need to avoid obstructions to air flow from existing buildings combined with the desire to minimize possible nuisance to the occupants from the noise of the machine. The third factor which had to be taken into account was the existence of turbulence created by the high and steep cliffs at either side of the island.

The site selected lies almost on the centre line of the island and close to the highest point, some 330 metres from the generator house. The site is as far from the cliffs as it is possible to get and air flow is unobstructed in every direction except for a very narrow sector due West occupied by the old lighthouse on the highest point of the island. There is a small but significant fall in the land towards the South-West, and the general lie of the land suggested that there would be a degree of augmentation to the prevailing south-westerly winds which, it was hoped would lead to enhanced output. The location was firm and dry and readily accessible from existing trackways. The ground consisted of some 250mm of top soil over granite.

The Electrical Supply System.

Electrical power is provided on an interrupted basis, the "guaranteed" period of supply being from 7 a.m. to 12 noon and 4 p.m. till midnight. The genera-ting plant driven by Lister diesel engines consists of three 6kVA single phase sets and one 27 kVA three phase unit. The original distribution system consisted of two separate single phase radial feeders and one single phase ring main distributing by underground cable from a switch-board in the generator house which allowed each circuit to be run individually with its own single phase set, or in groups, or from the three phase set. The run of these distribution cables is shown in Figure 2. The original supply had been provided for lighting, entertainment and services in the various holiday dwellings and those of the permanent staff and to power freezers for the storage of perishable foods Gas powered refrigerators are used for domestic purposes. The 27 kVA set, a recent addition, had been installed to allow a limited degree of heating for staff in the winter months in addition to the foregoing services. Space heating was provided by oil fired boilers or calor gas appliances. Diesel oil for the power station and for space heating and the gas cylinders all have to be handled from the beach to the various storage points on the island.

Wind Regime.

At the time of the initial investigation, there were no available wind records for the island although local opinion was that the winds were strong ! Average record for the area would indicate that the annual mean wind speed is 6 metres/second and that for about 30% of the year the wind speed would be less than 3 metres/second. During the period June to December 1981, a series of 99 daily wind readings were taken in the vicinity of the proposed aerogenerator site to provide more specific wind data for the island. These readings showed that there would be 74 days with some power, of which 60 had wind speed of at least 20 miles per hour, probably making the system self sufficient and, further, that on 25 of these days the machine would be delivering its full rated power. On the basis of this somewhat sparse information, it was estimated that an aerogenerator of 14 metre diameter would produce some 100,000 to 120,000 kWhrs per annum which compares to the estimated production from the diesel equipment at 20,000 kWhrs per annum. Operation of the aerogenerator would reduce the requirement for diesel generation to the order of 10,000 kWhrs or less. The balance of the wind energy would be used to produce hot water and space heating, leading to a proportional saving in other fuels.

The advantages of the aerogenerator system for the island were tabulated as follows : -

1. Longer hours of electrical production.

2. Less generator fuel consumed.

3. Less heating fuel consumed.

4. Less cargo commitment and handling labour required.

5. Dry space heating without condensation.

6. Reduced diesel running hours and less engine maintenance.

7. Hot water could be provided in all dwellings.

8. Space heating could be provided in all dwellings.

9. Reduced maintenance on building fabric, decor and furnishings.

10. The letting season could be extended possibly to the whole year.

11. System heating priorities were selectable.

The disadvantages were seen as : -

      1. High initial costs.
      2. The output varies with wind speed.

Foundations.

It was originally planned to construct the foundations using small concrete piers anchored to the granite mass using rock bolts but, when the site was excavated, it was found that the granite was badly fractured to a considerable depth and a mass concrete foundation cast in an excavation 1.5m deep was used instead. The twelve tower anchor bolts penetrate almost to the full depth of this foundation block where they are tied by a steel frame, and the whole block is steel mesh reinforced to ensure its integrity. Ducting was cast into the concrete to carry the power and control cables from the local control panel underground. The erection winch foundation was placed 14m to the West with the top of this block flush with the granite so that,when erection was complete, the anchor steelwork could be protected and the whole plinth concealed by turfing over.

Transport and Erection.

The shipment of the aerogenerator was organised in a similar manner to that described in the previous paper for the Fair Isle machine. The nature of the landing beach and the cliff road required special provision to be made on the island to allow the 3 largest packages to negotiate the cliff road and pass under the protected trees. To this end, a special low loading trailer was constructed on the island with a deck height of 0.5m.

Despite some trepidation about the beach landing, somewhat heightened by a wait of three days for the sea to subside, the actual landing proved to be relatively trouble free as conditions in the bay were nearly perfect for the four hours required to unload and land all the parts. Some minor difficulty was experienced with the two 6 metre tower sections and the 7 metre long wing cradle, all of which extended beyond the prow of the landing craft. These parts were handled by supporting them on the trailer and on the beach ramp of the landing craft while at sea, and skidding them into the correct position on the trailer on the beach before ascending the cliff road. An International B100 crawler-tractor with hydraulically powered front bucket was pressed into service to assist with the unloading of the trailers on site.

The erection procedure already described in the previous paper is illustrated diagrammatically in Figure 3. Unlike Fair Isle, the weather on Lundy after the landing was characterised by sustained high winds for the following twelve days, giving ample demonstration of the suitability of the site and causing some concern in judging the appropriate moment to lift the assembled structure into the upright position. The actual lift was undertaken in a southerly wind of 25 knots gusting to 35; a period of relative calm with forecast winds of increasing strength. This operation was completed by a team of enthusiastic volunteers in the time of 50 minutes.

Check out and commissioning of the machine in winds of 25 knots and above is somewhat nerve wracking when the machine goes straight to full power output as soon as it is up to speed, and a little extra care in triple checking that all the safety circuits are effective before running is justified. The machine produced 1,030 kWhrs in the first 24 hours of operation !

The control and load management system for the aerogenerator is substantially the same as used on Fair Isle and is described in detail in the preceding paper.

Wind Energy Utilization.

The aerogenerator with a nominal rating of 55kW has a considerable margin over the existing maximum demand on the island of some 20kW. To provide for effective utilisation of the surplus energy, all the habitable dwellings on the island system have been provided with frequency sensitive load control consumer units, each with three separate circuits rated at 1kW. A priority is given to providing hot water on the first available circuit subject to the overriding control of a change over contact thermostat which diverts this circuit to a storage heater when the water temperature has reached the set value. The remaining circuits are allocated to storage heaters for space heating.

Some considerable thought was given to the best form of storage heater to use, and CREDA model TSR 12 units were selected as they offered the capability of giving near instant heat during the initial charging period by virtue of a by-pass duct, controlled by a room thermostat built into the heater, in addition to the normal facility to select the amount of charge absorbed by the core for storage purposes. It was felt that the quick response when the machine came back on line after a period of calm would be subjectively better than the slow response of a normal core type unit. The degree of control also allows the unit to be operated at lower outputs to keep the buildings dry and in good decorative order when they are not occupied, allowing the available energy to be diverted to occupied dwellings as required. All these heater units have been fitted with two special 500 W elements. Despite the reduction in the energy input rate, it is still possible to fully charge the heater and obtain a thermostatic cut off, at which point the energy is automatically re-routed to the next unit line anywhere in the system according to the pre-set priorities.

The island continues to operate an interrupted service for the present in that it maintains the guaranteed hours for essential services. At periods when there is insufficient wind energy to meet the essential services demand, the network load will slow down the wind turbine to the point where it automatically disconnects the essential services and, if a guaranteed period is in force, the diesel is automatically started to provide the service. If the essential service load was relatively high when this occurred, an appreciable amount of energy would go to waste in the dump load but provision has been made to utilise this energy by providing direct connection from the wind turbine busbars to a number of suitably placed consumer unit circuits with a combined rating able to absorb the power likely to be generated in such a period.

To ensure an equitable distribution of energy over a daily cycle, the switching frequencies of these circuits are set with a lower priority than the others serviced through the main network.

The provision of this facility has involved the laying of a number of additional cables in the locality of the generator house. At present, this low wind distribution system is operating on a temporary basis into the staff accommodation. On completion of the current restoration and rebuilding programme involving the tavern, shop and three houses in the vicinity of the old hotel complex, this energy will be fed into a conventional wet central heating system shared through these buildings. To provide an adequate heat store in a reasonable volume, a phase change heat store system, using hydroscopic salts developed by Calor Alternative Technology Group, will be used. The phase change heat store will be charged by immersion heaters from the WTG system, the load being spread across all three phases and controlled by consumer unit system subject to generated frequency. During periods of calm weather, the water will be heated by an exhaust heat exchanger fitted to the 27 kVA diesel set, thus maintaining a guaranteed background heating supply.

All the island freezers are controlled by time switch to limit their operation to the normal guarantee period, a minor adjustment in setting ensuring a staggered sequence for starting. During the winter period, a number of de-humidifiers are also operated in the unoccupied letting accommodations and these are controlled by time delay relays in order to avoid the large starting surge which would be experienced were all these units to start simultaneously when the system was energised, so protecting the diesel and the aerogenerator from such surges and the associated problems. Time delay relays have been added to the freezers also to ensure sequential starting should a change over from W.T.G. to diesel occur during guaranteed hours. This provision makes it possible to sustain a supply on one 6 kVA set during daylight.

The consumer unit circuit dedicated to water heating in each accommodation is operated at a selected frequency such that it is possible for this circuit to be energised when the diesel set is in operation, thereby ensuring that hot water can be provided at all times, subject to the guaranteed hours, when there is no wind. In practice, the operation of this feature allows the diesel generator to be operated at a substantially constant load between 70 and 80% rated load. This means that the plant is operating at its best efficiency and a desirable service is provided with the most economical use of diesel fuel enhanced by the exhaust gas heat exchanger.

The manner in which the distribution system is managed on the island is illustrated in Figure 4 which shows the initial plan for distribution of the wind energy to the various locations This system is developing as the building work progresses towards completion, but the evolvement of the system on a day by day basis is presenting no problems as it is readily possible to re-programme the various control elements in the system to accommodate alterations as well as to harmonise the energy distribution with the use of the various dwellings as required.

The operation of the system on the 2nd March 1983 is represented by the consumer unit setting log shown in Figure 5, which allows rapid assessment of the loading and balance of the system and the existing priorities. This also shows some of the changes already introduced in that it has been decided to put in a three phase four core feeder to Millcombe House picking up the new agents house en route. The supply to the re-furbished buildings in the Old Hotel complex will also be three phase four core with a separate service cable. These additions, although not yet connected, will allow for maximum utilization of the available wind energy.

Performance.

During the period from 23rd November to 2nd March the aerogenerator was at operating speed for a total number of 2,060 hours, during which a total energy capture of 59,228 kWhrs was achieved. Some 13,714 kWhrs of this energy was dissipated in the dump resistors, the majority of this occurring during the initial period as the load system is being set up and the proportion being dumped is being progressively reduced as the system load is extended to its finished status.

During this period of 100 days, which includes 5 days outage whilst working on the aerogenerator, we have logged : -

Days with no wind generation . . . .   10 days

Days with partial wind generation . . . . . .   70 days

Days with no diesel generation   . . . .   30 days

Diesel fuel consumed   . . . .   390 gall.

Diesel power generation to network . . . . . . 6,303 kWhrs

Aerogenerator power to network   . . . . . .45,514 kWhrs

Aerogenerator at service speed   . . . . . . 2,060 hours

Mean output of aerogenerator in service. . . . 28.75 kW

Mean output of aerogenerator for period. . . . 24.68 kW

It is difficult to compare past and present fuel consumption with any significance as there have been a large number of contractors staff present on the island since September 1982, at times more than doubling the usual population for this period. In addition, the building works have required service power all day for some quite heavy power tools and site flood lighting, both morning and evening, during the short winter days.

The network load, appreciably higher, and subject to larger peak loads than normal service demand, has been greater than the available wind power on a number of observed occasions and caused a disconnection of service which would not have occurred on the normal service load. It is, therefore, reasonable to expect less diesel service time when the load on the network reverts to the normal pattern. Generator fuel consumption for the previous years operation was approximately 3,000 gallons, with no electrical water or space heating being provided as is now the case.

Consumption of other fuels is likewise difficult to assess due to the temporarily increased population but a section of staff quarters has shown a reduction is gas consumption from 9 to 3 cylinders over the period in question. It is also reported that the condition of the unoccupied dwellings is better maintained since heating was installed. The simultaneous installation of de-humidifiers will also have contributed significantly in this improvement.

It would appear that the aerogenerator will exceed its predicted production of energy in its first year of operation, and it seems certain that Lundy will make most efficient use of this facility when the utilization system is fully commissioned to the mutual benefit of the Trust, the staff and the holiday makers.

Acknowledgements.

The authors wish to thank the following organisations for information used in the preparation of this paper : -

Landmark Trust,
National Trust,
Northern Engineering Industries plc.
      NEI Clarke Chapman Ltd.
      International Research & Development Co. Ltd.*

* International Research & Development Co. Ltd. is a wholly owned subsidiary of Northern Engineering Industries plc. The responsibility for design, engineering, development and marketing small wind energy systems based on horizontal axis wind turbine technology for unit ratings up to 250kW was transferred from NEI Clarke Chapman Ltd. to IRD in July 1982 in order that the wider range of technical expertise within IRD would be readily available in support of this endeavour. The full range of manufacturing resources within the NEI organisation are available in support.

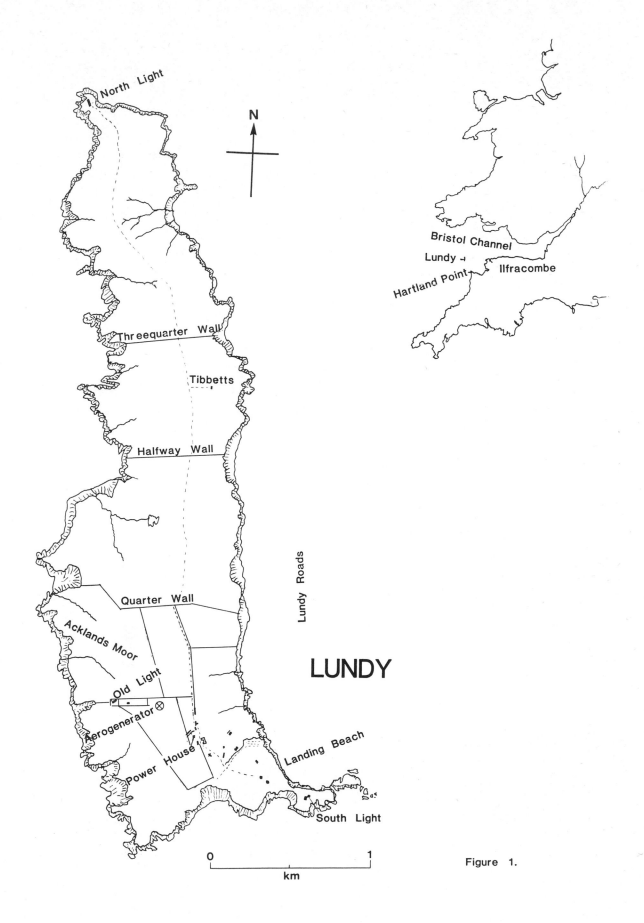

Figure 1.

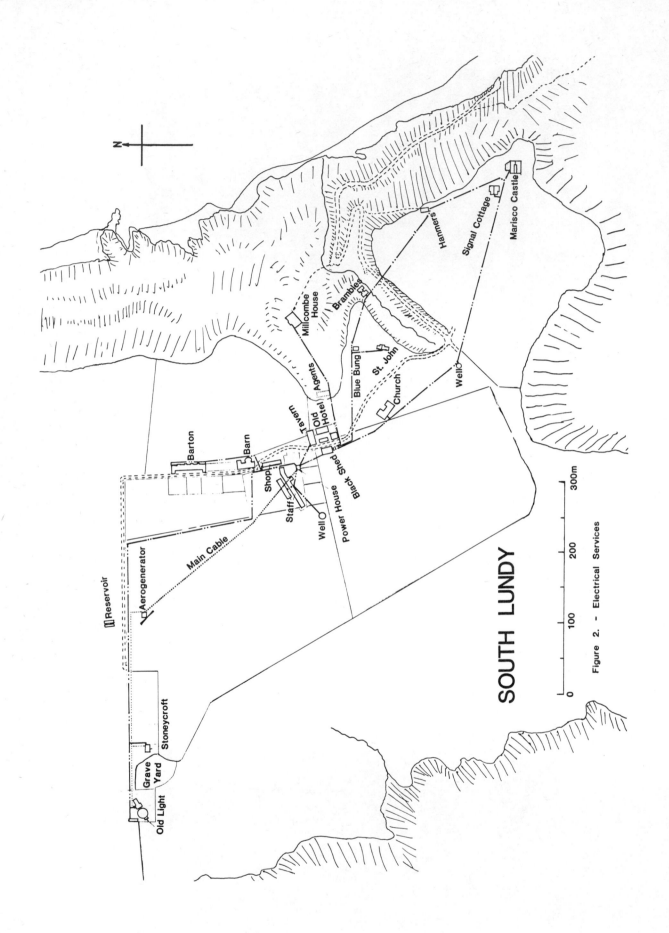

SOUTH LUNDY

Figure 2. - Electrical Services

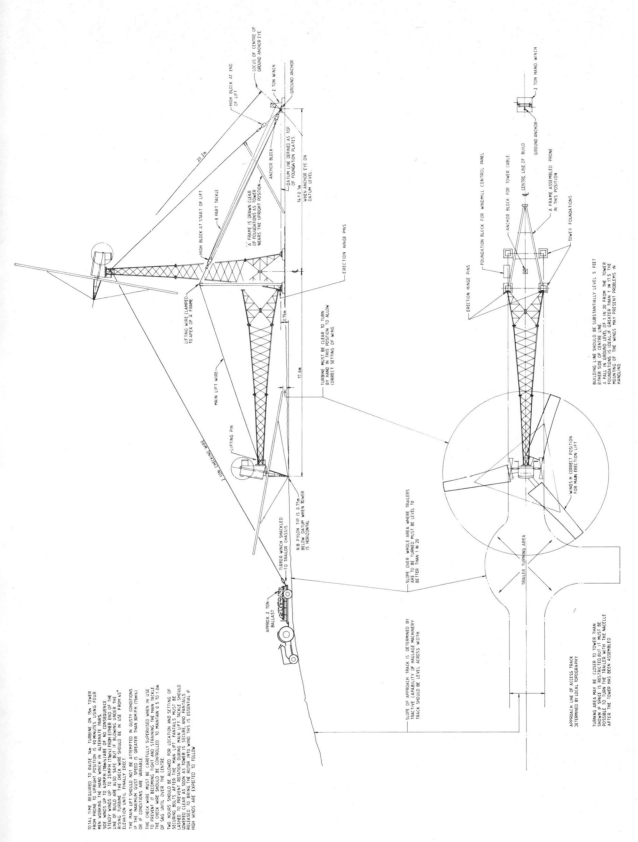

Figure 3 - Diagram of Aerogenerator Erection Procedure

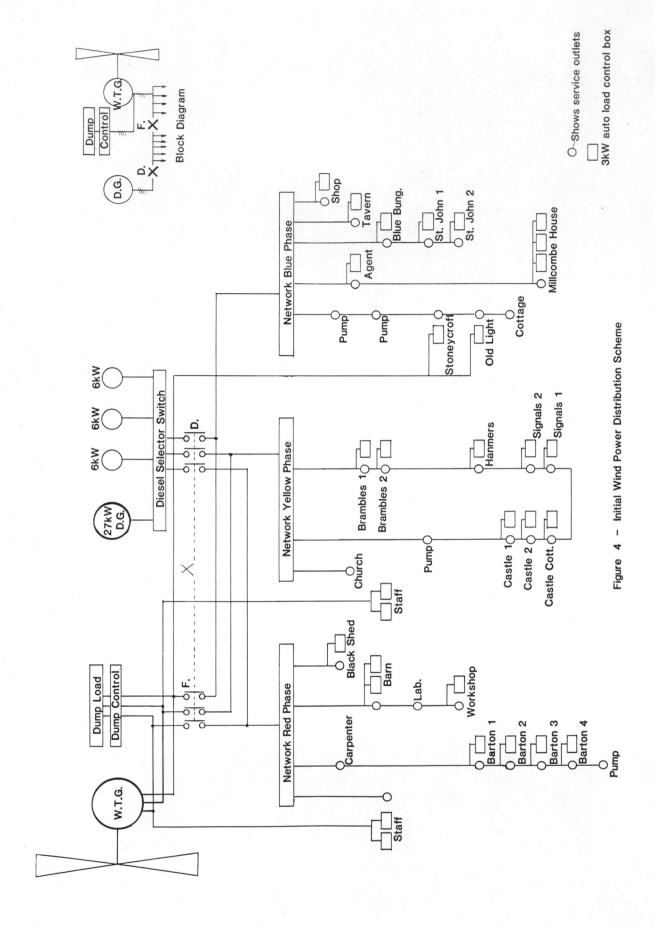

Figure 4 – Initial Wind Power Distribution Scheme

CONSUMER UNIT SETTING LOG

DWELLING		PHASE	SET TURN ON FREQUENCY HZ.																			
			50.1	50.2	50.3	50.4	50.5	50.6	50.7	50.8	50.9	51.0	51.1	51.2	51.3	51.4	51.5	51.6	51.7	51.8	51.9	52.0
BARTON 1		R	1			1	1																
BARTON 2		R	1			1	1																
BARTON 3		R		1	1			1															
BARTON 4		R		1	1			1															
BARN A		R	1					1	1														
BARN B		R						1	1	1													
MILLCOMBE	N/C	R																					
AGENTS	N/C	R																					
OLD HOTEL	N/C	R																					
PHASE LOAD	KW	R	3	5	7	9	11	15	17	18	18	18	18	18								
STAFF A		Y						1	2													
STAFF B		Y						1	2													
BRAMBLE 1		Y			1							1		1								
BLACK SHED		Y	1					1														
HANMER		Y					1		1	1												
CASTLE COTTAGE		Y					1		1													
CASTLE 1		Y				1			1	1												
CASTLE 2		Y				1			1	1												
SIGNALS 1		Y	1	1				1														
SIGNALS 2		Y			2			1														
MILLCOMBE	N/C	Y																					
AGENTS	N/C	Y																					
OLD HOTEL	N/C	Y																					
PHASE LOAD		Y	2	3	6	8	10	15	23	26	26	27	27	28								
BRAMBLE 2		B	1				1	1														
BLUE BUNGALOW		B	1		1																	
ST JOHNS 1		B		1	1																	
ST JOHNS 2		B		1		1																
STONEYCROFT		B				1		1		1												
OLD LIGHT	N/C	B																					
MILLCOMBE	N/C	B																					
AGENTS	N/C	B																					
OLD HOTEL	N/C	B																					
PHASE LOAD		B	2	4	6	8	9	11	11	12	12	12	12	12								
TOTAL LOAD			7	12	19	25	30	41	51	56	56	57	57	58								

DATE 2nd MARCH 1983

FIGURE 5

MEASUREMENT OF ATMOSPHERIC TURBULENCE
AN ASSESSMENT OF LASER DOPPLER ANEMOMETRY

R J Delnon   ERA Technology Ltd
Dr R Johnson   ERA Technology Ltd
C W A Maskell   ETSU
Dr J M Vaughan   RSRE

Abstract

Comparative measurements of atmospheric turbulence have recently been
made by ERA using different types of instrumentation. The instruments
included a Laser Doppler Anemometer (LDA), ERA gust anemometers, a
Vector Instruments cup anemometer and wind vane, and a tethered kite
anemometer (TALA kite).   The objective of the work was to compare the
outputs of the various instruments under reasonably strong wind conditions
and specifically to assess the suitability of using an LDA for measuring
turbulence for wind power studies.   The work was done with financial
support from the Department of Energy and in co-operation with RSRE of
Malvern.   The tests were carried out at the RSRE Airfield at
Pershore, which is in flat terrain offering reasonably homogeneous
wind conditions.

This is an intermediate note describing the measurement programme
undertaken and the method of data analysis.   The data have not yet
been fully analysed and it is hoped that detailed results will be
presented in a subsequent paper.

Introduction

The recent increase in interest and research into wind energy and wind
turbine generators (WTGs) has inevitably called for a greater knowledge
of the characteristics of the wind.   If a wind turbine is to be operated
safely and with a maximum efficiency it is important that the wind
characteristics at a potential wind turbine site are fully understood.
The important characteristics include both turbulence (for stress and
fatigue analysis) and mean wind values (for overall power output estimations).

The only reliable method of assessing the wind characteristics at a
particular site is to carry out direct and detailed wind measurements at
the site.   For two main reasons, the requirement for measured data can
present serious difficulties.

First, there is a tendency for increasingly large wind turbines to be
required, in pursuit of the economies of scale.   The dimensions of the
largest current wind turbines require wind data to be obtained at heights
in excess of 100 metres.   In practical terms, this is not possible
using conventional anemometry.   The erection of a 100 metre tower is
extremely costly and for many sites, the surrounding terrain and
restricted access facilities prohibit it altogether.   The most obvious
exclusion in this respect is wind measurement over areas of water.
The maximum tower height usually considered practical for these purposes
is around 50 metres.

Secondly, as the acceptance of wind energy technology grows the number of sites at which measurements are required is increasing. For serious consideration, all sites require an evaluation based on measured data. In order to carry out measurements at all sites, then either every one is instrumented separately or instrumentation is redeployed on a site to site basis. Costs alone, either of multiple sets of instrumentation and tower hardware or of continuous redeployment and re-erection of towers, make both of these alternatives impractical using present conventional techniques.

At present these factors result either in measurements being restricted to a few special sites with major project undertakings for each, or a substantial decrease in the quality and quantity of measurements taken, eg. mean wind speed measurements up to 20 metres instead of mean and turbulence measurements up to 100 metres.

The need to erect tall towers is therefore a major impediment to the comprehensive and widespread measurements of adequate wind data at potential WTG sites.

Two alternatives to conventional tower mounted anemometry are the Tethered Aerodynamic Lifting Anemometer (TALA) kite, and the Laser Doppler Anemometer (LDA). The TALA kite can be used to estimate the wind profile and turbulence intensities for the height range of interest. It has the advantage of being very portable and inexpensive. However, for detailed measurements it has limitations because of its inability to resolve wind velocity components, and over hilly terrain it is very difficult to measure a velocity profile at a particular site because the horizontal distance between the kite and the operator changes according to the height at which measurements are being taken.

The second of the two alternatives, The Laser Doppler Anemometer (LDA) is part of a group of measuring techniques generally known as Remote Sensing Techniques. A second type of device within this group is based on sound propagation and is usually referred to as Sound Detection and Ranging (SODAR). The counterpart LDA devices are referred to as Light Detection and Ranging (LIDAR).

Applications of LDA systems have been investigated over recent years both in full scale and in the laboratory. Full scale applications at present are mainly in the aviation field, where wind conditions ahead of aircraft, both in normal flight and in landing approaches, have been monitored (1).

Laboratory applications are relatively well established in experimental aerodynamics and in some instances LDAs are superceding the conventional thermal anemometry.

The full scale use of LDA systems is still in the research stage, but the benefit to be gained from developing and applying Remote Sensing Techniques justify the technique to be seriously considered for potential wind turbine site evaluations. This consideration however has to be based on the LDA's ability to obtain suitable data and this can only be assessed via carefully staged experiments ranging from initial brief measurement programmes to more realistic 'on-site' programmes.

A brief programme of measurements was sought by ETSU in order to make the preliminary assessment, and this has now been carried out by ERA using an existing LDA system.

The system used was a ground-based version of the Laser True Airspeed System (LATAS) developed by Royal Signals and Radar Establishment (RSRE) at Malvern for in-flight measurement of true airspeed and detection of wind shear and turbulence.   The tests were made in conjunction with the RSRE staff involved in the instrument development.

The objective of the work was to evaluate the performance of an LDA system with specific reference to its suitability for measuring boundary layer turbulence for wind power studies.   The experimental measurements were made on the airfield at RSRE, Pershore, with the intention of compiling comparable spectra along the mean horizontal wind and across it and (by elevating the instrument beam) at heights up to 100m above ground.

Instruments Used in the Experiment

The instrumentation deployed comprised:

- LDA system

- 1 cup generator anemometer and wind vane

- 3 ERA gust anemometers

- 1 TALA kite

The LDA used for the field experiment was the prototype single axis CO_2 system developed by RSRE.

The system uses a continuous wave laser operating in the infra-red to measure the radial velocity of air in the focal volume of a 150mm diameter aperture telescope with a Germanium lens.   The laser and telescope are mounted on a tripod and form a very compact and robust package.   The high efficiency of the laser (10%) and the low power output required (3W) means that the power supply is particularly compact.   The simplicity and compactness of the overall system is further improved by compressed gas cooling of the infra-red detector. The return signal is split by processing circuitry into a number of radial velocity bins which are refreshed at fixed intervals which can be as short as $50\mu s$, resulting in a distribution of radial velocities within the focal volume of the telescope.   In the equipment's present form one of three compressed types of output data can be selected; these are the velocity corresponding to the peak of the distribution, the centroid of the distribution and the width of the distribution.

The output data rate for these signals depends on atmospheric conditions but is typically 20Hz or more.

The response of the LDA is affected by its radial range resolution, which decreases the increasing range.   However, provided the integral turbulence length scales of interest are significantly larger than the radial range resolution of the LDA, the low frequency part of the spectrum which is part of the most interest for wind power studies, can be readily evaluated.

Comparison of probe volume length with expected range of values of the principal length scales along the mean wind direction (2 and 3), suggests that the resolution limit poses no great problems up to a range of 200 metres, which in any case is close to the maximum focal range of the system.

The system in its present form only directly provides radial velocity measurements, which limitation can be overcome by directing the beam at a shallow elevation angle (up to about 30°) along the mean wind direction. This enables measurements to be approximated by simple corrections to a horizontal velocity for heights of up to 100 metres.

For comparison purposes in the measurements at Pershore Airfield, three conventional types of instrument were deployed in addition to the LDA.

A conventional Vector Instruments cup generator anemometer and wind vane were mounted on a tower at a height of 10 metres. The outputs were recorded on an analogue tape recorder.

Three ERA gust anemometers were oriented in an orthogonal array and also mounted at a height of 10 metres. The outputs were recorded on an analogue tape recorder. The gust anemometers basically consist of a perforated sphere, of diameter about 30mm, connected via a slender tube to a small flexible but stiff plate which is strain gauged. Only the sphere and its support tube are exposed to the wind and as they are deflected under the influence of wind pressure a signal is obtained from the strain gauge circuit. The anemometers are unidirectional and previous calibration tests have shown them to have a reliable cosinusoidal response characteristic. Gust anemometers have normally been deployed in pairs in order to measure horizontal speed and direction. The inclusion of a third unit in this work was an exploratory attempt to measure the vertical component.

The TALA kite was a conventional manually read version manufactured by Approach Fish Inc of USA.

Description of Experiment

The general disposition of the instruments during the experiments at Pershore Airfield is shown in Fig.2. The tower which carried the additional instruments was positioned as conveniently as possible directly upwind of the LDA orientation direction to allow the LDA to record wind conditions as similar as possible to the conventional anemometers. Local constraints such as runways and 'unco-operative' wind directions rendered this virtually impossible in the event, but this was not considered serious since homogeneous wind conditions could safely be assumed because of the uniform nature of the surrounding terrain.

The experiment was carried out over a $1\frac{1}{2}$ day period in December 1982.

The LDA output was recorded for approximately 12 minutes or more for each of the heights.

The outputs from the additional anemometers at 10 metres were continuously recorded during the observation periods. The TALA kite data was recorded in the form of manually read values of line tension every 20 seconds, wind direction and kite elevation angle every 2 minutes.

Although wind strength was not a critical factor, the timing of the data collection was selected to coincide with the expected maximum winds.

Measurements were obtained for two separate observation periods:

Period 1 - 16 December 1982, 1600 hrs - 1644 hrs.

> The elevation of the LDA was set to its minimum value of 6° and the range adjusted to record data at 10 and 20 metres height.

> TALA kite data was obtained from the same heights. The kite data indicated a windspeed at 20 metres of about 9m/s.

Period 2 - 17 December 1982, 1147 hrs - 1310 hrs.

> The elevation of the LDA was set to its maximum value of 27° and data recorded at heights of 10,20,30,50 and 75 metres.

> TALA kite data was obtained for 50 and 100 metres. The kite data indicated a windspeed at 50 metres of about 7.5m/s.

Proposed Data Analysis

The chief comparison to be made using the data gathered during the experiment is that of estimates of the co-spectrum of longitudinal turbulence, as derived by the fixed anemometry on the tower, the LDA equipment, and that predicted by consideration of classical theory, using an estimate of roughness length based on mean wind profile and characteristics of the terrain up wind of the measurement station.

The mean windspeed profile as measured using the TALA kite and LDA measurements will be compared. Variation of the longitudinal spectrum (following that of integral length scale) with height cannot reliably be made using the kite, so the spectra calculated from LDA measurements will be compared with those predicted by theory.

The raw data gathered from the experiment requires considerable processing to obtain the necessary spectra and estimates of mean velocity profile and mean wind direction.

The gust anemometer outputs, from which turbulence spectra at 10m will be derived are proportional to the square of wind velocity components in their respective planes of sensitivity. The instruments also have a band of resonance around the 13Hz which must be filtered before the results can be analysed. The signal supplied by the LDA similarly requires filtering as the centroid output was refreshed every 50ms during each run.

The analog signals for the azimuth wind components measured by the gust anemometers will then be passed through an A/D converter and converted to wind velocities from the calibration curves of the anemometers.

Mean windspeed and direction for each (nominally) 15 minute run will be calculated using the outputs of the fixed anemometry, and this mean wind direction will be used to confirm that the alignment of the LDA system was correct for the period of the run. The TALA kite results will give a second check for the levels in excess of 10m.

Power density spectra along the mean wind direction will be calculated using Fast Fourier Transform from the components supplied by the gust anemometers and the laser. These will be compared directly for 10m height and for qualitative differences as the height of the laser search volume is increased. Further comparison based on estimates for roughness length and boundary profiles and using the classical spectrum (Ref.4) will be made for all measured heights.

The analysis will be completed shortly and a final report prepared for the Department of Energy.

Acknowledgments

This work is being undertaken under contract to ETSU and the Department of Energy.

Mr R Foord and Mr P H Davis of RSRE prepared the LDA equipment and operated it during the field experiment.

Permission from the Directors of ERA Technology Ltd to publish the paper is gratefully acknowledged.

References

1.   Applied Physics Newsletter
      Royal Signals and Radar Establishment, Malvern, May 1982.

2.   Foord, R et al 'Measurements with CO_2 systems'.
      Symposium on Long Range and Short Range Optical Velocity.
      Measurements.   Institution Saint Louis, September 1980.

3.   ESDU Data Sheet, Item 75001.
      July 1975.

4.   'Turbulence Spectra, height scales and structure parameters
      in the stable surface layer'.
      J C Kaimal Boundary Layer Meteorology, Vol. 4.
      (1973) pp 289-309.

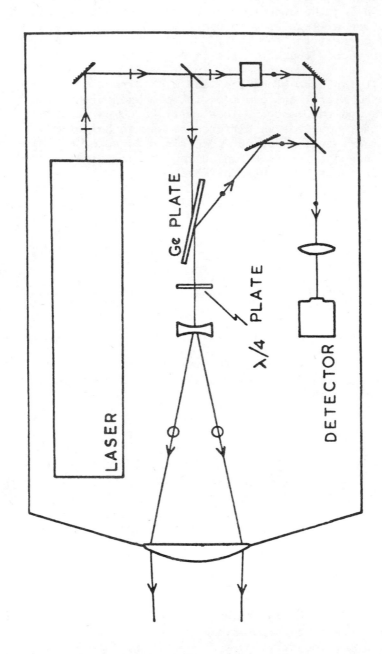

Fig.1: Outline of Prototype Single Axis CO_2 Laser Doppler Anemometer Developed by RSRE

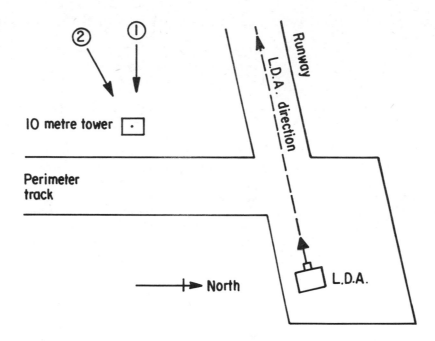

Fig.2a : ORIENTATION OF INSTRUMENTS

Distance between tower and LDA = 82 metres

1. Period 1 wind direction = 270°
2. Period 2 wind direction = 240°

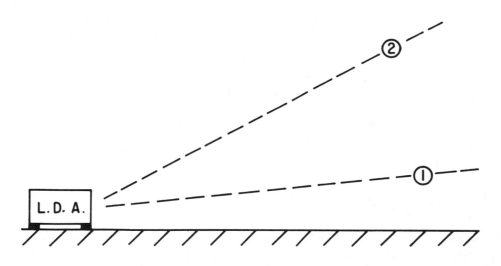

Fig.2b : SIDE VIEW ELEVATIONS OF LDA

Period 1 = 6°   Period 2 = 27°

THE DISTORTION OF LARGE-SCALE TURBULENCE BY A WIRE GAUZE DISC

E.J. Fordham
Cavendish Laboratory
University of Cambridge

Abstract

Measurements of the distortion of turbulence upstream of a circular wire gauze simulator have been made in large scale grid turbulence. It is found that significant, but not dramatic distortion effects occur with the radius to length scale ratio used. The main effect is the suppression of the low frequency components of the turbulence close to the gauze due to the dominance of the surface 'source' effect over the vortex-stretching effect in the distortion. The relevance of these results to wind turbines and moving reference frame turbulence models is discussed.

Introduction

Current efforts to understand the fluctuating forces, moments and stresses due to atmospheric turbulence acting on large wind turbines have so far recognized the importance of describing the turbulence in a moving reference from following the motion of a blade, and have resulted so far in moving frame turbulence models discussed by Fordham and Anderson (5), Connell (14), Kristensen and Frandsen (19) and others. All of these authors based their work on that of Rosenbrock (15) and assume homogeneous isotropic turbulence. Apart from the approximations involved in such a assumption when referring to atmospheric boundary layer turbulence, all the papers cited ignore the distortion of the turbulence structure by the presence of the wind turbine itself, which will reduce the wind speed at the rotor by a factor of about 1/3, for an ideally loaded ideal rotor. As noted by Vermeulen (1), this problem is regularly referred to in the literature (e.g. by Jensen and Frandsen (2), Raab (3), and de Vries (4)), but has not hitherto been investigated in the wind energy community. This paper describes an experimental first approach to the problem, using discs of wire gauze to simulate the wind turbine. A turning model was not used, since with a model of realistic size, it would be necessary to consider turbulence of larger integral scales than can be generated in all but the largest wind tunnels, in order to model realistically the length scale to diameter ratios appropriate for wind turbines operating in the natural wind. The moving-frame turbulence models referred to above all recognize the importance of turbulence integral scale. The relevance of the present approach to an actual wind turbine will be discussed in more detail below.

Rapid Distortion Theory (RDT)

Theoretical approaches for calculating changes in turbulence in distorted flows have been used in the past to investigate flows round bluff bodies, notably Hunt's theory (8) of flow round a circular cylinder tested experimentally by Britter et al. (11), and Graham (7) has used similar methods to investigate flow through a porous plate. These papers use generalisations of the rapid distortion theory of Batchelor and Proudman

(9). RDT is applicable to flows where the following conditions hold:

1) the turbulence is weak i.e. $\alpha = u_1'/\overline{U}_1 \ll 1$

2) the timescale l_x/u_1' for the non-linear interactions between the energy containing eddies (the "turnover" timescale or Lagrangian timescale) is large compared with the timescale for the distortion of order $a/\overline{U}_1$

   i.e. $u_1'a/\overline{U}_1 L_x \ll 1$. This is the 'rapid distortion' condition.

3) the body and turbulence Reynolds numbers are large

   i.e. $a\overline{U}_1/\nu \gg 1$ and $\beta = u_1'L_x/\nu \gg 1$.

Notation and values of critical parameters used in the present experiment are given in Table 1, from which it can be seen that the assumptions are well-satisfied in the present experiment. In atmospheric turbulence, intensities may well reach 0.15 or greater in rough terrain, so that condition 1) may be only marginally applicable; however the basic physical processes involved in the distortion are probably very similar.

Subject to the above conditions, the vorticity transport equation:

$$\frac{\partial \underline{\zeta}}{\partial t} + (\underline{u} \cdot \nabla)\underline{\zeta} = (\underline{\zeta} \cdot \nabla)\underline{u} + \nu\nabla^2\underline{\zeta} \tag{1}$$

where $\underline{u}$ is the flow velocity, and $\underline{\zeta} = \nabla \times \underline{u}$ can be shown (e.g. Hunt (8)) to reduce to:

$$\frac{\partial \underline{\omega}}{\partial t} + (\underline{U} \cdot \nabla)\underline{\omega} = (\underline{\omega} \cdot \nabla)\underline{U} + 0(\alpha,\beta) \tag{2}$$

where $\underline{\omega}$ is the turbulent vorticity, and $\underline{U}$ is the mean flow velocity.

This is the basic equation of rapid distortion analysis and its importance stems from the fact that it is of the same form as (1); hence the interpretation of (1) at large Reynolds numbers that 'vortex lines move with the fluid' can be applied to (2) in the sense that the vortex lines of the turbulence are stretched and rotated by the mean flow. (Fig. 7 illustrates).

The change in turbulent vorticity can then be calculated using Couchy's solution of (1) applied to (2):

$$\omega_i(\underline{x},t) = \omega_j(\underline{a},0)\partial x_i/\partial a_j \tag{3}$$

where $\underline{a}$ is a Lagrangian co-ordinate (i.e. $\underline{x} = \underline{a}$ at $t = 0$). The details of the calculation are set out by Hunt (8) for the case of a circular cylinder. Having calculated the distorted turbulent vorticity in this way, the turbulent velocity can be found as the Biot-Savart velocity field of the turbulent vorticity, plus a gradient of a fluctuating potential such that the necessary boundary conditions at the obstacle's surface are satisfied; in the case of a solid object, this will be that the normal component of velocity is always zero at the surface. There are two physical effects to consider in understanding distorted turbulence; firstly the stretching, rotation and 'piling-up' of vortex lines around an obstacle, and secondly the so-called "blocking" or source effect caused by the surface boundary condition. The former acts in a region up to a few radii away from the body, and the second acts in a region within distances of order L_x from the body. The relative importance of the two effects thus depends on the radius to length scale ratio a/L_x of the obstacle and the turbulence; when a/L_x is <u>small</u>, the source effect is dominant; when a/L_x is <u>large</u>, the vortex stretching effect is dominant except within distances of order L_x from the body.

Considering the present experiment, the vortex stretching mechanism will tend to amplify the streamwise component of turbulence measured here, whilst the source effect will tend to reduce it. Obviously the boundary condition at a gauze is not as severe as the condition at a solid surface, but the resistance of the gauze does produce a similar suppression of the turbulence close to the gauze. The details of the boundary condition for a gauze are discussed by Graham (7). The case of a wind turbine is discussed below.

Experiments with gauze discs

In the present study, simulators consisting of a wire gauze fixed to a thin brass hoop with a small diffuser angle (Fig. 3) were used in the wind tunnel of the Dept. of Applied Mathematics and Theoretical Physics, Cambridge; a very coarse square mesh grid was used to generate turbulence of probably the largest scale obtainable with reasonable homogeneity in a tunnel of this size (0.45 m x 0.45 m square section). Relevant parameters are given in Table 1; some comparison is made with the results of Britter <u>et al.</u> (11); the same tunnel was used in their experimental test of Hunt's theory (8). A large length scale was important in order to be able to model the a/L_x ratio expected for large wind turbines operating in atmospheric turbulence; it was also intended to make two-point correlation measurements and examine the way in which the lateral correlation was changed as the flow approached the gauze. Meaningful measurements of this type could not be done unless the lateral correlation was significantly greater than zero between points separated across the rotor disc. A moderately dense gauze was chosen, with a resistance coefficient determined by directly measuring the pressure drop across the gauze in a duct; fair agreement with the predictions of ESDU (20) for the gauze porosity was obtained. The mean flow field of such a gauze can be calculated on the virtual source theory of Taylor (10); iso-speed contours and some velocity vectors are shown in Fig. 1 for k = 2.0. The accuracy of this theory has been tested on the gauze centre-line (see Fig. 2) in both turbulent and low turbulence flow; good agreement is obtained, although it is not possible to check the value of k accurately by this method.

Turbulence measurements were made using DISA type 55P01 hot wire probes and TSI type 1054 B linearised constant temperature anemometers; the probes were mounted and traversed through small holes punched in the gauze. Streamwise components of turbulence only were measured. The turbulence signals were AC coupled with a time-constant of 2.2 seconds to buffer amplifiers and a 7-track FM tape recorder with a bandwidth to beyond 2 kHz. The recordings were later digitised and recorded on magnetic tape for analysis on the University of Cambridge IBM 3081 to obtain the probability density, auto-correlation and spectral density functions of the signals, and in the case of two-point measurements, the cross-correlation function.

The homogeneous turbulence was also measured without a gauze sample; although the scales, intensities and single point spectra were available from the work of Britter et al. (11) (Figs. 5, 6), lateral correlation measurements had not been made, which are of interest here. The results of the lateral correlation measurements are shown in Figs. 4 and 5. A typical spectrum measurement is shown in Fig. 13, showing agreement with Britter et al. (11). The form of the longitudinal spectrum can be approximated by the von Karmàn expression:

$$\hat{\phi}_{11}(\hat{\kappa}_1) = \frac{1}{\pi} \{1 + (1.34\hat{\kappa}_1)^2\}^{-5/6} \tag{4}$$

where $\quad \hat{\phi}_{11}(\hat{\kappa}_1) = \phi(n)\overline{U}_1/4\pi u_{1\infty}'^2 L_x \tag{5}$

and $\quad \hat{\kappa}_1 = \kappa_1^* L_x = 2\pi u L_x/\overline{U}_1.$ Here n is the ordinary cyclic frequency.

The turbulence was not isotropic in several respects (see Table 1 for u_2'/u_1' and u_3'/u_1' ratios and Fig. 6 for the departure of ϕ_{22} and ϕ_{33} from isotropy at low $\hat{\kappa}$) but was approximately so; in particular the lateral correlation of the u_1-component did not show great deviations from isotropy, unlike measurements in the atmosphere (e.g. Fordham and Anderson (5), Tennissen (12), Counihan (13)).

Results obtained

1)   Spectra: spectra obtained on the centre line of both gauzes, and at points off axis for the smaller gauze, are shown on Fig. 9 to 12 for various axial positions. The main feature is the expected reduction in energy at low frequencies as the gauze is approached; slight amplification is observed at intermediate and low frequencies at the larger distances from the gauze. The spectra have normalised according to (5) with respect to upstream values of U_1, u_1' and L_x. The behaviour at high frequencies was not very clear, but there appeared to be slight suppression of the high frequency components close to the gauze. The overall r.m.s. fluctuation velocities are plotted in Fig. 17. Again the main feature appears to be the reduction in intensity close to the gauze, with slight amplification at the larger distances measured, compared with the free turbulence measurements, presumably a reflection of the integrated spectral energy which shows a slight increase in these cases over the intermediate energy containing range around $\hat{\kappa}_1 = 1$. (The von Karmàn spectrum (4) is shown plotted in area preserving form in Fig. 14).

The physical mechanisms at work here have been discussed in a previous section; as expected, close to the gauze, the source effect is dominant for the low frequency components, although not as severe as would be expected on the grounds of 'blocking' alone. Vortex-stretching is also occurring; however the source effect dominates close to the gauze for this a/L$_x$ ratio. For smaller a/L$_x$ ratios, (larger scales) the dominance is expected to become more pronounced. Furthermore the range of frequencies over which the source effect dominates increases as the gauze is approached.

2) Autocorrelations: autocorrelation functions corresponding to the above spectra were obtained, and extreme cases plotted in Fig. 16, normalised with respect to local variance. The time-scale is significantly but not dramatically reduced as the gauze is approached; the spectrum is becoming more like white noise as its flat range is extended.

3) Cross-correlations: cross-correlation functions, with time-lag, were obtained for various points across the rotor disc, and typical results are shown in Fig. 19 and 20. Zero-lag correlation coefficients for other cases are shown in Fig. 18. The results all show a definite reduction in correlation coefficient, even though normalised with respect to local variance, which shows a reduction as the gauze is approached. This is presumably because of the reduction in energy of the large wavelength components of the turbulence which are those responsible for the lateral correlation.

4) Probability distributions: probability distribution functions for the turbulence were also examined; flatness factors and skewness factors for the distributions were evaluated. Unfortunately a very small number of erroneous 'outliers' produced by the digitiser prevented accurate estimates of these parameters; however the qualitative behaviour was clear, showing significant increases in flatness and non-zero skewness as the gauze was approached, i.e. the turbulence became markedly non-Gaussian. Britter et al. (11) also noticed a similar effect in measurements near a circular cylinder and proposed an explanation in terms of the orientation of vortex lines expected for large fluctuations.

The relevance of gauze discs to wind turbines

This study has demonstrated that two effects are important in the distortion of the turbulence 1) the vortex-stretching by the mean flow field, and 2) the source effect due to the surface boundary condition. With regard to 1), whilst the gauze does not model such features as the periodicity in the flow close to the rotor, the flow field plotted in Fig. 1 is very similar to the field calculated by Anderson (16) from a vortex ring wake model for corrections to field test data. The flow reduction is assumed to be uniform on Taylor's virtual source theory (10), which compares well with the induced velocity profile shown in Fig. 21, at least for the optimum blade. For an untwisted blade, the agreement is less good, but the upstream flow may be expected nevertheless to be qualitatively similar.

The relevance of the surface source effect to a wind turbine is less clear, and must be considered conjectural until similar experiments can be done with turning models or real turbines. However it should be noted that the blade passing frequency for a notional tip speed ratio 8 rotor at this scale is much larger than the frequencies of the turbulence components which are most affected by the source effect; this suggests that the representation by a static gauze is at least plausible.

The other important parameter is the ratio a/L_x; it was not possible in the present experiments to examine a wide range of a/L_x ratios, but the behaviour is expected to be sensitive to this ratio. However it is clear from known length scales in atmospheric turbulence that large wind turbines will be operating in turbulence with scales of the same order as their diameter. The values of a/L_x used here are therefore in a sensible range for modelling a real wind turbine. It is not clear from these experiments whether there will be any significant effects due to the marked anisotropy of the atmospheric wind, with lateral to longitudinal scale ratios significantly smaller than the isotropic ratio.

Implications for moving reference frame models

The moving reference frame models of Fordham and Anderson (5), Kristensen and Frandsen (19) and Connell (14) so far use isotropic turbulence theory. However they can be extended to accommodate anisotropy and distortion, provided that the lateral correlation functions with time lag are known. The difficulty with accommodating distortion is predicting how the lateral correlations and autocorrelations are likely to change. The details of such a calculation have yet to be done, but the results of the present study indicate that timescales of the turbulence may be reduced, and that lateral correlations may also be reduced. This would lead to moving-frame spectra which are rather more broad-band in character than the ones so far published. The total r.m.s. turbulence is expected to be reduced, but the probability distribution may become significantly non-Gaussian, with a high flatness factor.

Conclusions

This study has demonstrated that significant, if not dramatic, turbulence distortion effects occur in front of a wire gauze disc in large-scale grid turbulence. The mean flow field of a wind turbine is believed to be well-modelled by a gauze, as a first approximation, so that vortex-stretching effects will be similar in the two cases. The relevance of the gauze's source effect is conjectural, but plausible in view of the high blade-passing frequency of a notional rotor of the same scale. The main effects, which are expected to be stronger as a/L_x i.e. further reduced, are that the time scales of the autocorrelations become shorter, and that lateral cross-correlations are reduced. The implications for moving reference-frame turbulence models are that the rotationally-sampled wind is likely to show a more broad band spectrum than would otherwise be predicted.

Acknowledgements

The financial support of the SERC and CEGB and the University of Cambridge is acknowledged. The hot-wire apparatus was loaned by CERL, Leatherhead, and the FM tape recorder and digitiser by the Fluid Mechanics Group of the Cavendish Laboratory. The measurements were made in the Dept. of Applied Mathematics and Theoretical Physics, Cambridge. Dr. John Mumford gave much assistance in setting-up the experiments and provided some of the analysis software, and Mr Ray Flaxman made up the gauzes.

References

1. Vermeulen, P.E.J.; 'Definition of the concept of turbulence in relation to wind turbine design'. TNO Report 81-09061, (July 1981), Apeldoorn, Netherlands.

2. Jensen, N.O. and Frandsen, S.; 'Atmospheric turbulence structure in relation to wind generator design'. 2nd Int. Symp. on Wind Energy Systems, Amsterdam. BHRA (1978).

3. Raab, A.; 'Combined effects of deterministic and random loads in turbine design'. 3rd Int. Symp. on Wind Energy Systems, Copenhagen. BHRA (1980).

4. de Vries, O.; 'Fluid dynamic aspects of wind energy conversion', AGARD-AG-234, (July 1979).

5. Fordham, E.J. and Anderson, M.B.; 'Analysis of the results of an atmospheric experiment to examine the structure of the wind as seen by a rotating observer'. Proc. 4th BWEA Workshop, Cranfield, Bedford, Pub. BHRA (1982).

6. Milborrow, D.J. and Ainslie, J.F.; 'Calculation of the flow patterns and performance of wind turbines using streamline curvature methods'. Proc. 2nd BWEA Workshop, Cranfield, Bedford (April 1980).

7. Graham, J.M.R.; 'Turbulent flow past a porous plate'. J. Fluid Mech., 73, 3, pp. 565-591 (1976).

8. Hunt, J.C.R.; 'A theory of turbulent flow round two-dimensional bluff bodies'. J. Fluid Mech., 61, 4, pp. 625-706 (1973).

9. Batchelor, G.K. and Proudman, I.; 'The effect of rapid distortion of a fluid in turbulent motion'. Q. J. Mech. and Appl. Math., 7, 1, (1954).

10. Taylor, G.I.; 'Air resistance of a flat plate of very porous material'. ARC Reports and Memoranda No. 2236 (Jan. 1944).

11. Britter, R.E., Hunt, J.C.R. & Mumford, J.C.; 'The distortion of turbulence by a circular cylinder'. J. Fluid Mech., 12, 2, pp. 269-301 (1979).

12) Tennissen, H.W.; 'Structure of mean winds and turbulence in the planetary boundary layer over rural terrain'. Boundary Layer Meteorology 19, pp. 187-221, (1980).

13. Counihan, J.; 'Adiabatic atmospheric boundary layers'. Atmospheric Environment 9, pp. 871-905 (1975).

14. Connell, J.R.; 'The spectrum of wind fluctuations encountered by a rotating blade of a wind energy conversion system'. Battelle Pacific Northwest Laboratories Report PNL 4083 (1981).

15.   Rosenbrock, H.H.;   'Vibration and stability problems in large wind
      turbines having hinged blades'. Electrical Research Association
      Report C/T 113, (1955).

16.   Anderson, M.B.,; Ph.D. dissertation, University of Cambridge, 1981.

17.   Powles, S.J.R., Anderson, M.B., Wilson, D.M.A. and Platts, M.J.;
      'Articulated wooden blades for a horizontal axis wind turbine;
      design, construction and testing'. Proc. 4th BEWA Workshop,
      Cranfield, Bedford, March 1982.   (BHRA 1982).

18.   Rayner, J.M.V.; J. Fluid Mech., 91, pp. 697-764 (1979).

19.   Kristensen, L. and Frandsen, S.;   'Model for power spectra of the
      blade of a wind turbine measured from the moving frame of
      reference'. Jnl. Wind Engineering and Industrial Aerodynamics,
      10, pp. 249-262 (1982).

20.   Engineering Sciences Data Unit; 'Pressure losses for round wire
      gauzes normal to the flow Data Item no. 72009, (Fluid Mechanics,
      Internal Flow seris, Volume 3).

21.   Batchelor, G.K.; 'Thoery of homogeneous turbulence', C.U.P.,
      (reprinted 1982).

Table 1

Grid mesh size		11.4 cm
Bar to mesh ratio		1:4
Turbulence intensity	$u_1'/\overline{U}_1$	0.064
(c.f. Britter et al. (11)		0.062)
Lateral turbulence ratio	u_2'/u_1'	0.84
Vertical turbulence ratio	u_3'/u_1'	0.85
Length scale	L_x	5.2 cm
(c.f. Britter et al. (11)		6.0 cm)
Mean wind speed	U_1	10 ms^{-1}
Gauze resistance coefficient	k	~ 2.2
Diameter of gauze samples	$2a_1$	7 cm
	$2a_2$	10 cm
Blockage area ratios for gauze discs	$a_1 = 3.5$ cm	2%
	$a_2 = 5.0$ cm	4%
Radius to length scale ratio	a_1/L_x	0.67
	a_2/L_x	0.96
Gauze diameter Reynolds numbers	$2\overline{U}_1 a_1/\nu$	4.7×10^4
	$2\overline{U}_1 a_2/\nu$	6.7×10^4
Turbulence Reynolds number	$u_1' L_x/\nu$	2.1×10^3
Gauze wire Reynolds number	$d\overline{u}_1/\nu$	1.2×10^2
Timescale for non-linear interactions	L_x/u_1'	84 ms
Timescale for mean flow distortion	$a_1/\overline{U}_1$	7 ms
	$a_2/\overline{U}_1$	10 ms
Blade passing frequency for notional two-bladed rotors at $\lambda = 8$	ν_b ($a_1 = 3.5$ cm)	730 Hz
	($a_2 = 5.0$ cm)	510 Hz
Reduced blade passing frequencies	$\hat{\kappa}_b = 2\pi(2\nu_b)L_x/\overline{U}_1$	24
	(for a_1, a_2)	17

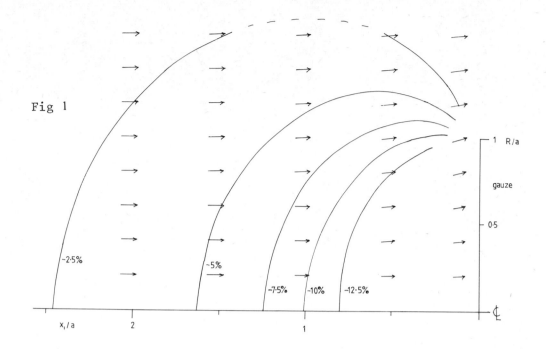

Fig 1

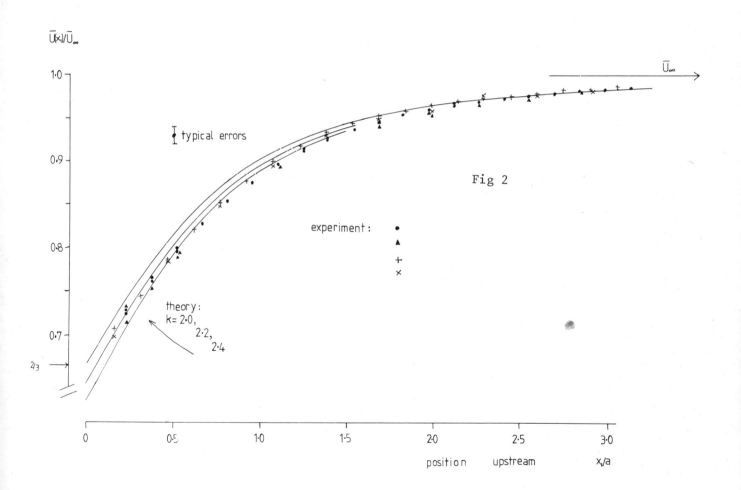

Fig 2

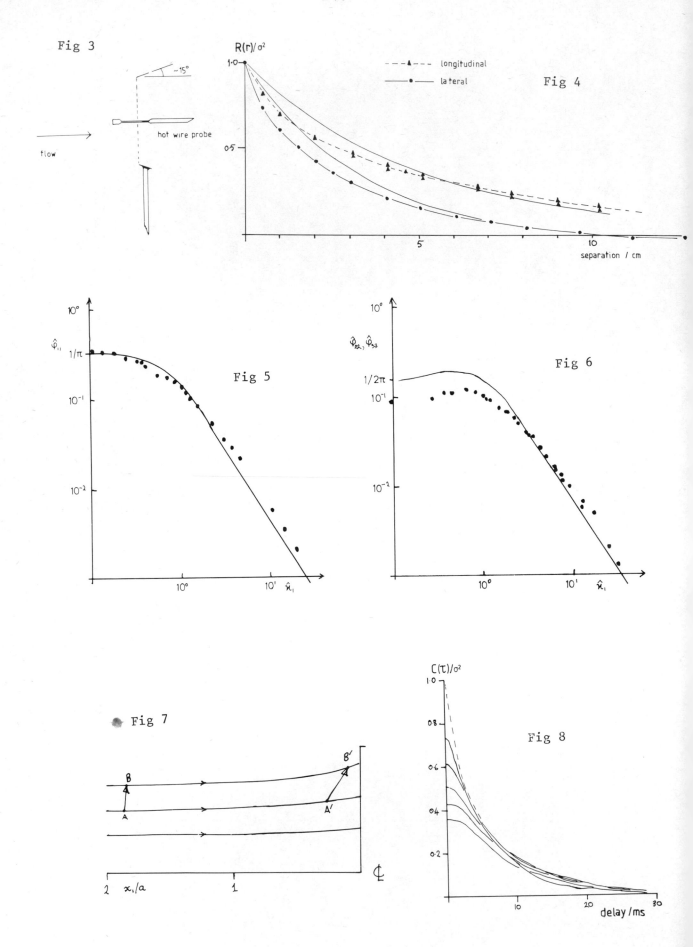

Fig 3

flow

hot wire probe

~15°

Fig 4

$R(r)/\sigma^2$

1·0

0·5

- - -▲- - - longitudinal
——●—— lateral

5

10

separation / cm

Fig 5

$\hat{\varphi}_{11}$

10^0

$1/\pi$

10^{-1}

10^{-2}

10^0

10^1

$\hat{\varkappa}_1$

Fig 6

$\hat{\varphi}_{22}, \hat{\varphi}_{33}$

10^0

$1/2\pi$

10^{-1}

10^{-2}

10^0

10^1

$\hat{\varkappa}_1$

Fig 7

B

A

B'

A'

₵

2   x_1/a

1

Fig 8

$C(\tau)/\sigma^2$

1·0

0·8

0·6

0·4

0·2

10

20

30

delay /ms

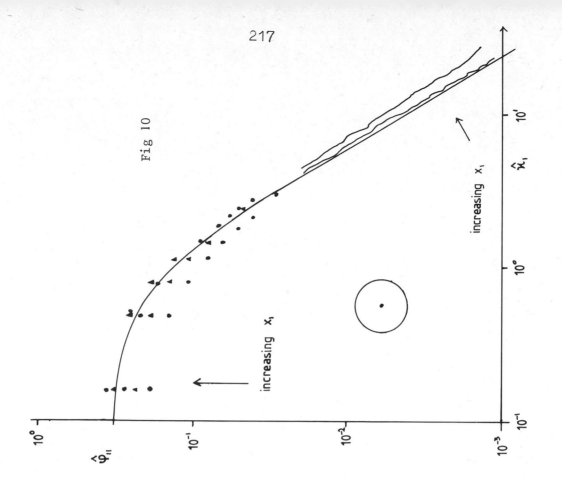

Fig 10

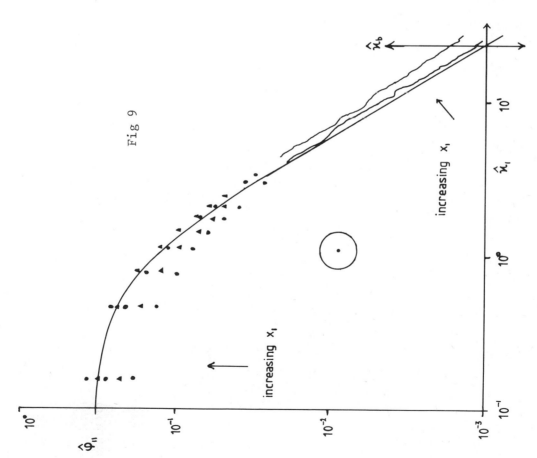

Fig 9

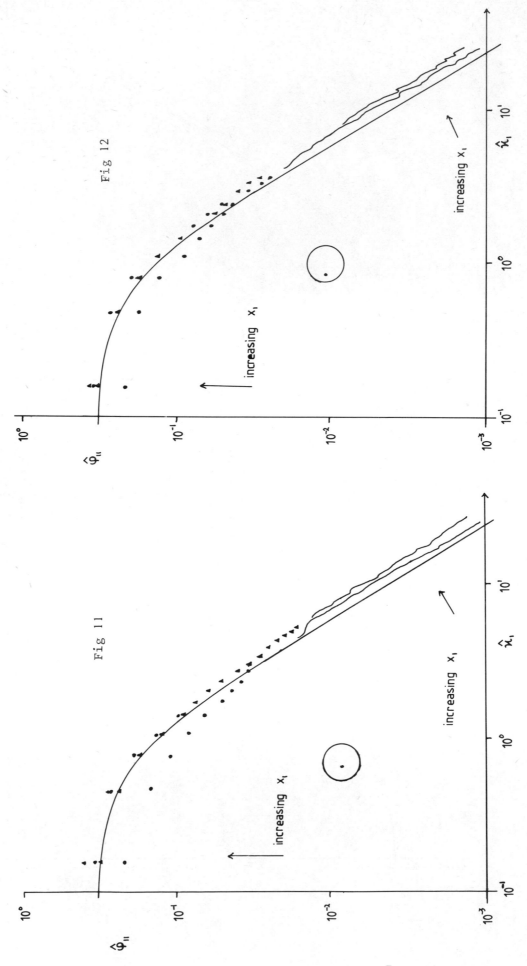

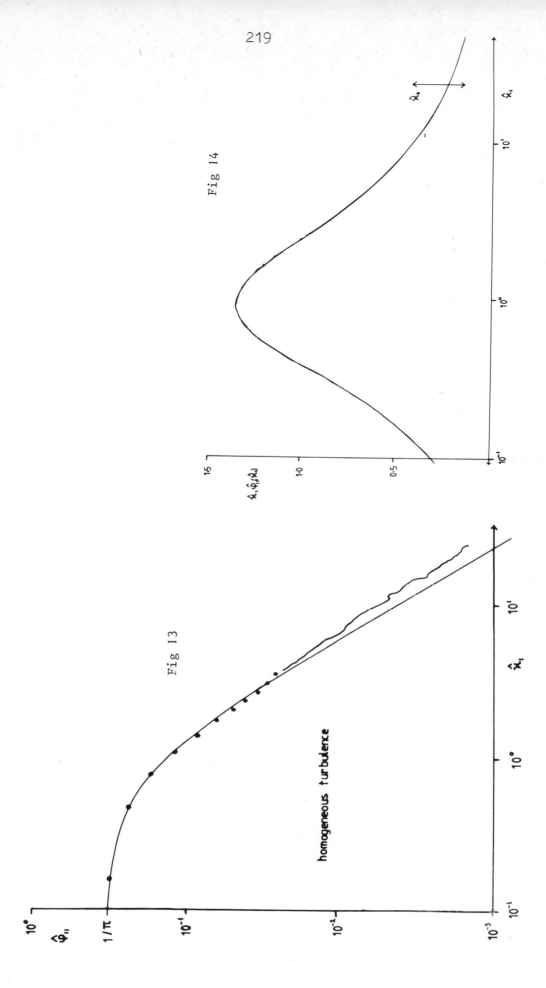

Fig 14

Fig 13

homogeneous turbulence

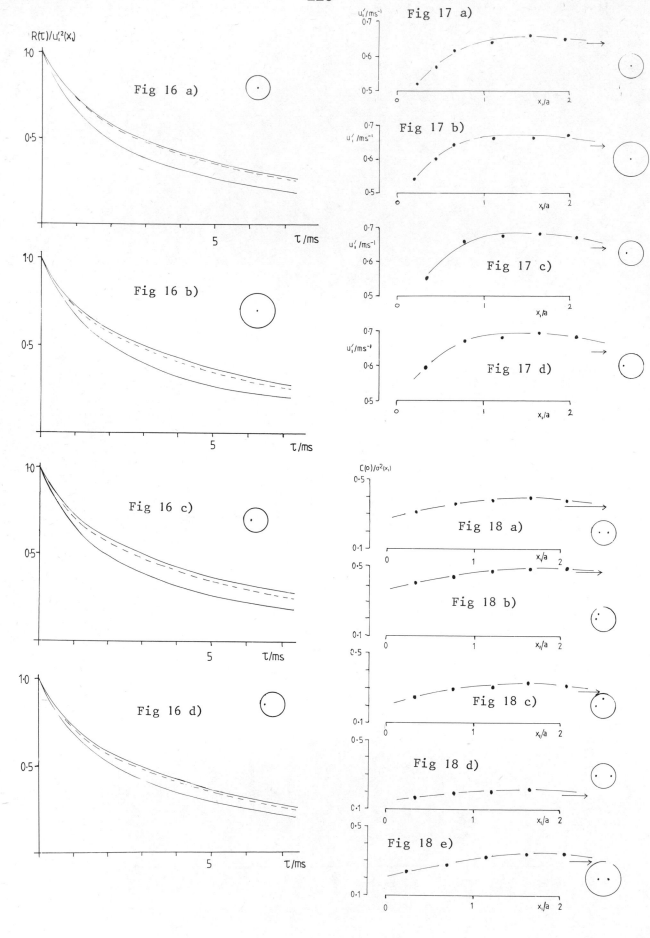

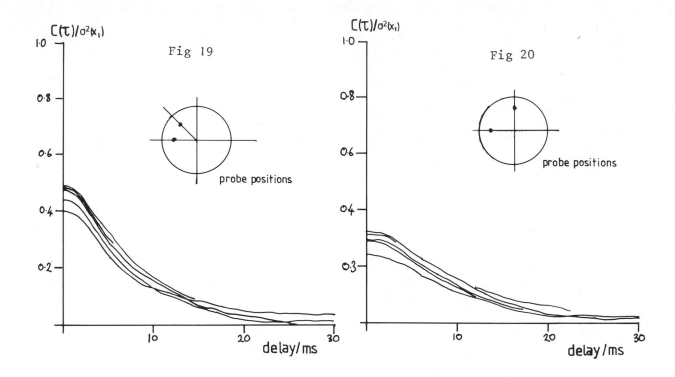

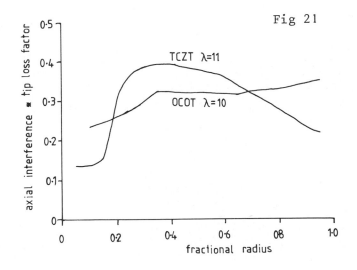

Notes on the figures

1) Iso-speed contours (percentage reductions indicated) and sample flow vectors for flow upstream of a circular gauze.

2) Mean velocities on gauze centre line; theoretical curves for k = 2.0, 2.2, 2.4; experimental points for 7 cm dia. gauze; low turbulence ($\bullet$), in grid turbulence ($\blacktriangle$); for 10 cm gauze, low turbulence ($+$), and in grid turbulence ($\times$).

3) Sketch of gauze discs used and probe disposition.

4) Correlation structure of the turbulence used. Longitudinal correlations inferred from auto-correlations by the use of Taylor's hypothesis; lateral correlations measured directly. Continuous lines show exponential model with length scale of 5 cm.

5) Power spectrum of longitudinal component obtained by Britter et al. (11)

6) Power spectra of lateral components obtained by Britter et al. (11)

7) Schematic illustration of stretching and rotation of vortex lines by the mean flow field.

8) Cross-correlation functions of the undistorted turbulence with time-lag. The zero-lag values are the first five lateral coefficients in Fig. 4. Typical auto-correlation curve shown dashed for comparison.

9) Distortion of the turbulence spectrum on the centre-line of the 7 cm gauze. Only selected points plotted for low $\hat{\kappa}_1$; extreme cases only plotted as continuous lines for high $\hat{\kappa}_1$. Typical reduced blade passing frequency $\hat{\kappa}_b$ indicated. Solid line is equation (4).
Symbols (for increasing upstream distance): $\bullet$ $x_1/a = 0.23$
$\blacktriangle$ $x_1/a = 0.45$
$\bullet$ $x_1/a = 0.66$
$\blacktriangle$ $x_1/a = 1.10$
$\bullet$ $x_1/a = 1.97$

10) As Fig. 9, but for 10 cm dia. gauze.
Symbols (increasing upstream distance): $\bullet$ $x_1/a = 0.20$
$\blacktriangle$ $x_1/a = 0.45$
$\bullet$ $x_1/a = 0.66$
$\blacktriangle$ $x_1/a = 1.11$
$\bullet$ $x_1/a = 1.99$

11) As Fig. 9, but probe position at radius 1.3 cm.
Symbols (increasing upstream distance): $\bullet$ $x_1/a = 0.34$
$\blacktriangle$ $x_1/a = 0.78$
$\bullet$ $x_1/a = 1.21$
$\blacktriangle$ $x_1/a = 1.65$

12)   As Fig. 9, but probe position at radius 2.3 cm.
      Symbols (increasing upstream distance):   ● $x_1/a = 0.34$
                                                   ▲ $x_1/a = 0.78$
                                                   ◕ $x_1/a = 1.65$
                                                   ▲ $x_1/a = 2.08$

13)   Power spectrum of longitudinal component of turbulence obtained
      in the present experiments. c.f. Fig. 5.

14)   Von Kàrmàn spectrum (equation (4)) plotted in area-preserving
      form as $\hat{\kappa}_1\hat{\emptyset}_{11}(\hat{\kappa}_1)$ v. log $\hat{\kappa}_1$. Note position of notional blade-
      passing frequency $\hat{\kappa}_b$.

16)   Extreme case of auto-correlation curves corresponding to the
      spectra shown in Figs. 9 to 12. Upper curve: $x_1/a \simeq$ =; lower
      curve: $x_1/a \simeq 0.2$ ((a) and (b)), $x_1/a \simeq 0.3$ ((c) and (d)).
      Normalised with respect to local variance. Typical auto-
      correlation curves for undisturbed turbulence shown dashed.

17)   Variation of r.m.s. turbulence velocity upstream of gauze. Probe
      radial positions indicated by symbols at right (corresponding
      to a) b) c) d) in Fig. 16, and to Figs. 9-12 respectively).
      Turbulence velocity of 0.64 ms^{-1} for undisturbed turbulence
      indicated by arrow.

18)   Zero-lag coefficients for various pairs of probe positions across
      the gauze disc, as function of upstream distance. Probe positions
      indicated by symbols at right. Correlation coefficients for these
      separations in undistorted turbulence indicated by arrows. Note
      that these coefficients are normalised with respect to the local
      variance.

19)   Cross-correlation functions with time-lag for probe positions
      indicated (7 cm dia. gauze). Zero-lag values correspond to
      Fig. 18b).

20)   Cross-correlation functions with time-lag for probe positions
      indicated (7 cm dia. gauze). Zero -lag values correspond to
      Fig. 18c).

21)   Axial interference factor times tip loss factor as function of
      non-dimensional blade radius for two of the Cambridge machines;
      the optimum-twist blade used by Anderson (16) and the tapered
      chored, zero-twist blade described by Powles et al. (17)

AN ANALYSIS OF THE AERODYNAMIC FORCES ON A
VARIABLE GEOMETRY VERTICAL AXIS WIND TURBINE

M. B. ANDERSON
SIR ROBERT McALPINE & SONS LTD

ABSTRACT

A theoretical model is developed, based on the two-disc multiple streamtube approach, to predict the aerodynamic forces on a variable geometry vertical axis wind turbine. The model includes the effect of the lift and drag produced by inclined struts, interference losses and parasitic drag. Results are presented, for a 25 metre diameter turbine, in terms of the overall performance and the harmonic content of the aerodynamic forces on the blade and those applied to the tower.

INTRODUCTION

The variable geometry vertical axis wind turbine was proposed as an alternative to the Darrieus wind turbine under development in Canada, the U.S.A. and other parts of the world. It retains all the advantages of vertical axis operation while eliminating the complex blades of the Darrieus machine, replacing them with straight, untapered, untwisted aerofoils. In addition, the variable geometry allows constant speed operation without any of the problems associated with cyclical stall.

Variable geometry as a means of shedding the excess power at high wind speeds was first proposed by Musgrove (1). The 25 metre diameter prototype turbine discussed in this paper, whilst employing a slightly modified method of changing the blade geometry (Fig. 1), uses the same underlying principle of operation.

One of the inherent characteristics of the vertical axis wind turbine is that the aerodynamic forces in both the rotating and non-rotating frames of reference are periodic. The dynamic response of the rotor, tower and drive train to these forces is an important consideration in terms of fatigue and stability.

This paper presents a model which has been developed to accurately predict the aerodynamic performance and loads by extending the two-disc multiple streamtube model (2,3,4) to include the complex geometry, inclined struts and parasitic drag associated with the crossarm and blade/strut junctions of the 25 metre turbine. The effects of interference losses at the blade/strut junctions is also discussed.

Results will be presented in terms of a) the overall performance (i.e., power coefficient) and b) the amplitude of the Fourier components of the aerodynamic forces on the rotor and those transferred to the supporting structure.

AERODYNAMIC MODEL

An important aerodynamic difference between vertical and horizontal axis wind turbines is that, even under ideal conditions, the flow with respect to the vertical axis aerofoil is unsteady. During one revolution the flow direction and relative velocity are periodic for a vertical axis wind turbine whereas they are steady for a horizontal axis wind turbine. This periodicity is a serious drawback, when formulating an aerodynamic model, as it means that it is impossible to relate the time averaged torque to the change in angular momentum of the flow, as is usual in horizontal axis wind turbines. Momentum considerations, therefore, only deal with velocity components in the direction of the undisturbed flow. It is possible to avoid this limitation by using a vortex model (5), however the additional complexity when compared with the small improvement in the predicted loads over conventional streamtube theories does not usually warrant its use.

Blade-element theory which is the basis of a streamtube model, calculates the forces on a blade due to its motion through the air (combination of wind velocity, induced velocity and rotational velocity of the blade), to determine the aerodynamic loading of the entire rotor.

Basically, blade-element theory is lifting line theory applied to a rotating blade. It assumes that each blade-element behaves like a two-dimensional aerofoil to produce aerodynamic forces (lift and drag), with the influence of the wake and the rest of the rotor contained entirely in the induced velocity at the element. The aerodynamic forces at each element are equated to those derived from momentum considerations enabling the induced velocity, in the downwind direction, to be determined. This is based on the underlying assumption that each element is independent of every other. For a finite number of blades, of finite length, the application of lifting line theory is not strictly valid near the blade tip as cross-flow will reduce the pressure difference between the upper and lower surfaces. This effect has been modelled, in horizontal axis wind turbines, by the inclusion of a tip loss factor. Based on experimental results Stacey (6) has shown that the lift distribution, on a vertical axis wind turbine when the blades are vertical, is approximately rectangular and therefore tip loss effects are negligible and that it is an adequate approximation to use two-dimensional aerofoil theory.

A number of streamtube models, which are essentially adaptations of the original Templin (7) model, have been proposed for determining the induced velocity and hence the aerodynamic loading. These range from the single streamtube to the multiple streamtube with two-discs including unsteady effects. The aerofoil data used in these models ranges from straight line approximations to interpolating tabulated data spread over a wide range of Reynolds number and angles of attack. It is generally agreed (2,3,4,8) that the multiple streamtube model, in particular the two-disc version, agrees with experimental evidence more closely than the single streamtube theory.

The two-disc multiple streamtube model, which will be used in this paper, treats the turbine as two non-interacting, discrete actuator discs. One representing the upwind blade and the other the downwind blade. It assumes that the flow between the discs is non-expanding, that atmospheric pressure is attained at some point inside the turbine on each streamline and that the wind velocity approaching the second disc corresponds to the wake velocity of the first disc. The effect of this is that the flow velocity at the upwind disc is greater than the downwind disc, thus introducing asymmetry into the loading; as observed experimentally. It has been shown (3) that this method of formulation results in half the flow retardation taking place within the turbine.

A number of authors (2,9) have introduced unsteady aerofoil data into streamtube models but at present there is insufficient evidence to support its inclusion and it has therefore been ignored.

The two-disc model, in its present form, does not allow one to include the effects of any lift producing members other than the blades. For the 25 metre turbine (Fig. 1) and indeed a number of Darrieus turbines the rotor contains several inclined struts whose effects must be included when determining the induced velocities.

a)   Inclined Struts

For the 25 metre turbine there are eight inclined struts (Fig. 1). Each one can be considered to be an inclined blade, producing both lift and drag, and thus can be thought of as an 'additional blade' whose radial position and inclination is different from the main blades (in general its chord and sectional properties will also be different). To incorporate these 'additional blades' it is necessary, for each streamtube, to include the contribution of the aerodynamic forces, acting in the downwind direction, from each of them before equating with the force derived from momentum considerations.

b)   Interference Losses

At the junction between the blades and struts, i.e., hinge points, losses will occur due to flow separation. This will result in a local reduction of the blade lift and increased blade drag which will consequently degrade the overall performance of the turbine. Recent tests (10) at R.A.E. Farnborough on a full-scale model of the 25m blade/strut area have shown that it is possible to design fairings for the blade/strut junctions which minimises interference losses to such an extent that they can be neglected (except for parasitic losses which are described below).

c)   Parasitic Drag Losses

For the 25m turbine these losses will occur in three areas:-

i)   Along the entire length of the cross-arm.
ii) At all the blade/strut and cross-arm/strut junctions.
iii) At the swinging link situated at each end of the cross-arm.

It can be shown (11) that the effect of the parasitic drag losses on the downwind force is small and therefore can be neglected when calculating the induced velocity. They, however, must be included when evaluating the overall performance. It may be shown that the performance loss due to parasitic drag is approximately proportional to the cube of the tip speed ratio. For this particular design the constant of proportionality is between 0.00024 and 0.00050 depending on reef angle (blade inclination).

RESULTS AND DISCUSSION

A computer program has been developed, based on the previously discussed aerodynamic model, to analyse the 25m turbine. The relevant specifications of the turbine are:-

Rated power (at generator)	130 kW
Rotational speed	27 r.p.m.
Shutdown windspeed (at 25m height)	30 m/s
Rated windspeed (at 25m height)	11 m/s
Turbine diameter (with blades upright)	25 m
Blade length (tip to tip)	18 m
Blade chord (aerofoil section)	1.25 m (NACA 0015)
Main strut chord (aerofoil section)	1.07 m (NACA 0018)
Actuator strut chord (aerofoil section)	0.4 m (NACA 0031)
Reef angle	0-70 degrees

The aerofoil data, used in this analysis, is based on tabulated data presented by Sheldahl and Klimas (12) and covers a wide range of angles of attack and Reynolds number. A four-point bivariate interpolation routine is used to determine the lift and drag coefficient for the appropriate angle of attack and Reynolds number. The data presented by Sheldahl and Klimas unfortunately does not include the NACA 0031 and therefore the sectional characteristics of this aerofoil were modelled using data for a NACA 0025 but with a modified minimum drag coefficient.

Presented in Fig. 2 is the predicted power coefficient as a function of tip speed ratio for the 25m turbine with and without struts. For both cases parasitic losses have been included. It can be seen that the peak power coefficient, for both cases, is in excess of 0.4 which compares well with horizontal axis wind turbines. Including the struts has two effects; firstly, it shifts the peak of the performance curve to the left reflecting the increased solidity and secondly, reduces the efficiency at high tip speed ratios due to increased drag. The rapid decrease in efficiency for tip speed ratio less than three is due to blade stall occurring simultaneously along the entire length of the blade.

The effect of reefing the blades (increasing the blade inclination) on the performance is shown in Fig. 3. The performance coefficient has been normalised with respect to the swept area when the blades are upright and parasitic losses have been included. As the reef angle increases both the efficiency of the turbine and the tip speed ratio decreases. In addition the effects of stall become less dramatic indicating that it is not occurring simultaneously along the entire length of the blade.

With a knowledge of the efficiency of the rotor, drive train and generator it is possible to determine the reefing strategy as a function of wind speed (Fig. 4) for constant electrical power output above the rated wind speed. From this relationship it is possible to determine harmonic content of the predicted cyclically varying aerodynamic forces as a function of wind speed. Fig. 5 shows the variation of the harmonic content (1P = rotational frequency : 0.45 Hz) of the blade normal force (i.e. perpendicular to the blade and in a radial plane) with windspeed. As would be expected there is a contribution from all the harmonics. The amplitude of these harmonics decreases with increasing frequency. The results below the rated windspeed are in general agreement with those predicted and experimentally verified by Stacey and Musgrove (13). As the windspeed increases the blades progressively reef and the 1P component increases, reaches a maximum and then decreases. The higher harmonics, especially the 2P, increase with increasing windspeed reflecting the increased asymmetry in the upwind and downwind passes of the blade.

For both the aerodynamic torque and tower forces only the even harmonics (2P, 4P, ..etc.) appear (Figs. 6 and 7). In both cases the 2P components increase with windspeed, reach a maximum and then decrease reflecting the reduction in peak loads due to reefing. For the 4P components the variation with windspeed is slightly more complex with a minimum occurring at approximately 21 m/s. This minimum is due to blade stall not occurring simultaneously along the entire length of the blade. The subsequent increase in the 4P components with windspeed is due to the innermost portion of the blade stalling, however it is still small compared with the 2P components.

The mean aerodynamic torque and the mean downwind tower force are not presented in Figs. 6 and 7 as they are approximately equal to the amplitude of their respective 2P components. The mean sideways force is zero.

CONCLUSIONS

A theoretical model has been developed to predict the aerodynamic forces acting on a variable geometry vertical axis wind turbine. The model has been used to analyse the 25 metre diameter prototype from which the following conclusions can be drawn:-

a)   The inclusion of struts reduces the efficiency at high tip speed ratios and moves the point at which the peak efficiency occurs to a lower tip speed ratio.

b)   The analysis has shown that a peak power coefficient in excess of 0.4 is possible.

c)   Reefing of the blades not only limits the power output but also blade and tower forces.

d)   Reefing prevents simultaneous stalling along the entire length of the blade and thus minimises the effects of cyclical stall.

ACKNOWLEDGEMENTS

The work described in this report has been supported by Sir Robert McAlpine & Sons Ltd. Acknowledgement is also given to the Department of Energy who have provided support funding for the associated development programme. It is also my pleasure to thank I. D. Mays, P. J. Musgrove and G. Stacey for their contributions to this work.

229

REFERENCES

1. Musgrove, P.J., 1976. 'The variable geometry vertical axis windmill', Proc. of the International Symposium on Wind Energy Systems, Cambridge.

2. Stacey, G. and Musgrove, P.J., 1982. 'Developments on vertical axis wind turbine streamtube theories', Proc. of the Fourth BWEA Wind Energy Conference, Cranfield.

3. Read, S. and Sharpe, D.J., 1980. 'An extended multiple streamtube theory for vertical axis wind turbine', Proc. of the Second BWEA Wind Energy Conference, Cranfield.

4. Paraschivoiu, I., 1981. 'Double-multiple streamtube model for Darrieus wind turbines', Wind Turbine Dynamics, NASA Conf. Publication 2185, Cleveland, Ohio.

5. Strickland, J.H., Webster, B.T., and Nguyen, T., 1979. 'A vortex model of the Darrieus turbine : An analytical and experimental study', Trans. of ASME Journal of Fluids Engineering.

6. Stacey, G. 1983. Private communication.

7. Templin, R.J., 1974. 'Aerodynamic performance theory for the NRC wind turbine', National Aeronautical Establishment, Ottawa. Report LTR-LA-160.

8. Sharpe, D.J., 1977. 'A theoretical and experimental study of the Darrieus vertical axis wind turbine', Kingston Polytechnic Research Report.

9. Masse, B., 1981. 'Description de deux programmes d'ordinateur pour le calcul des performances et des charges aerodynamiques pour des eoliennes axe vertical', Institut de Rescherche de l'Hydro-Quebec. Report IREQ-2379.

10. Sibley, N., 1983. Private communication.

11. Musgrove, P.J., 1981. 'The variable geometry vertical axis wind turbines, Phase II Report', Vertical Axis Wind Turbines Consortium.

12. Sheldahl, R.E., and Klimas, P.C., 1981. 'Aerodynamic characteristics of seven symmetrical airfoil sections through 180 degrees angle of attack for use in aerodynamics analysis of vertical axis wind turbine', Sandia Laboratories, Report SAND80-2114.

13. Stacey, G. and Musgrove, P.J., 1982. 'Frequency analysis of aerodynamic forces on a straight bladed vertical axis wind turbine', Proc. of the Fourth International Symposium on Wind Energy Systems, Stockholm.

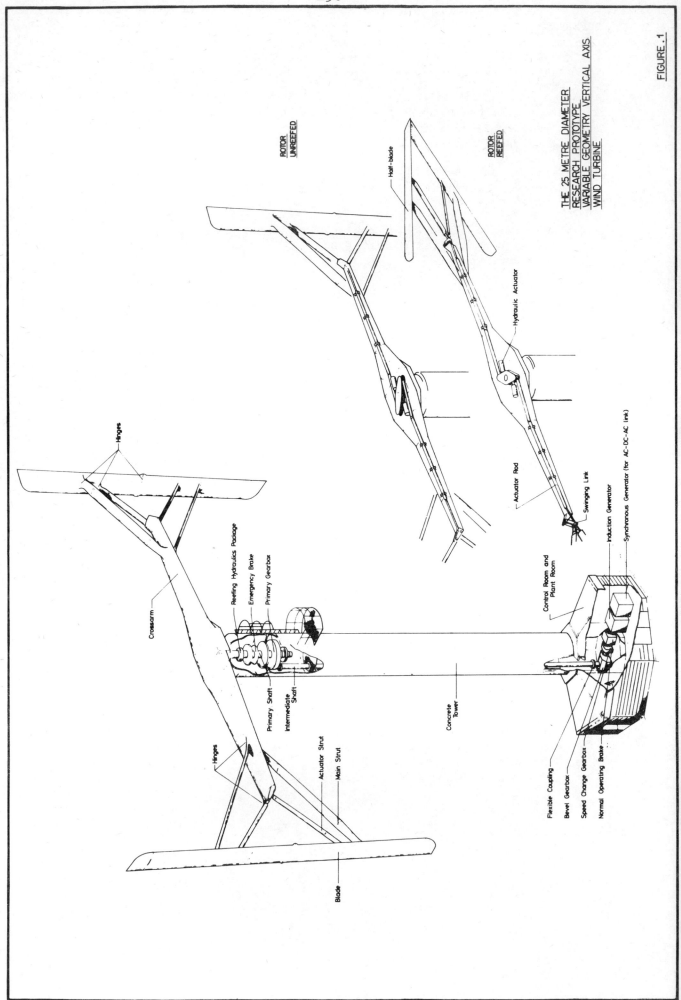

Hinges

Crossarm

Reefing Hydraulics Package
Emergency Brake
Primary Gearbox

Primary Shaft

Intermediate Shaft

Hinges

Actuator Strut

Main Strut

Blade

Concrete Tower

Flexible Coupling

Bevel Gearbox

Speed Change Gearbox

Normal Operating Brake

Control Room and Plant Room

Induction Generator

Synchronous Generator (for AC-DC-AC link)

ROTOR UNREEFED

Half-blade

Hydraulic Actuator

Actuator Rod

Swinging Link

ROTOR REEFED

THE 25 METRE DIAMETER
RESEARCH PROTOTYPE
VARIABLE GEOMETRY VERTICAL AXIS
WIND TURBINE

FIGURE .1

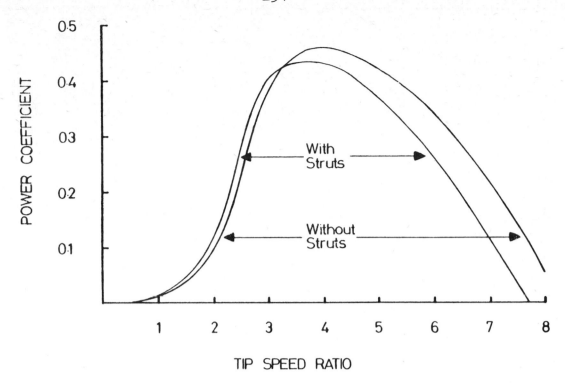

Figure 2.   The effect of struts on the power coefficient of the 25m
diameter turbine.

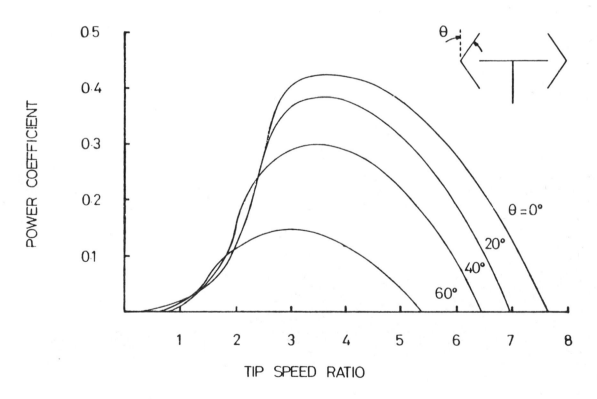

Figure 3.   The effect of reefing on the power coefficient of the 25m
diameter turbine.

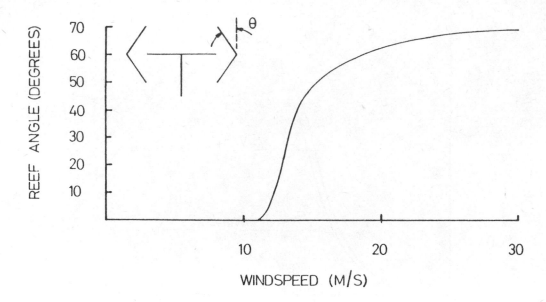

Figure 4.   Variation of reef angle with wind speed for constant
electrical power output above the rated wind speed.

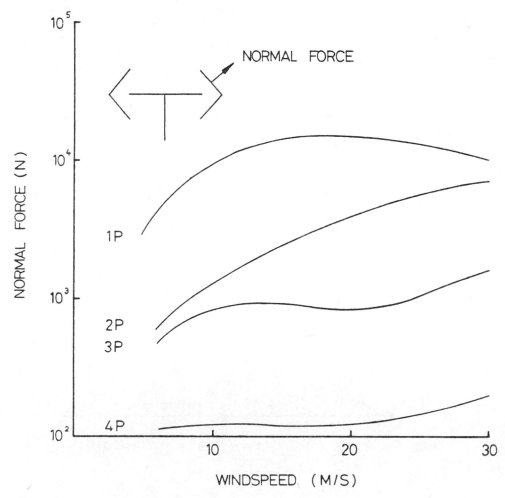

Figure 5.   Variation of the amplitude of the Fourier components of
the normal force with wind speed.

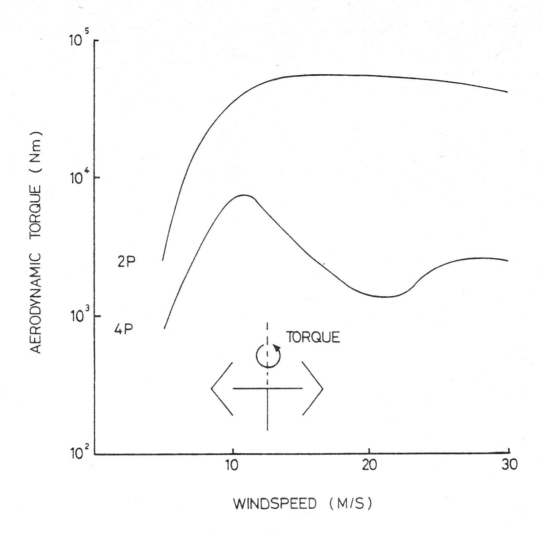

Figure 6.   Variation of the amplitude of the Fourier components of the torque with wind speed.

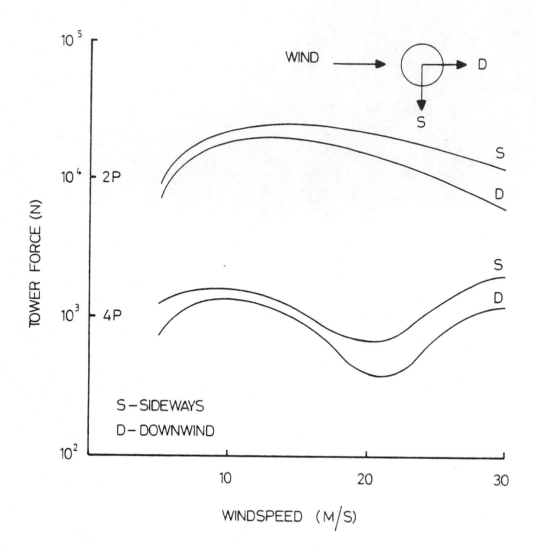

Figure 7.   Variation of the amplitude of the Fourier components of
the tower forces with wind speed.

SURVEY OF MATERIALS SUITABLE FOR USE IN MW SIZED AEROGENERATORS

L.M. Wyatt, U.R. Lenel and M.A. Moore,
Fulmer Research Laboratories, Stoke Poges,
Slough, Berks.

Abstract

A comprehensive survey of materials of construction for MW sized aerogenerators has been carried out, revealing that it should be possible to build aerogenerators to operate reliably for thirty years, using existing materials technology. However, certain shortcomings in materials information have been identified and appropriate research is recommended where this will allow the more efficient and confident use of materials of construction.

This paper outlines the factors affecting the use of materials for construction of aerogenerators and indicates the areas where materials information is inadequate or inconsistent. Appropriate actions to improve confidence in materials performance, and hence confidence in aerogenerator performance, are suggested.

Introduction

The cost-effective generation of electricity from wind requires the installation and operation of aerogenerators which:

* are rated at several (3-7) MW

* can be amortized over a life not less than 30 years

* are sufficiently reliable (provided scheduled maintenance is carried out) that no significant outage occurs.

These requirements pose significant engineering and materials problems. The purpose of this survey was to identify the shortcomings in existing materials information which might prevent the confident and most cost-effective use of materials of construction in meeting the above requirements. The survey involved examination of performance of existing aerogenerators, performance of materials in comparable environments and fundamental properties of materials.

It has been established that there is no fundamental reason why an aerogenerator of the required size, reliability and life should not be built. Failures which have occurred have been due mainly to unsatisfactory design or engineering, or malfunction of control systems, and further investigation of critical engineering aspects would be advantageous[1].
It has been found that components of an aerogenerator other than the rotor can be constructed using conventional technology, and the requirements for further materials research for these components are relatively minor[1]. The major materials problems occur in the rotor, and it is these aspects which are discussed in this paper.

Designs of aerogenerator currently under consideration include the horizontal axis aerogenerator, the vertical axis curved blade aerogenerator,

(Darrieus) and the vertical axis straight blade aerogenerator (with arrowhead furling).   The stress conditions, and hence the materials requirements, vary from one type of machine to another.

In a horizontal axis machine, loading is imposed by wind, gravity and inertia.   The maximum loads are imposed by wind alone (maximum gust and hurricane) and the stresses induced depend on design only and not on material.   Changes in wind speed and gusting cause fluctuating stresses in the blade, and changes in local wind speed due to wind shear and tower shadow cause cyclic stresses at the frequency of rotation.   Gravity loads also induce cyclically alternating stresses at the rotational frequency, and the stresses induced depend on the weight of the blade.   Loads due to centrifugal force do not vary cyclically, but depend on the weight of the blade and the speed of rotation.   Inertia loads occur during start-up and shut down.   In order to operate satisfactorily an aerogenerator blade must be able to withstand the maximum load imposed during the maximum gust, of which the statistical probability of occurrence is only once during blade life.   The blade must also withstand fatigue stresses caused by fluctuating loads (due to start up, shut down, gravity, gusting, wind shear and tower shadow) over a period of thirty years.

Stresses in vertical axis blades result from centrifugal, gravitational and aerodynamic loads.   Centrifugal loads produce bending moments in a straight vertical blade, and depend on blade mass and speed of rotation. The troposkein shape of the curved Darrieus blades eliminates most of the bending moment, and centrifugal loads are manifested mostly as a tensile stress.   Small bending moments remain due to imperfections in the troposkein shape and gravity.   In both types of vertical axis machine, aerodynamic loads alternate as the blades go through each cycle, leading to fatigue stresses.   Start-up and shut-down lead to varying inertia loads.   Maximum loads occur at the highest wind speeds (maximum gust and hurricane).

The relative magnitudes of the various stresses in each type of machine can be modified by design parameters such as for instance the number and configuration of blades, blade taper and thickness and provision of support struts.   The following discussion of blade materials assumes that optimum designs are chosen where possible.

Blade Materials

There are two forms of blade construction.   In one the aerofoil, or part of it, also carries the stress.   In the other, stress is withstood by a stress member (the spar) to which is attached a separate aerofoil.

Carbon steel, GFRP, wood, stainless steel, CFRP, aluminium, titanium and prestressed concrete have been proposed or used for the manufacture of blades.

Carbon steel is excellent and GFRP has many favourable properties for the stress bearing member.   GFRP is also pre-eminently suitable for the aerofoil, while wood, in the form of wood resin composite, has favourable properties for a stressed skin construction.   Aluminium is susceptible to fatigue, and is thus only suitable for the curved blade vertical axis machine in which cyclic bending moments are absent.   Economic, supply or engineering considerations, or lack of experience in utilisation, eliminate the other candidates.

Steel

Steel combines the advantages of a high elastic modulus and a high yield strength which together resist buckling, a high fatigue strength and ready availability. Its drawbacks are a high density which magnifies gravity or inertia generated fatigue stresses, a requirement for very high quality welding and uncertainties in the effect of environment.

The requirement to resist cyclic bending moments in horizontal axis and straight bladed vertical axis machines necessitates sophisticated design to minimise stresses and almost immaculate fabrication to limit the size of defects and the magnitude of locked in stresses.

For a horizontal axis machine, the most effective design would probably be a D spar or an ellipse tapering in section and wall thickness from root to tip, but such designs are complicated by the need to twist the aerofoil. This introduces problems in fabrication and a better compromise is probably the use of an elongated hexagon, formed by longitudinal welding of bent steel plate, on which can be mounted a twisted aerofoil. Increase of the root section reduces stress, but is restricted by the need to limit the thickness/chord ratio of the aerofoil. The resulting compromise produces a spar that may be fatigue critical along its whole length and requires a very high standard of fabrication and inspection and very accurate fatigue calculations.

For a straight bladed vertical axis machine, the high centrifugal force ensures that one side of the blade never goes into tension, and the blade, being straight, has a well defined neutral axis. The most effective design would therefore locate closing welds and welds to which access is difficult in positions where fatigue is not a serious problem.

There are essentially two methods of assessing the fatigue performance of a welded steel spar.

One utilises fatigue "N" line curves such as are published in "DD 55 1978 fixed Offshore Structures"[2], and sums fatigue damage by Miner's law. This may not be completely satisfactory for aerogenerator design for two reasons. The first is that the curves are empirical. The values correspond roughly to second lower standard deviations of the ESDU fatigue data - that is they correspond to a failure probability of one in a hundred fatigue specimens. The failure probability of a structure with many metres of weld is greater than this and for a structure lacking redundancy, such as an aerogenerator blade spar, it would be wiser to take the third lower standard deviation. Secondly this design procedure assumes that locked in tensile stresses at the toe of the weld (where failure usually starts) are such that the stress cycle is always in tension. It is not clear that it would be possible to guarantee the fatigue performance of a steel aerogenerator spar under these conditions for thirty years.

The second method of assessing fatigue performance, giving a greatly improved certainty of performance, is fatigue crack growth calculations. After welding, surface dressing and stress relieving, the welds in all highly stressed positions are inspected to reveal the size and position of the largest defects. A limiting value for the life of the component can then be determined by calculating crack growth rates.

It should be possible by modern techniques to weld material such as BS.4360:1979, Grade 50 D to a maximum defect size of 1.25mm on the surface

or 2.5mm internal. Boeing claim to be able to weld a twisted "D" spar
to this standard. This, combined with good design should ensure that
the stress intensities which cycle at rotational stresses are below the
crack growth initiation level at the commencement of life. At later
stages during the life of the component, start-up, shut-down and maximum
gust stresses which occur at much lower frequency will have lengthened
the crack so that crack growth occurs even at the lower cyclic stresses.

The resulting crack growth calculation is complex but even so Boeing[3]
have devised a procedure which they have verified, experimentally, which
demonstrates that a twisted "D" spar which they have manufactured will
have a minimum life of 30 years.

Effect of Offshore Environment on Steel

The procedure outlined above refers to fatigue behaviour in air. The
high humidity and salt spray to which aerogenerator components will be
subjected in an offshore environment may adversely affect fatigue
performance but no immediately relevant investigations have been carried
out.

The effects of immersion in sea water have however, been extensively
studied. The influence of some of the more important parameters is well
illustrated in Figures 1 and 2[4]. The steel tested (X65) is higher in
tensile strength than is recommended for an aerogenerator blade and this
will almost certainly have exaggerated the effect of the environment.
Investigators on steels comparable with Grade 50D find lower accelerations
[5,6]. The acceleration of fatigue crack growth is very dependent both
on frequency and stress intensity. At frequencies of 0.3Hz and below,
corresponding to alternating stresses at the rotational frequency and
lower (such as start up and shut-down), the acceleration of fatigue crack
growth can be very large at stress intensities in the middle of the range
studied. There is however, little or no effect at low stress intensities
near the initiation level.

No work appears to have been carried out on the effects of sea spray but
the results in Figure 3[7] on the effect of alternate wetting and drying,
point to the conclusion that the acceleration is probably intermediate
between that in air and that immersed.

The conclusion to be drawn is that unless design, fabrication and
inspection can keep the stress intensity at the tip of a defect to below
the level of initiation or until information on the effect of salt spray
atmospheres on fatigue crack growth becomes available, steel strength
members in aerogenerator blades operating offshore must be protected.

It may be possible to do this, using a zinc epoxy or zinc silicate coat
on the spar and covering it completely with a fibre glass aerofoil. The
effect of salt spray atmospheres on the fatigue crack growth in
constructional steel requires to be evaluated to underwrite design and to
make it possible to assess the effect of defects in the protection.

G.F.R.P.

G.F.R.P. has advantages of low density, high resistance to tensile and alternating stresses, resistance to environment and ready availability. These properties make it an optimum material for use as an aerofoil and it is used for the strength member of a number of horizontal axis designs.

The main problem for horizontal axis and for straight bladed vertical axis machines is the low modulus which results in very large deflections and renders it very difficult to design to withstand buckling under large wind forces (for horizontal axis aerogenerators) and large centrifugal forces (for straight bladed vertical axis aerogenerators). The requirement to avoid buckling effectively imposes a limiting design strain of 0.2%. This is where possible achieved by optimizing the orientation of the fibre reinforcement.

Spars or complete aerofoils may be manufactured by filament winding, by pre-impregnated tape winding or by hand lay up. Winding procedures require very large rotating machinery. Filament winding can be used to incorporate very long fibres, but it is difficult to lay fibres at an angle close to the axis. Tape winding can lay fibres of any required orientation but the fibre length is limited by the width of the tape. Optimal properties can be obtained by a combination of the two. A typical load bearing member would have about 70% of the filaments orientated axially, 10% hoop and 20% at $\pm 45^{\circ}$.

Hand lay up permits the incorporation of fibres running the whole length of the blade. However, hand layup requires a longitudinal joint the complete length of the blade and it is essential to ensure that this lies on the neutral axis.

The fatigue of G.F.R.P. has been extensively studied by helicopter blade manufacturers[8]. Some of the data produced is shown in Figures 4 and 5. Fatigue of G.F.R.P. composites does not spread by crack propagation as in a metal but takes the form of a disintegration of the matrix which is dependent on the applied strain. This is well illustrated in Figure 6[9], which indicates that limiting strain at high values of N tends to that of the matrix material. The effect of long exposure to marine atmosphere of a helicopter blade on the fatigue properties of the material is illustrated in Figure 4, in which the values marked Δ were taken from a blade which had flown for 2,600 hours in a marine atmosphere.

There is a substantial margin over the design limit of 0.2% strain so that a correctly manufactured blade is unlikely to fail, even assuming a deterioration in properties due to moisture absorption greater than has been observed.

It is possible to achieve the 0.2% strain criterion for horizontal axis wind turbine blades but this may require an increase in the wall thickness and possibly also the thickness of the aerofoil such that the blade is actually heavier than the equivalent steel blade would have been. The much greater deflections than occur with a blade with a steel strength member are easily accommodated in a down wind configuration aerogenerator, but can lead to severe problems with upwind configuration.

In VAWTs the requirement to limit drag when the blade is in the "no lift" position restricts aerofoil thickness and makes it impossible to design a G.F.R.P. blade to 0.2% strain.

A substantial effort on fatigue and environmental affects would be required before a fatigue life of 30 years in a marine environment could be guaranteed at a strain of 0.4%.

These problems could be overcome by the incorporation of a proportion of higher modulus fibres such as carbon or steel. The use of steel fibres however, increases the weight of the blade, and the incorporation of carbon fibres must await the availability of this material at an economic price.

Joints between blade or hub, or between sections of a blade usually involve the attachment of the G.F.R.P. body to a steel flange. A well developed technology of attachment has been developed for helicoptors. Models of typical root end fixings have been subjected to fatigue testing[10] and have withstood loads well in excess of those likely to be experienced in service.

G.F.R.P. Aerofoils

G.F.R.P. has optimal properties for the aerofoil of a blade for any type of aerogenerator. The aerofoil may be integral with the stress member, or may be mounted on a stress member constructed of any of the materials considered in this paper. The attachment of fibre glass to any of the other materials is well understood.

G.F.R.P. is resistant to water droplet erosion at velocities below approximately 80m/sec which will be exceeded only by the largest projected designs. Fibre glass aerofoils will normally be protected by a non-load bearing layer of impregnated woven glass fibre which can be repaired on site, or at a local maintenance depot. If erosion seems likely to present a problem it can be overcome by cementing on a layer of titanium sheet. The polymers (typically epoxies) used as matrices for GFRP are degraded by ultra violet radiation and ultra violet additives or coatings, such as for example polyurethane, must be provided.

Wood

Its lightness, good specific strength, cheapness and ready availability make laminated wood a promising material for aerogenerator blades.

Two laminating techniques, "Wood Resin Composite" employed by Gougeon Brothers in the U.S. and Structural Polymer Systems in the U.K., and "Compressed Wood" employed by Permali are available. Both techniques will provide blades with favourable properties but the hand lay-up process so far used for compressed wood has made it prohibitively expensive for aerogenerators, although service experience with large fan blades has been uniformly favourable. Wood resin composite construction, which involves impregnating veneers of soft woods with epoxy resin, laying up the veneers in a female mould, curing under vacuum bag and glueing the resultant two halves together, is, once the cost of the expensive mould has been amortised, cheaper than any material (except possibly reinforced concrete).

Present design allowables for wood laminates are based on those used for aircraft propellers in the 1914-1918 war and are probably conservative. The mechanical properties of wood laminates are highly anisotropic, and shear and cross grain tensile stresses must be minimised. The individual plies and blocks are therefore orientated so far as is possible in the tensile direction, the laminae are partially impregnated by using a resin of controlled viscosity and the wood is enveloped in a fibre glass sheath, which resists splitting and shearing. Where there are high concentrations of stress which may lead to cleavage it may be advisable to incorporate additional fibre glass plies within the material.

Wood construction is generally considered to have a high resistance to fatigue. This concept has arisen because the design procedures usually place a conservative stress ratio on design for steady stress, and ensure that a cyclic stress is lower still. The available experimental data on the fatigue properties of wood are as a result very meagre and Figures 7[11] and 8[12] for laminated and compressed woods respectively summarise the information available.

If the possible use of wood as a blade material for a programme of aerogenerator construction is contemplated, an extensive programme of fatigue testing of wood resin composites is required. The influence of fibre glass layers in restraining shear and cleavage should be studied and torsional fatigue tests which are most likely to induce these should be included.

The major influence of environment on wood is that caused by the absorption or loss of moisture. These effects are much reduced by the partial impregnation with epoxy and the outer layer of fibre glass. It is important however, to ensure that the moisture content of a wooden component remains sensibly constant, and parts should be fabricated in an environment resembling the operating environment as closely as possible.

As in the case of G.F.R.P., wooden blade sections usually require to be bonded to a steel flange at one or both ends and the attachment will usually be the most highly stressed point of the blade. Most of the aerogenerators so far built depend on stud fittings, on which fatigue tests have demonstrated adequate margins over operational loads[10].

No problems appear to prevent the manufacture of wooden blades for horizontal axis aerogenerators, although to obtain the required rigidity it may be necessary to use a higher thickness to chord ratio aerofoil than would be necessary for an aerofoil with a steel strength member. The operational performance of wooden blades, for instance in the Gedser and Mod-OA machines, has been highly satisfactory and no deterioration of the wood has been detected.

Straight bladed vertical axis blades must have a low thickness to chord ratio and a satisfactory design for wood with the required rigidity has yet to be produced. An optimum design might well utilise a combination of steel and wood.

Conclusions

A survey of materials suitable for use in construction of MW sized aero-generators has indicated that it should be possible to build aerogenerators capable of operating reliably for thirty years provided design and engineering are carried out to high standards. The requirements for further materials research for components other than the rotor are relatively minor, but a number of investigations on rotor blade materials would enable more confident and efficient use of materials. The following research programmes are recommended:

1. The effect of a salt spray environment and protective coatings on fatigue crack growth in welded steel.
2. If designs utilising GFRP with a strain limit greater than 0.2% are to be considered, a programme of fatigue testing, including the effect of moisture absorption, should be carried out.
3. A comprehensive test programme on mechanical properties and fatigue performance of laminated wood composites.

Acknowledgements

The work reported here was funded by the U.K. Department of Energy.

References

1. Wyatt, L.M., Lenel, U.R., and Moore, M.A.
   Survey of Materials Suitable for Use in MW Sized Aerogenerators.
   Fulmer Research Institute R948/2/September 1982. (Work Commissioned by the Department of Energy, Energy Technology Support Unit Harwell.   (Unpublished).

2. British Standards Institution,
   Draft for Development DD55 1978.

3. Boeing Mod-2 Wind Turbine System Concept and Preliminary Design Report US Dept. En. Solar Technology Division, Washington DC, 20545, 1974.

4. Vosikovsky O.   Fatigue-Crack Growth in an X-65 Line-Pipe Steel at Low Cyclic Frequencies in Aqueous Environments.   Trans. ASME October 975 pp. 298-304.

5. Benson, J.M., and Novak, S.R.
   Subcritical Crack Growth in Steel Bridge Members,
   Highway Research Board, Nat. Acad. Sci. Sept. 1974.

6. Kitigwa, H.
   A fracture Mechanics Approach to Ordinary Corrosion Fatigue of Unnotched Steel Specimens.   Conference - University of Connecticut.
   June 1971.

7. Watanabe M. and Mukay, Y.   Corrosion Fatigue Properties of Structural Steel and its Welded Joints in Seawater.   Conference on Welding in Offshore Constructions.   The Welding Inst. Newcastle. February 1974.

8. Och. F.
   Fatigue Testing of Composite Rotor Blades (MBB)
   51st Meeting of AGARD Structures and Materials Panel, Sept. 1980.
   Aix-en-Provence, Sept. 1980.

9. Dharan, C.H.K.
   Fatigue Failure in Graphite Fibre and Glass Fibre Composites.
   J. Mat. Sci. 10 pp. 1665-1770, 1975.

10. Faddoul, J.R. and Sullivan, T.L.
    Structural Fatigue Tests for Large Wind Turbine Blades,
    NASA Lewis Research Centre, Cleveland, Ohio, 44135.

11. Sitzhof and Nordhoff.
    Int. Journ. of Fracture 16 609-616. Alph An den Rhyn.
    The Netherlands, 1980.

12. Messrs. Permali Ltd., (Gloucester),
    Private Communication.

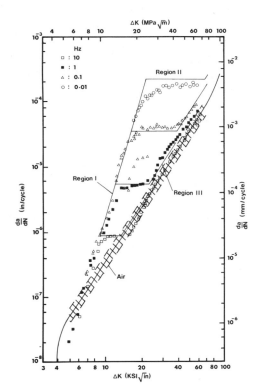

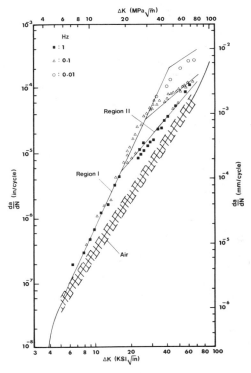

Figure 1. Fatigue-crack growth
rates in 3.5 percent salt water
at cathodic potential and four
frequencies (R = 0.2). (The 0.1 Hz
results fall into two groups
because two series of tests were
carried out using different wave
forms.)

Figure 2.   Fatigue-crack growth rates in
3.5 percent salt water at free corrosion
potential and three frequencies (R = 0.2).

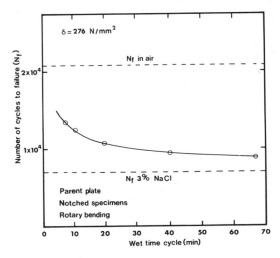

Figure 3.   Results of Cyclic wet-dry Tests
on mild steel.

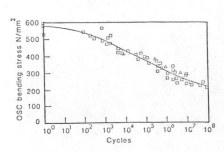

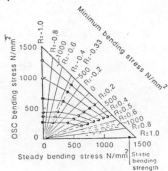

Figure 4.   Bending Fatigue S/N   Figure 5.   Bending Fatigue Haigh
Curve.   Diagram.

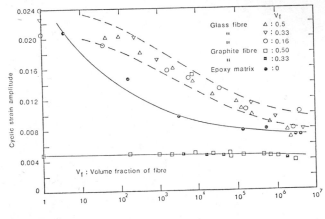

Figure 6.   Reversals to failure
($2N_F$ cycles).

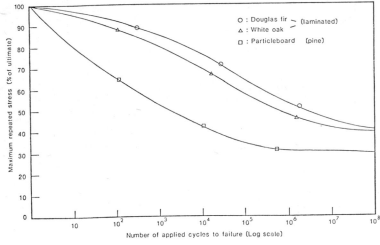

Figure 7.   Bending S-N curves for
wood and wood composites.

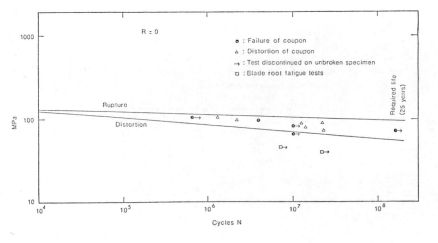

Figure 8.   Results of Fatigue Tests or
Compressed Wood (and blade
roots).

AN APPRAISAL OF STRAIGHT BLADED
VERTICAL AND HORIZONTAL AXIS WINDMILLS

K. McAnulty*
Energy Technology Support Unit
Harwell.

Abstract

Vertical axis (VA) wind turbine designs are now being presented as alternatives to the traditional horizontal axis (HA) pattern. Claims for the superiority of one type over the other have been made from time to time, and an attempt is made to put the claims in perspective. Theoretical and experimental results are compared with a view to establishing any fundamental differences. This comparison is then extended to some engineering and operational considerations which might have a bearing on the potential for further development of each type.

Introduction

A comparison is sought between 'optimum' HA and VA windmills. The VA type is the straight bladed design developed from the original Darrieus patent and known as the Vertical Axis Variable Geometry (VAVG) wind turbine. The curved blade Darrieus type of VA machine is not considered here, although some conclusions apply to both.

Obviously, no windmill can be optimum with respect to all parameters, and a comparison based on one aspect of performance may be contradicted by a comparison with some other characteristics. In the following, the two types of windmill are examined under nine headings which encompass major aspects of design and operation.

Performance in a Uniform Wind

Figure 1 shows the variation of power coefficient with tip speed ratio for VA and HA machines. The factors which influence the point at which maximum C_p is achieved are design dependent, but the actual peak values shown, and the general shape of the curves are characteristic. It will be noticed that two curves are given for the VA type. This is because the proportions of the straight bladed VA machine which correspond to optimum power coefficient are not well defined, and rotors having a given solidity ratio may be constructed with different blade proportions. For HA machines this is not the case, and it is possible to derive a condition which fixes the aspect ratio of the rotor blades once the aerofoil operating point is chosen (Milborrow (1)).

Practical VA windmill rotors will have aspect ratios of about 15, so that the curve for infinite aspect ratio must be seen as a theoretical limit that cannot be approached.

---

* On secondment from EASAMS Limited, Frimley.

It will be noted that the curves for the VA machines in Fig. 1 appear rather peaked towards the lower end of the tip speed ratio range. Lest this feature be interpreted as disadvantageous, Fig. 2 shows the same data plotted against the ratio of tip speed ratio to tip speed ratio for optimum power coefficient. The differences between the characteristics for HA and VA machines can be seen to be relatively slight. It appears, therefore, that differences between the performance of the machines based on theoretical estimates of power coefficient are not significant, although practical VA machines may have lower peak values than similar HA ones.

Performance in Free Air

Figures 1 and 2 apply to windmills immersed in uniform winds. Real windmills operate in the atmosphere where neither wind speed nor direction are constant. Since the HA machine must be aligned with the wind direction to produce optimum performance, while the VA machine has no directional properties, it might be supposed that the latter would be at an advantage in free air.

Figure 3 shows the results of free air tests on three configurations of a VA machine. The results of a comparable test on an HA machine of similar size are given in Fig. 4. It can be seen that there is little to choose between the two types with respect to performance. The freedom from the cost and complexity of wind direction sensors and yaw control equipment is a significant factor in favour of the VA design.

Control

The control of both HA and VA windmills is primarily by alteration of the angle of attack of the blades. (Other methods are available but their use is not widespread.) Theoretical curves for the variation in power coefficient with tip speed ratio for various reef angles (and thereby angles of attack) for the VA variable geometry windmill are shown in Fig. 5 (Sir Robert McAlpine and Sons Ltd. (2)). A similar set of characteristics for a HA rotor are given in Fig. 6 (Griffiths, R.T. and Wollard, M.G. (3)). Using this mode of control, the operation of both types of machine can be extended into regions of low tip speed ratio (i.e. high wind speed for a given rotational speed), thus allowing operation well beyond the rated wind speed. The advantages of this feature lie in the minimisation of the number of times the windmill needs to be stopped from full power, and are available to both HA and VA machines.

Many advanced HA machines, however, do not incorporate variable pitch over the entire length of the blade, and operational flexibility is balanced against complexity of rotor design. The VA machine designer would appear to have no choice in this regard, since, if stalling the entire blade twice per revolution as tip speed ratios approach the lower end of the range is to be avoided, some form of control of the angle of attack is essential.

It is appropriate, however to observe that the need to achieve adequate control incurs a cost in performance. The power coefficient curves of Figs. 1, 2, 3 and 4 show peaks of certain values of tip speed ratio. The decrease in power coefficient as tip speed ratio is reduced past the maximum is due largely to stall. Given the variability of the wind, it is not possible to operate at maximum power coefficient continuously, and, if stall is to be avoided, the rated tip speed ratio will be somewhat higher than that at which the power coefficient peaks. The choice of the rated tip speed ratio involves a compromise between protection from stall due to gusts, and decreased power output as a result of operating at a tip speed ratio other than the optimum. Both types of machine must be designed with this compromise in mind, and there is, as yet, no indication that one suffers more than the other.

Cyclic Stall and Pitch Control

The occurrence of stalling in VA windmills presents one of the most striking differences between this type of machine and the HA windmill. For a straight bladed VA machine with its blades in the vertical position, stalling and flow reattachment will occur twice per revolution as tip speed ratios fall below that for which power coefficient is a maximum.

Under these conditions the aerodynamic force on the blades becomes uneven, so that the excitation of higher vibratory modes of the turbine structure becomes a possibility (British Aerospace (4)). Operation under such conditions for prolonged periods would be inadvisable, so that the provision of a blade pitch change mechanism, which relieves the entire blade of this experience, becomes not a matter of choice but of necessity.

The position with the HA machine is somewhat different. The rotor may be designed in such a way that stall, when it does occur, is progressive, usually extending from the hub outwards. The effects of stall on the whole windmill depends on the details of the design. Rigidly mounted, three-bladed fixed pitch rotors may be operated successfully in such a way that blade stall plays a role in regulation. With other rotor configurations, e.g. teetered hub, the consequences of stalling a substantial portion of the blades are to be avoided, and the rapid response of the pitch change mechanism is necessary for rotor stability.

With the Musgrove machine, cyclic stall is avoided by reefing the blades, which are hinged in the middle and are supported by articulated struts. At high reef angles (high wind speeds), aerodynamic forces are large and support-strut geometry unfavourable so that we have a control mechanism which must support loads considerably in excess of their normal operating loads. By way of contrast, the pitch control mechanism of a HA machine is required to react only against the aerodynamic moments of (usually part) of the blade. The control mechanism is therefore lighter, possibly more reliable and requires less maintenance.

Rotor Configuration and Load Patterns

The most obvious external difference between the two types of machine is the shape of the rotor, and one might reasonably look in this direction for differences in effectiveness (left undefined). Figure 8 shows the pattern of major loads on both types of rotor.

In both types the power generating surface intersects an air stream in a cross-section generated by the cyclic motion of that surface. The total length of structure needed to allow the blade to generate the cross section, however, differs markedly. For the HA machine it is simply the diameter of the rotor. If we assume VA rotor proportions which maximise the area enclosed with respect to perimeter, (i.e. a rotor whose diameter equals the blade length), we find that we need 2.66 times the length of structure.

For the VA machine, blade bending stresses will be induced by inertial (steady) and aerodynamic (reversing) loads. Very light blades capable of carrying aerodynamic loads should mean low inertial stresses and lead to light rotors. Practical considerations however (material selection, methods of fabrication, cost) have influenced blade design with the result that the ratio of inertial to aerodynamic loads is about 8. The blades of currently projected designs carry bending loads due to centrifugal forces which swamp the aerodynamic loads, so that rotors tend to be somewhat heavier than one might have expected. Cross arm stresses will be predominantly gravity induced bending, which being unidirectional avoids the fatigue life complications that accompany the fluctuating gravity of HA rotors. It must be remembered that the weight of the reefing mechanism must be added to the weight of the blade/cross arm structure, and that, for reasons given above, this could be significant.

This is perhaps an area for closer study, since the total tower head weight appears to influence total cost of windmills strongly. In the meantime we cannot assume that the VA rotor will be lighter than an HA rotor of the same swept area.

Output Shaft Orientation

Another consequence of the difference in rotor configurations is the direction of the power output shafts. Broadly speaking, the VA machine provides a high torque, low speed output, directed vertically. The 25 m diameter prototype uses this feature to site the generator and some of the speed increasing gearing at the base of the tower, driven by a long, intermediate speed shaft.

There appears to be no overwhelming advantage in this arrangement. Siting the generator and some of the gearing on the ground certainly results in reduced tower head weight, but the VA machine, with its low (by a factor of 2 or so) speed output requires additional gearing, and the transmission of large amounts of mechanical power over long distances by rotating shafts is not a straightforward matter. There is, in any case, no reason why HA machines cannot, by the use of a direction changing stage of gearing at the tower head, be built with a similar arrangement.

Blade Shape

Until recently, HA windmill blades have been built with variable chord, and twist. When combined with an undercambered aerofoil section which may also vary along the length of the blade, these features make for a component whose shape may cause manufacturing difficulties and hence increased cost.

In contrast, the blades of current VA machines are of uniform symmetrical section. Whether or not these differences could account for a significant cost difference depends largely on the methods of manufacture and the number of items to be constructed from one set of tooling.

Although no firm cost figures are available to allow a valid comparison, it is likely that for production of small numbers of machines, the blades for VA windmills would be cheaper. This does not, however, imply that the entire rotor would be cheaper.

Dynamic Characteristics

Due to the relative novelty of the Musgrove type of VA windmill, no large body of well validated dynamic analysis exists. Such experimental work as has been done however, suggests that unexpected resonances can occur even in simple prototypes (see Pretlove, A.J., *et al* (5)) and it would be premature to conclude that the VA machine offered any advantage in this area.

Potential for Range Scale Development

Comprehensive analysis of the design of HA windmill blades has led to the identification of possible size limiting factors, (Milborrow, D.J. (1)). While it appears that both stresses induced by both aerodynamic and inertial loads in VA machines are independent of size (British Wind Energy Association (6)) the need to accommodate the reversing aerodynamic loads while maintaining reasonable fatigue life can result in rapid growth in blade weight with size, especially for metal structures of conventional design. Whether or not this will impose limits on the maximum size of this type of machine is not yet clear.

The growth of cost with size is more difficult to quantify and must be expected to depend on many factors. Studies in the USA discussed in Boeing Engineering and Construction (7), show a minimum in the busbar energy cost at 2.5 MW, corresponding to a 100 M rotor in a 6.26 ms wind, for land based machines. More recent work by Douglas, R.R. (8), shows the trend of cost against machine size reaching a minimum at about 7.2 MW, with a rotor diameter for the wind speeds considered, of 130 M. This suggests, but does not prove, that extreme size may be of academic interest only.

Interests in the potential for large scale development has been heightened by estimates of substantial savings in the costs of preparation of foundations for machines sited offshore, (British Wind Energy Association (6)). The saving of £300/kW in foundation preparation in going from 3.7 MW to 20 MW indicates the order of the economies, which, if the promise held out by the VA variable geometry concept is fulfilled, could be a major factor in the future exploitation of wind power.

Conclusions

- Free air performance for both types is similar.

- The absence of yaw control mechanism is a significant factor in favour of the VA type.

- Pitch control over the entire blade allows both types to operate in high winds.

- Cyclic stall in high wind speeds makes extensive (whole blade) pitch control essential for VA machines, and incurs an unquantified penalty in terms of rotor complexity and tower head weight.

- There appears to be no favoured output shaft arrangement.

- The simple blade shape of VA machines should result in cost savings.

- Considerable uncertainty surrounds the dynamic characteristics of VA machines, in contrast to a well developed and validated analysis for the HA type.

- Large scale development promises to bring with it substantial economic advantages. Provided light weight rotors can be built for VA machines (utilising advanced, high specific strength materials), the possibility of their development to large sizes deserves attention.

References

1   MILBORROW, D.J. The Analytical Design of Horizontal Axis Wind Turbine Blades and their Size Limits. Central Electricity Generating Board. CERL October 1982.

2   SIR ROBERT McALPINE AND SONS LTD. Design Study of 25 M Diameter Variable Geometry Vertical Axis Wind Turbine. August 1981.

3   GRIFFITHS, R.T. and WOLLARD, M.G. Performance of the Optimal Wind Turbine. Applied Energy (4) 1978.

4   BRITISH AEROSPACE, (WEYBRIDGE-FILTON DIVISION). Wind Tunnel Report WT 722. Wind Tunnel Tests in a 3 Metre Diameter Vertical Axis Windmill. July 1979.

5   PRETLOVE, A.J., *et al*. Cross Arm Vibrations on the Rutherford Laboratory VGVAWT. Energy Group Report 81/3 December 1981. Rutherford Laboratory and Reading University.

6   BRITISH WIND ENERGY ASSOCIATION. Wind Energy for the Eighties. BWEA 1982.

7   BOEING ENGINEERING AND CONSTRUCTION. 250 kW Wind Turbine System for Electric Power Generation - Mod-2 Project. Seattle, Washington, December 1978.

8   DOUGLAS, R.R. Conceptual Design of the 7 megawatts MOD-5B Wind Turbine Generator. 5th Biennial Wind Energy Conference and Workshop, Washington, DC. October 1981.

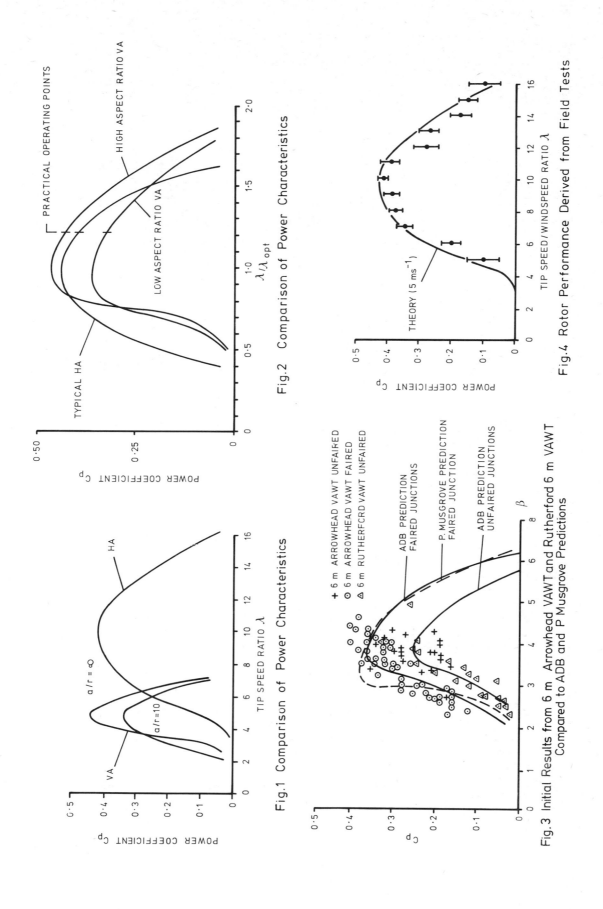

Fig.1 Comparison of Power Characteristics

Fig.2 Comparison of Power Characteristics

Fig.3 Initial Results from 6 m Arrowhead VAWT and Rutherford 6 m VAWT
Compared to ADB and P Musgrove Predictions

Fig.4 Rotor Performance Derived from Field Tests

252

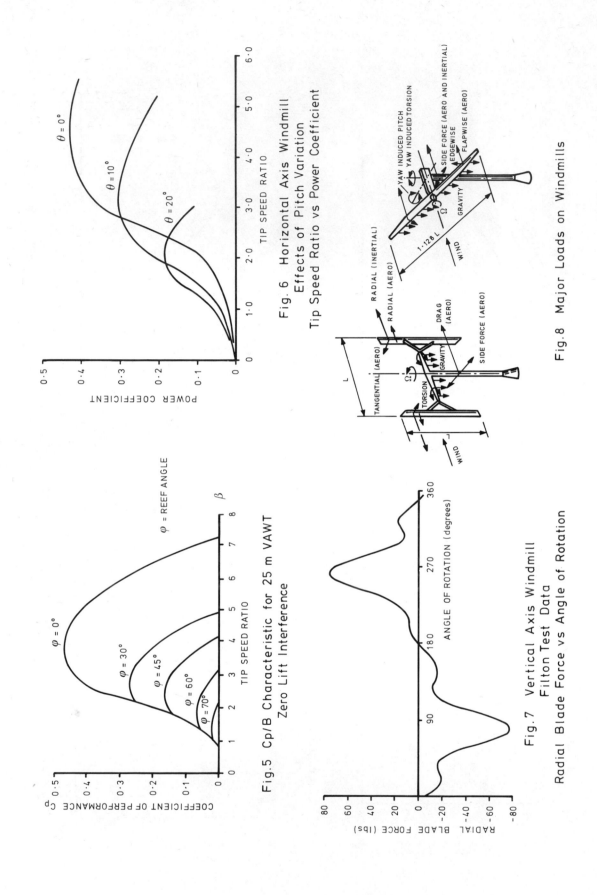

Fig.5 Cp/B Characteristic for 25 m VAWT
Zero Lift Interference

Fig. 6 Horizontal Axis Windmill
Effects of Pitch Variation
Tip Speed Ratio vs Power Coefficient

Fig.7 Vertical Axis Windmill
Filton Test Data
Radial Blade Force vs Angle of Rotation

Fig.8 Major Loads on Windmills

THE 3 MW ORKNEY WIND TURBINE

P B Simpson, D. Lindley, A S Lee, A Caesari
Wind Energy Group
Greenford, Middlesex

ABSTRACT

On August 30, 1982, the Wind Energy Group signed contracts with the North of Scotland Hydro Electric Board and the Department of Energy to design, manufacture and construct a 3 MW horizontal axis wind turbine on Mainland Orkney. By the end of March 1983 the modified design of the 60m diameter wind turbine will be substantially complete. In May the design will be submitted to the Department of Energy and the North of Scotland Hydro Electric Board for approval for manufacture and construction.

This paper outlines development of the 3 MW turbine since completion of the original design in December, 1979 and describes the alternative arrangements and design modifications that have been considered. The principal features of the machine that is now proposed for construction on Orkney are described.

INTRODUCTION

At the end of August, 1982, the Wind Energy Group (WEG) comprising Taylor Woodrow Construction Ltd, British Aerospace PLC and GEC Power Engineering Ltd, signed contracts with the North of Scotland Hydro Electric Board and the Department of Energy to design, manufacture and construct a 60 m diameter, 3 MW wind turbine on Burgar Hill, Orkney. The 60 m wind turbine is the major element in the Orkney Project which incorporates a number of further elements, the principal of which is a 20 m diameter, 250 kW wind turbine. The history of the Orkney Project and details of the other elements have been described by Lindley and Stevenson (1), Lindley (2), Lindley et al (3) Lindley and McPhie (4), Hassan, Henson and Parry (5) and Hassan and Lindley (6). The 20 m machine has been further described by Armstrong, Ketley and Cooper (7), Garrad (8) and Cooper (9). This paper describes development of the design of the 60 m machine and reviews current status and programme.

DESIGN OBJECTIVES

A reference design for a 60 m diameter machine was completed in 1979 and a report was issued by British Aerospace Dynamics Group to the Department of Energy (10). That design had its origins in 1976 at the beginning of the UK wind energy development program. The design philosophy that had been adopted then was one of simplicity and ruggedness, for installation on hill top sites in Scotland, connected to the main electricity supply network. During 1980 the development of the design program took a new direction. The planned location of the machine was transferred from Bennan Hill in Ayrshire, South Scotland to Burgar Hill on Orkney. This relocation was the catalyst for a change in design philosophy, insofar as the 'weak' grid of the islands required the machine to be equipped with a synchronous rather than an induction generator and consequently control of power quality was introduced as an essential performance criterion. Independently, but closely associated with the need for power quality control, the change in design philosophy was also driven by recognition of developments in wind turbine technology world wide, which pointed clearly to the future of wind turbines lying in the development of rather lighter installations than the original simple, rugged, UK approach would permit.

Redesign of the 60 m wind turbine was therefore undertaken on the basis of the following objectives:

* cost optimised rotor

* compact and light transmission

* effective control of power quality

* light nacelle structure

* ease of maintenance, inspection and repair

* elegant and pleasing appearance.

The revised design has been based on a specification in which functional requirements are stated. The requirements that have influenced the design most materially are as follows:

* nominal output rating of 3 MW

* operation in parallel with diesel generation, with total system demand ranging from 25 MW down to the minimum compatible with stable automatic operation of the combined system

* torque trimming by variation of pitch angle of the blade tips, driven by hydraulic actuators

* attenuation within the drive train of torque peaks arising from wind turbulence

* a mechanical failsafe brake, capable of stopping the rotor from an overspeed condition in extreme wind conditions.

DESIGN DEVELOPMENT

Development of the new design took place in the following stages:

* teetering hub study

* concept study

* parametric studies

* trade off studies.

Teetering Hub Study

A study was undertaken to examine in general the implications for the design of large wind turbine generators of teetering the rotor hub, with particular reference to the 60 m machine (11). The conclusions were that normally, during operation when teetering is unrestrained, cyclic tower head loads are reduced by about 90% and cyclic out-of-plane blade loads by about 80%.

In view of the clear advantage in the reduction of rotor hub and tower head loads, particularly benefitting tower, nacelle and yaw mechanism design, it was concluded that the new design of the 60 m machine would be based on a teetered rotor.

Design Concept Study

In the initial stage of the concept study a full range of design options was reviewed. Some of the options considered and the choices that were made are shown in Table I.

Power Train and Nacelle

In the second stage of the concept study, various alternative design features were combined in three separate arrangements of power train and nacelle, and these were then considered in some detail. The arrangements were known as Schemes A, B and C. Basic outlines of the power train arrangements are shown in Figure I and a summary of their essential features is given in Table 2. The particular combinations of features incorporated in each of the three arrangements were not the only ones considered, but an assessment of the alternatives in three specific combinations made the study more practicable.

A brief summary of the conclusions concerning the advantages and disadvantages of the alternative power train and nacelle arrangements is given in Table 3. The principles of Scheme B were adopted as the basis for further design.

Tower

Consideration was given to various geometries and types of construction for the tower. Comparisons were drawn between the appearances of various forms with varying proportions and between alternative steel and concrete structures. A number of the alternative geometries considered is shown in Figure 2. The range of proportions was governed predominantly by consideration of system first mode natural frequency. In principle, the geometry (v) in Figure 2 was preferred.

Comparison of alternative structures in prestressed concrete and steel established that the use of neither material offered a clear advantage. Comparisons indicated that concrete offered a cost advantage, with respect to both construction and, to a lesser extent, maintenance. It was established that the proportions of concrete and steel structures would be similar and it was judged that there would be little to choose between them as far as appearance was concerned.

It was concluded that the process of construction for a concrete tower is more suited to the Burgar Hill site and that, on balance, concrete offered a cost advantage. Further design studies were therefore centred on prestressed concrete structures.

Erection and Plant Handling

Alternative methods of erection and provision for maintenance and repair were also considered. The basic options may be described as:

1. Full self-erect and built-in plant handling capability.

2. Partial self-erect and built-in plant handling capability, with heavy/high lift equipment required at first erection.

3. Minimal self-erect and built-in plant handling capability, with a high degree of reliance on repeated use of lifting equipment.

The earlier design of this machine had provided option 1. As this option is incompatible with a light nacelle structure it was rejected for the revised design. Option 3 was considered to be inappropriate for a remote site such as Burgar Hill where, initially at least, only one machine is to be erected. In principle, Option 2 was therefore chosen and a number of alternative schemes were considered. The method that has been adopted is outlined later in the paper.

Parametric Studies

Parametric studies were carried out relating to:

* optimisation of aerodynamic design of rotor blades

* optimisation of structural design of rotor blades

* sensitivity of system first mode natural frequency.

An evaluation was also made of increase in construction costs against gain in value of energy captured with increasing hub height.

Rotor Blades - Aerodynamic Design

For the purposes of this parametric study the 1980 reference design was defined as the "Base-Line Standard". The objective of the study was to strike a sound economic balance between the reduction in manufacturing cost on the one hand and the reduction in energy capture on the other that would result from simplification of the base-line standard blade geometry. The geometry had been optimised in the earlier phases of the project and it is instructive to note that if an optimum power ratio of 0.45 could be maintained at all windspeeds up to rated power, given the site data on which the study was based, annual energy capture would only increase to just over 4% above that achieved with the base-line design.

The following parameters relating to blade geometry were considered:

* chord width

* taper ratio

* thickness/chord (t/c) distribution

* blade angle setting or variation (twist)

* aerofoil section

* radial position of blade/root juncture.

Changes were made to the base-line standard affecting chord width (inboard), twist, t/c distribution and blade/root juncture. The net effect of the changes was to simplify blade manufacture considerably. The net reduction in annual energy capture was calculated to be just over 2%.

The single change causing the largest simplification in blade geometry related to twist. Optimum, base-line and modified blade angle setting/variation are shown in Figure 3. In the modified design, the twist which extended over the whole length of the base-line design is replaced over the fixed pitch span of the blade between 15% and 70% radius by a constant setting of 4°. Over the variable pitch tip the twist corresponds with optimum, giving an improvement on the base-line standard. The net effect, however, is a reduction in energy capture of 0.5%.

Rotor Blades - Structural Design

Initial studies were made of straight, tapered hexagonal box beam spars of constant plate thickness around the section, but varying thickness along the spar (Figure 4a). These studies determined a preference for a spar box occupying 60% of the aerofoil chord width. As for the blades of the 20 m diameter machine, the spar would be housed within the aerofoil section profile, which twisted around it along its length. The aerofoil section would then be built up with ribs, blocks and stiffening members, covered by a glass fibre reinforced plastic (GFRP) skin.

In parallel with the parameter study of aerodynamic performance, alternative structural arrangements were examined in which profiled upper and lower flanges directly form a substantial part of the aerofoil section, with GFRP leading and trailing edges completing it (Figure 4b). This arrangement corresponded with the proposal to eliminate twist over the fixed pitch span of the blades. Consultations with a number of potential blade fabricators confirmed that significant savings would be achieved in blade manufacture with this arrangment and the principle was therefore adopted.

Further development of the structural design led to the webs of the spar box section being moved inwards. This moved the welded joints between flanges and webs from the highly stressed corner locations and also reduced the unsupported width of the flanges which otherwise would require stiffening.

The structural arrangement incorporating a box spar within an aerofoil section formed with GFRP has been retained for the twisted, variable pitch tip blades.

System First Mode Natural Frequency

The first mode natural frequency (f_n) of a horizontal axis wind turbine generator is governed principally by the following parameters:

    i)   hub height
    ii) tower head mass
    iii) 2nd moments of area of sections through tower
    iv) distribution of mass of tower and contents
    v) elastic modulus of tower material
    vi) elastic behaviour of foundation.

During the concept design studies it had been assumed that in the revised design f_n would be roughly 1.6 x rotational frequency (P) i.e. the structural system would be soft. (By comparison, in the reference design for a rigid hub machine, f_n had been 3.4 P and the structural system was stiff.) Parametric studies were made of both steel and concrete alternatives, to establish the sensitivity of f_n for both types of structure to variations in tower head mass and tower geometry. At a later stage sensitivity to hub height and variation in the elastic modulus of concrete, E_c, were also investigated. Given the stiff nature of the materials underlying the Burgar Hill site, there was no need to consider variation in the elastic behaviour of the foundation.

The investigation of variation of f_n with E_c and hub height established that with the stiff aggregate found on Orkney and correspondingly high values of E_c, a choice existed between a soft system with hub height at 65 metres (f_n 1.7 P) and a stiff system with hub height at 45 metres (f_n 2.5 P). A trade off study, described below, was therefore undertaken to investigate the merits of the alternative, higher configuration.

Trade Off Studies

Trade off studies were carried out to investigate:

    a) the relationship between increase in construction costs and gain in value of energy captured with increasing hub height, and

    b) the relative complexity, cost and performance of alternative systems of power quality control.

Hub Height

Comparable outline designs were prepared for prestressed concrete towers suited to hub heights of 45, 55, 65 and 75 metres. The dimensions of these are shown in Figure 5. As indicated on the Figure, the 'soft' towers for hub heights of 45 m and 55 m are not suited to the Orkney site. The estimated values of the elastic modulus of concrete made with local aggregates resulted in the first mode natural frequency with these hub heights being too high for a soft system. At the time of the design study, including uncertainty bands, the range of E_c values was 36 kN/mm^2 at 1 year to 60 kN/mm^2 at 30 years. Subsequent further testing of concrete and aggregate specimens has reduced this range to 42 - 54 kN/mm^2.

Governed by consideration of f_n , as described above, the choice lay between 45 m stiff and 65 m soft towers. The difference in basic construction costs (shuttering, reinforcement and concrete) for these structures was calculated to be greater than £90k, whilst the value of the increased energy capture at the greater height was calculated to be about £11.5k per year.

In view of the fact that additional cost increase associated, for example, with crane rigging at first erection and permanent provision for plant handling would also be incurred with increase in hub height, it was concluded that for a prototype machine the value of the increased energy capture was insufficient to justify the greater hub height and to offset the increase in capital expenditure. The original hub height of 45 m was therefore retained, giving a stiff structural system.

Power Quality Control

The basic requirement of a system providing control of power quality from a wind turbine generator connected in combination with diesel generation to a weak grid is that fluctuations in the power absorbed by the rotor from the turbulent wind are substantially smoothed within the drive train. There are a variety of ways of achieving this and a number of options were considered in detail for the Orkney 60 m machine.

Following an assessment of relative cost, effect on blade pitch and overall control, and of performance and reliability, a system has been devised which offers considerable advantages with respect to compactness and reliability, whilst providing very effective and flexible means of control.

DESIGN FEATURES OF THE LSI

The machine concept that has resulted from the design prpocess described above is illustrated in Figure 6. It is known as the LSI. The full technical specification is given in Table 4. An artist's impression is shown in Figure 7. Within the limitations of this paper, the following description can be given.

Rotor

The rotor is in five major sections:

* hub section, comprising casting and inner blades

* two outer blades

* two tip blades.

Hub Section

The centre hub steel casting is of banjo configuration. The casting carries the centre trunnion, attaching rotor to main shaft, mounted on elastomeric, teeter bearings. The casting spans 3.6 m and the inner blade spars, extending to 30% radius, are butt welded to it.

The hub casting incorporates the mountings for:

- teeter locks/stops/dampers
- pitch change system equipment
- monitoring equipment
- spinner attachments
- erection slinging points for the complete rotor.

The inner blade steel box spar occupies 60% of the chord width and forms part of the aerofoil profile. Leading and trailing edges are manufactured from GFRP and are bolted to the edge of the spar.

The working aerofoil section ends at 15% radius. Closure pieces in GFRP form a transition to the spinner, which is also manufactured from GFRP.

Outer Blades

The outer blades extend from 30% to 70% radius. They are attached to the inner blades by bolted joints. The spar, leading and trailing edges are of similar construction to those of the inner blades. A typical section is shown in Figure 8a.

Tip Blades

The tip blades are just under 9 m long. A steel box spar is housed within the aerofoil section profile, which is twisted around it. The aerofoil section is built up with closely spaced, GFRP covered marine ply ribs bolted to the spar. Profiled foam blocks are located between the ribs, wood stiffening members are located along the leading and trailing edges and the whole assembly is covered by a GFRP skin. A typical section is shown in Figure 8b.

The tip blades are mounted on the outer blades with rolling element bearings, and pitch change is driven by single hydraulic jacks.

Drive Train

The drive train is made up of a primary gearbox mounted in the nacelle and a secondary gearbox mounted on the generator, located in the top section of the tower. The arrangement is shown schematically in Figure 9.

A mechanical brake is mounted on the rear of the primary gearbox casing. The hydraulic pump serving the pitch change control system is mounted on an extension to the main shaft, protruding through the rear of the gearbox.

Primary Gearbox

The primary gearbox carries the rotor on the input shaft. It is fitted with 3-point mountings to apply statically determinate loads to the casing and supporting nacelle structure. It contains two parallel epicyclic gear stages with 69% of power transmitting through both stages and 31% transmitting through one stage only. A bevel gear providing a vertical output drive down the axis of the tower forms a third stage.

Secondary Gearbox

The secondary gearbox is driven by a flexibly coupled shaft. It consists of a single epicylic stage with input to the planet carrier and output taken from the annulus gear.

Power Transmission Control System

The power transmission control system comprises:

* variable blade tip pitch control

* power quality control

* mechanical brake.

The main functional requirements of this system are:

i) to regulate the average power output, primarily restricting this to rated power at wind speeds above 17 ms^{-1},
ii) to minimise short term output fluctuations arising from turbulance, and
iii) to limit rotor speed.

Blade tip pitch control fulfils the first and partially fulfils the third requirement, a newly devised system fulfils the second, whilst the mechanical brake provides emergency back-up to fulfil the third requirement.

Blade Tip Pitch Control

Blade tip pitch control is achieved with an hydraulic system arranged as follows:

* 10 kW electric motor mounted on the aft end of the primary gearbox

* Constant pressure pump mounted on an extension to the main shaft through the aft end of the primary gearbox

* Hydraulic circuit to the hub through the primary gearbox and main shaft

* Low pressure reservoir mounted on the rotor hub

* High pressure lines out to blade tip junction with outer blades at 70% radius

* Servo amplifiers and valves, and double acting jacks adjacent to 70% radius.

Power Quality Control

A system has been designed which will smooth the whole spectrum of power fluctuations in turbulent wind to within a range that will be fully acceptable in terms of power quality. The response of this system to an idealised gust input is shown in Figure10.

Brake

The brake comprises two units of standard design, mounted back to back on the aft end of the primary gearbox casing. It is designed to stop the machine from overspeed, following loss of electrical load and failure of the blade tip pitch change control system, which would normally be used to brake the rotor. Each of the units contains three discs, splined to an extension of the primary gearbox second stage output shaft. Stationary friction surfaces consist of pads of organic friction material. The braking force is applied to the pads by springs. These are normally compressed by pneumatic pressure, holding off the pads, providing a failsafe system.

Nacelle

The nacelle main structure is a girder frame which transfers the primary gearbox loads to the tower head at discrete points, via individual components of the yaw bearing. The design is such that the primary vertical plane loads are carried by triangulated tension and compression members. Torque reaction is also provided by vertical plane forces at the primary gearbox rear supports. The girder frames extend from the gearbox mounting points to the rear end of the nacelle, joined by horizontal wind bracing members. Outriggers are attached to the main frames to support the sheeting rails of the exterior cladding. A travelling hoist gantry supported on the main girder frames covers the length of the nacelle and a separate hoist for installation and removal of medium weight componets from and to the ground is accommodated in the rear of the nacelle.

A crane jib capable of lifting and lowering the complete rotor is built into the front section of the nacelle. This jib can also be deployed both to lift the primary gearbox into the nacelle at first erection and to remove and replace it, if necessary, during the operational lifetime of the machine.

The procedures involved in lifting gearbox and rotor are shown schematically in Figure 11.

Yaw Mechanism

The design of the yaw mechanism is based on the following requirements:

* discrete bearings at structure load application points

* low cost consistent with reliability and maintenance free life

* replacement of wearing parts and limited life components with minimum down time.

The main vertical loads are supported on pad type bearings sliding on a steel ring fitted to the tower head. These may be changed in situ by jacking the main frame locally off the tower head. Horizontal loads are taken by rollers in contact with the inner edge of the ring. Uplift forces arise at one or two points of support only under conditions of extreme loading when the machine is inoperational and these are carried by fixed restraints.

The yaw drive is provided by two electrically driven reduction gears mounted in the nacelle and driving pinions which engage teeth on the steel ring at the tower head. The teeth consist of hardened steel round pins, which may be removed and replaced in situ as required.

STATUS AND PROGRAMME

The contracts relating to the Orkney 3 MW machine provide for review of the design before the construction phase is entered. The current programme schedules submission of the proposed design for review by the North of Scotland Hydro Electric Board and the Department of Energy at the end of May, 1983. Subject to approval of the design, completion of final details and manufacture will commence in the third quarter of 1983 and commissioning should follow in the second quarter of 1985.

ACKNOWLEDGEMENTS

The design and construction of the Orkney 3 MW Wind Turbine Generator is jointly funded by the Department of Energy and the North of Scotland Hydro Electric Board and the authors are grateful for their permission to publish this paper. The authors also acknowledge the support given to the Project by the Boards of Taylor Woodrow Construction Limited, British Aerospace Public Limited Company and GEC Power Engineering Limited. They also acknowledge the work carried out by many of their colleagues in developing the design described in this paper.

REFERENCES

1.   Lindley, D. and (a) "Orkneys horizontal axis wind turbine project" Modern
      Stevenson, W.G.   Power Systems, Dec, 1981 pp 37 - 41.

                   (b) "The horizontal axis wind turbine project on Orkney",
                       Proc. of the Third BWEA Wind Energy Conference,
                       Cranfield, UK, April 1981, pp 16 - 32.

2.   Lindley, D.   "A programme to integrate large scale wind turbine
                           generators into the Orkney Mainland Grid"
                           Heliotechnic Educational, London, May 1981.

3.   Lindley, D.   "The Orkney Wind Turbine Project - A Progress Review",
      Armstrong, J.R.C.   Proc. of the Fourth BWEA Wind Energy Conference,
      Hassan, U.   Cranfield, UK, April 1981, pp 34 - 50.
      Simpson, P.B.
      Stevenson, W.G.

4.   Lindley, D.   "Horizontal Axis Wind Turbine Project on Mainland
      and   Orkney, United Kingdom", presented at the Fourth
      McPhie, J.   Conference on Electric Power Industry, Bangkok, Nov 1982.

5.   Hassan, U.   "Design of a minicomputer based data acquisition and
      Henson, R.C.   processing system for the Burgar Hill, Orkney, Wind
      and   Turbines", Proc. of the Fourth International Symposium
      Parry, E.T.   on Wind Energy Systems, Stockholm, Sweden, Sept 1982.

6.   Hassan, U. and   "Preliminary results of a survey of the Burgar Hill,
      Lindley, D.   Orkney, Wind Turbine Site", ibid.

7.   Armstrong, J.R.C.   "The 20m diameter wind turbine for Orkney", Proc, of
      Ketley, G.R. and   the Third BWEA Wind Energy Conference, Cranfield, UK,
      Cooper, B.J.   April 1981, pp 54 - 62.

8.   Garrad, A.D.   "Wind Turbine Transmission Systems", ibid, pp 40 - 47.

9.   Cooper, B.J.   "The Control System for the 20 m 250 kW Wind Turbine
                           Generator for Orkney", paper presented to IEE
                           Colloquium "Energy Generating Systems for Wind Power",
                           IEE London, April 1982.

10.   British Aerospace   "Report on the Work Carried out in Phase II of the 60 m
      Dynamic Group   Horizontal Axis Wind Turbine Generator Study and
                           Recommendations for Phases III & IV", Feb 1980 (unpublished).

11.   Wing Energy   "Teetering Hub Design Study - Part II: 60 m Study"
      Group   WEG 60-6000, February 1982 (unpublished).

Table 1   REVIEW OF DESIGN OPTIONS

Item	Design Options	Advantages	Disadvantages	Action/Choice
1. Blade Construction (main load carrying structure)	Steel	Established manufacturing capability.	Fatigue vulnerable	
	GRP	Durability	Development of manuf. capability required. Transport	Steel
	Wood			
2. Blade Geometry	Simple	Established manufacturing capability	Lower performance	Parametric Study
	Optimised	Testing of manufact. capability required	Higher capital cost	
3. Aerodynamic Control	Partial Span Pitch Control	Self-start capability	Speed of response limited by induced loads	Partial span pitch control driven hydraulically
	Ailerons	Fast response	Noise, higher loads	
	Mechanical	Reliable	Less adaptable	
	Hydraulic	Widely adaptable	Integrity, maint.	
	Electrical	Minimum link problems	Bulky	
4. Rotor	Rigid	Simple, robust	Transmits full aerodynamic loads	Teeter Hinge
	Teeter Hinge	Reduces loads	Engineering Development required	
	Δ_3	Assists optimisation of structural dynamics	Additional complexity	
5. Rotor Axis	Inclined	Blade clearance	-	Inclined
	Horizontal	Periodic loads lower	-	
6. Mechanical Transmission	Rigid	Simple, robust	Requires electrical smoothing of power spikes	Soft mounted epicyclic
	Quill shaft	Simple	Adds length to nacl.	
	Soft mountings	Compact, tuneable	-	
	Parallel shaft	Simple, robust	Bulky-less compatible with soft mountings	
	Epicyclic	Compact	UK supply limited	
7. Braking	Frictional	Additional security	Bulky	Include friction brake
	Aerodynamic only	Minimum cost	Less secure	

Item	Design Options	Advantages	Disadvantages	Action/Choice
8. Synchronous Generator	Single speed	Simple	-	Review, considering inclusion of variable speed option for practical investigation at full scale
	Variable speed	Performance gain. Potential for eliminating soft transmission	Additional cost in prototype, at least	
9. Nacelle Configuration	Truss framework	Adaptable	Fatigue sensitive	Review
	Bed girder	Simple, ease of installation & repair	-	
	Structural shell	Economy of materials	Machinery replacement more difficult	
10. Yaw Bearing	Slewing Ring	Virtually off-shelf	Repair & replacement difficult	Review
	Other	Can be designed to carry high BM's	-	
11. Electrical Links	Slip Rings	Compact	Maintenance	Slip rings
	Twisting Cables	-	Bulky, unproven at required duty	
12. Tower Stiffness	Stiff	Rugged	Massive	Soft
	Soft	Matches strength requirements, slender	-	
	Soft-soft	Materials economy, slender	Inadequate strength? Restricted access	
13. Tower Material	Reinforced concrete	Ease of construction, maintenance free.	-	Review
	Prestressed concrete			
	Structural steel	Slender	Transport, erection, site welding	
14. Erection	Self-erect	Minimum cost over life of machine	Complexity	Self-erect, using tower as basic support with demountable aids as required
	Craneage	Simpler	Probably too costly	

Table 2: <u>POWER TRAIN ARRANGEMENT OPTIONS</u>

<u>Scheme A</u>

The principal feature of this scheme is a large ring gear of around 4m diameter which carries the hub. Other features are as follows:

- Planetary secondary gearbox with rotating planet for soft transmission
- Bevel drive to vertical generator
- Rotor supported on primary gearwheel
- Space frame supporting structure
- Rotary pumps and accumulator for soft transmission
- Ferrobestos yaw bearing pads
- Hydraulically actuated walking type yaw drive
- Transmission brake between primary gear and rotor.

<u>Scheme B</u>

This scheme has features similar to the MOD-5A design and has the rotor mounted directly off an epicyclic gearbox. The main features include:

- Integral gearbox with planetary primary stage
- Double helical stage and bevel drive to vertical generator
- Rotor supported on planet carrier
- Soft transmission arranged by rotating ring gear driving hydraulic pistons supplying accumulator for soft transmission
- Box beam supporting structure
- Ferrobestos yaw bearing pads
- Hydraulically actuated walking type yaw drive
- Transmission brake between primary gear and rotor.

<u>Scheme C</u>

This scheme is essentially a scaled-up 20m design with in-line power train. The principle features are:

- Independent roller bearings supporting the rotor
- Integral 3-stage planetary gearbox supported on rotor shaft overhang
- Gear casing torque reaction on springs for soft transmission
- Horizontal generator
- Box beam supporting structure
- Slewing roller type yaw bearing with geared yaw drive
- Transmission brake on rotor shaft.

Table 4 : TECHNICAL SPECIFICATION

TECHNICAL SPECIFICATION - Wind Turbine Generator LS1

ROTOR	
Number of blades	2
Diameter	60m
Speed	34rpm
Direction of rotation	clockwise looking downwind
Cone Angle	0°
Tilt Angle	5°
Location relative to tower	upwind
Hub type	teetered
Teeter angle	± 5°
Spinner	teeters with rotor

BLADE	
Material	steel + GFRP
Aerofoil	NACA 44xx series
Pitch mechanism	pitch of outer 30% span
Control	hydraulic
Power coefficient, max	0.43
Tip speed ratio at Cp max	8

TRANSMISSION	
Primary gearbox	2 stage epicyclic + bevel
Ratio	31.3
Brake	caliper/6 disc
High speed connection	differential gear + flexible couplings

GENERATOR	
Type	synchronous
Rated power	3750 Va
Voltage	11kV
Speed	1500 rpm
Frequency	50Hz

CONTROL SYSTEM	
Overall control	supervisory microprocessor
Blade tip control	closed loop
Up to synchronisation	blade tip pitch control
Operating mode	power limited

YAW DRIVE	
Type	electric motors
Rotation rate	109/minute
Braking	friction reaction

TOWER	
Type	prestressed concrete
Height to rotor hub	45m
Diameter of cylinder	3.8m

PERFORMANCE	
Rated power	3MW
Wind speeds (hub height)	
- cut-in	7 m/s hourly mean
- rated	17 m/s hourly mean
- cut-out	27 m/s hourly mean
- survival	70 m/s 3-sec gust
Annual Energy with 11 m/s AMWS at hub height	9 GWh

DESIGN LIFE	
All components	20 years min

INSTALLATION	
Burgar Hill, Orkney, 1985	

Table 3: COMPARISON OF POWER TRAIN AND NACELLE ARRANGEMENT OPTIONS

DESIGN CONCEPT STUDY - SUMMARY CONCLUSIONS

FEATURE	SCHEME A	SCHEME B	SCHEME C
Short power train	+	+	-
Light power train components	+	+	o
Energy storage in power train	o	-	-
Simple soft transmission arrangement	-	+	+
State of development	o	o	+
Access to centre of rotor	+	+	o
Rotor bearings replaceable in situ	+	-	o
Low rotor axis height	o	+	+
Main brake adjacent to rotor	+	+	o
Light nacelle structure	+	o	-
Reduction of nacelle size	o	+	-
Simple yaw bearing and drive	+	o	o
Main current slip rings eliminated	+	+	-
Simple lubrication system	o	+	o
Consequential			
Easier to erect	o	+	-
Easier to maintain	+	o	o
Cost of transmission	-	+	-
Cost of nacelle and yaw mechanism	+	o	-
Bottom line			
Weight	+	+	-
Cost	o	+	-

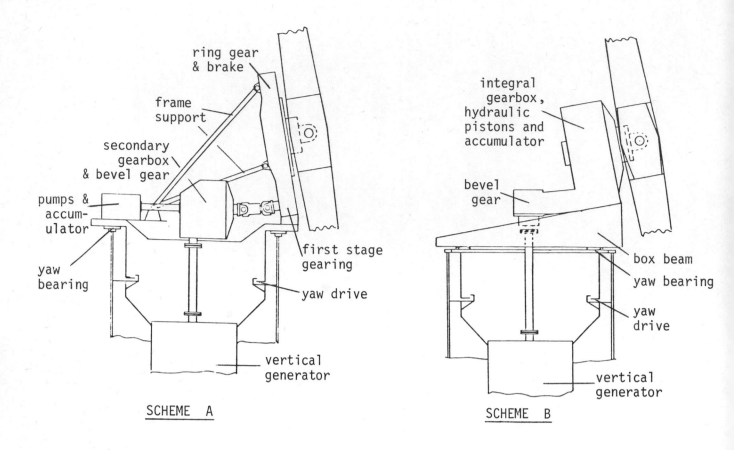

ring gear
& brake

frame
support

secondary
gearbox
& bevel gear

pumps &
accum-
ulator

yaw
bearing

first stage
gearing

yaw drive

vertical
generator

SCHEME A

integral
gearbox,
hydraulic
pistons and
accumulator

bevel
gear

box beam

yaw bearing

yaw
drive

vertical
generator

SCHEME B

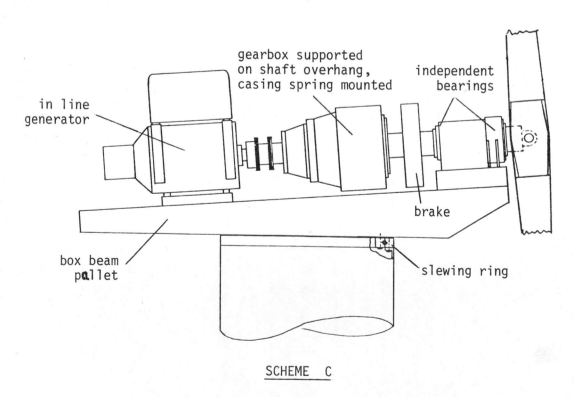

gearbox supported
on shaft overhang,
casing spring mounted

independent
bearings

in line
generator

brake

box beam
pallet

slewing ring

SCHEME C

Figure 1 : BASIC OUTLINES OF ALTERNATIVE ARRANGEMENTS OF POWER TRAIN

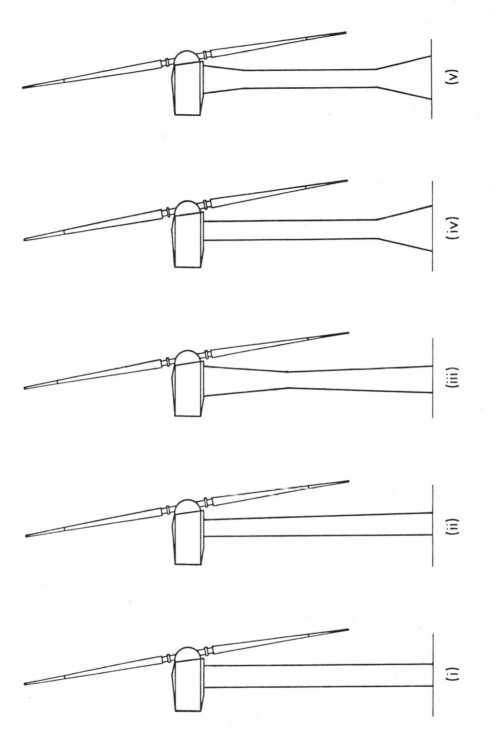

Figure 2: ALTERNATIVE TOWER GEOMETRIES

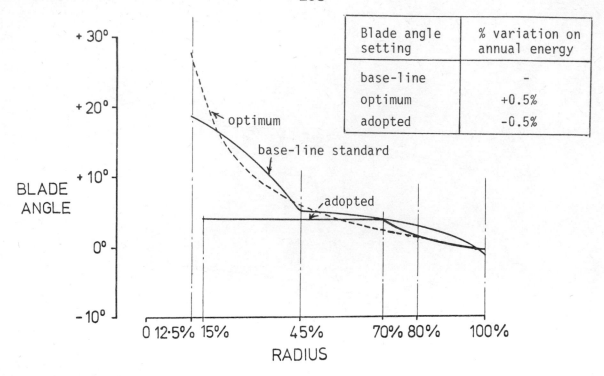

Blade angle setting	% variation on annual energy
base-line	–
optimum	+0.5%
adopted	-0.5%

Figure 3 : <u>EFFECT OF BLADE ANGLE SETTING ON PERFORMANCE</u>

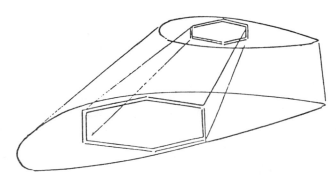

(a) Hexagonal Box Beam Within the Aerofoil Section

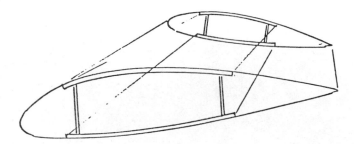

(b) Profiled Box Forming Part of the Aerofoil Section

Figure 4 : <u>ALTERNATIVE BLADE STRUCTURAL ARRANGEMENTS</u>

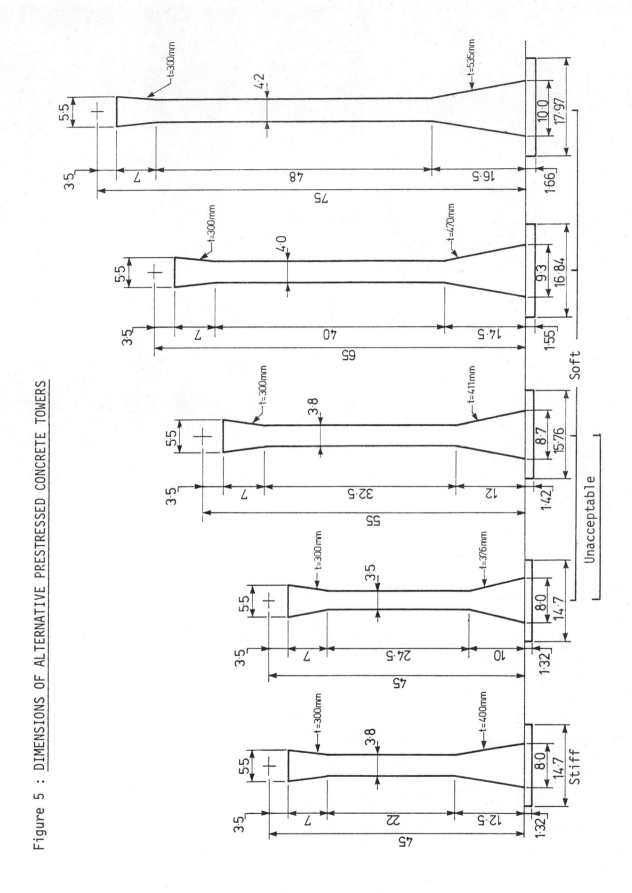

Figure 5 : DIMENSIONS OF ALTERNATIVE PRESTRESSED CONCRETE TOWERS

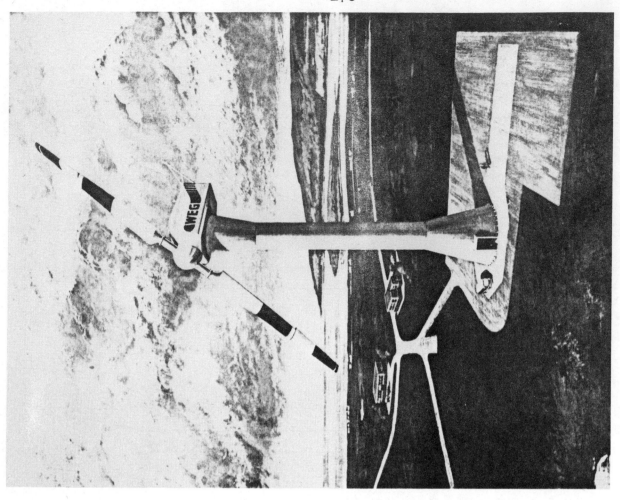

Figure 7 : <u>ARTIST'S IMPRESSION</u>

PARTIAL SPAN
PITCH CONTROL

HYDRAULIC
ACTUATOR

TEETERED
HUB

INCLINED
ROTOR AXIS

MAIN SHAFT BEARINGS
INTEGRAL WITH
PRIMARY GEARBOX

BOX STEEL SPAR
with fibreglass
l.e. + t.e.

INTERNAL STEEL SPAR
with fibreglass
envelope

30% partial span
control

LOCAL CONTROL

REMOTE OPERATION

DISC BRAKE AT
LOW SPEED

NACELLE SIZE
REDUCED TO MINIMUM

SECONDARY
GEARBOX

VERTICAL
GENERATOR

CONCRETE TOWER
3.8 m DIA

RACK AND PINION
LIFT

Figure 6 : <u>MACHINE CONCEPT</u>

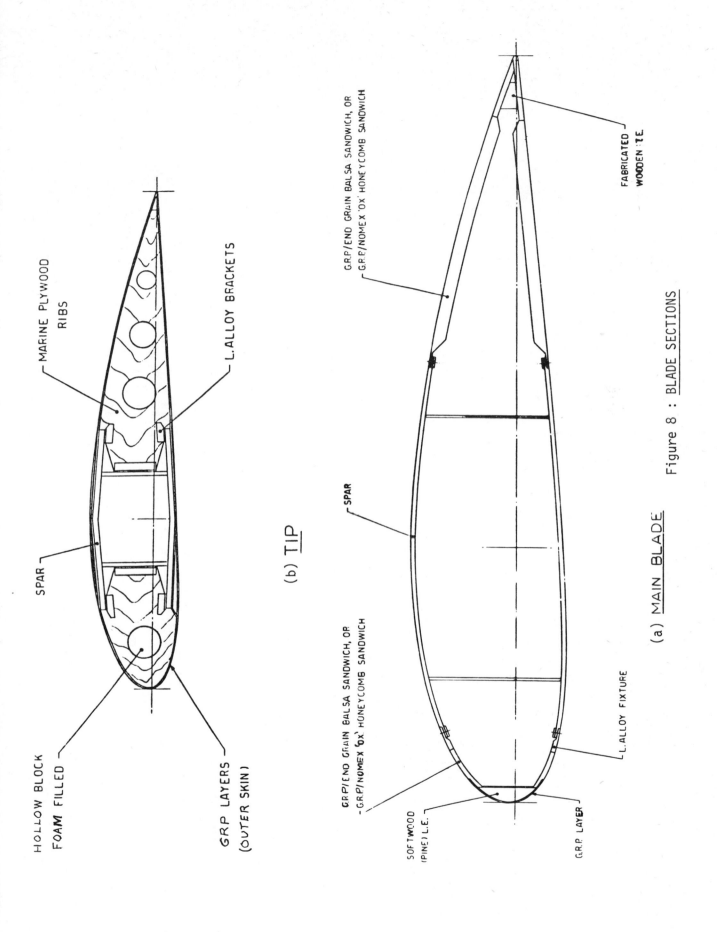

HOLLOW BLOCK
FOAM FILLED

MARINE PLYWOOD
RIBS

SPAR

L. ALLOY BRACKETS

GRP LAYERS
(OUTER SKIN)

(b) TIP

G.R.P./END GRAIN BALSA SANDWICH, OR
G.R.P./NOMEX 'OX' HONEYCOMB SANDWICH

SPAR

G.R.P./END GRAIN BALSA SANDWICH, OR
-G.R.P./NOMEX 'OX' HONEYCOMB SANDWICH

SOFTWOOD
(PINE) L.E.

G.R.P. LAYER

FABRICATED
WOODEN T.E.

L. ALLOY FIXTURE

(a) MAIN BLADE

Figure 8 : BLADE SECTIONS

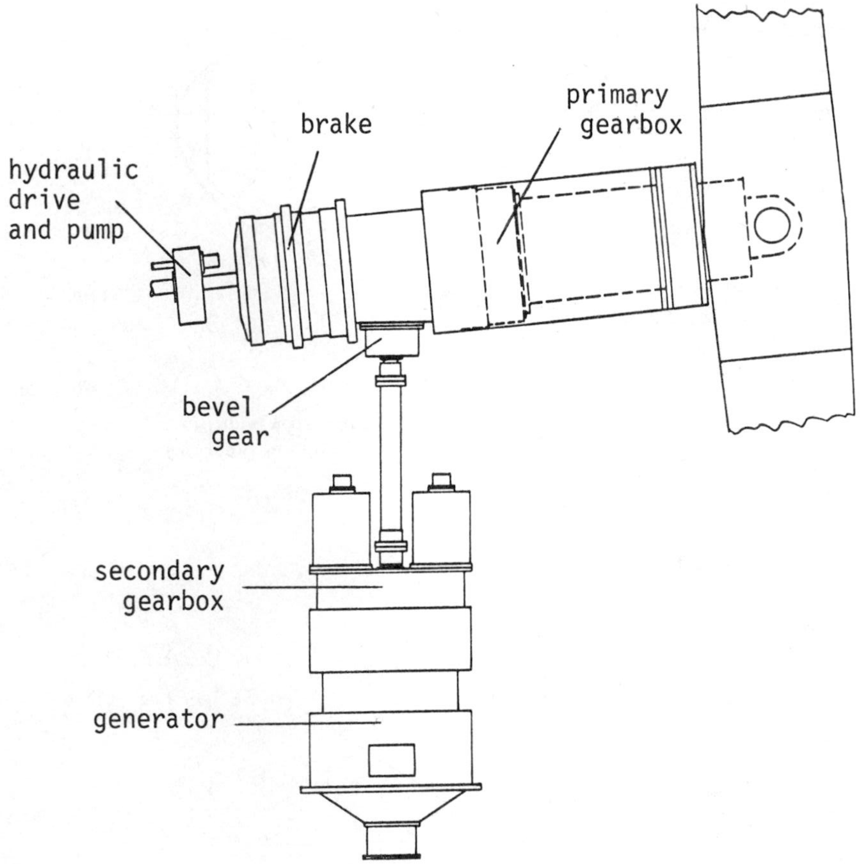

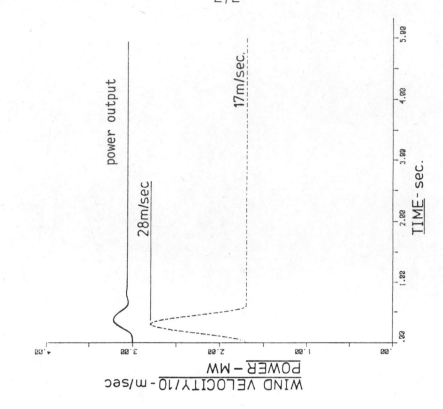

hydraulic
drive
and pump
brake
primary
gearbox
bevel
gear
secondary
gearbox
generator

Figure 9 : SCHEMATIC ARRANGEMENT OF DRIVE TRAIN

WIND VELOCITY/10 - m/sec
POWER - MW
power output
28m/sec
17m/sec.
TIME - sec.
.00
1.00
2.00
3.00
4.00
5.00

Figure 10 : GUST POWER RESPONSE WITH POWER QUALITY CONTROL SYSTEM

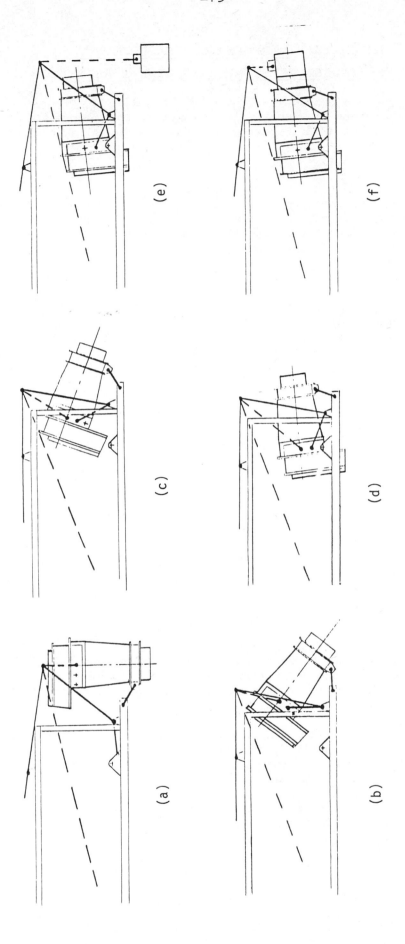

Figure 11 : PRIMARY GEARBOX AND ROTOR INSTALLATION

LABORATORY PERFORMANCE MEASUREMENTS
OF A MODEL VERTICAL-AXIS WIND TURBINE

S Gair, W S Bannister, M Millar
Departments of Electrical & Electronic Engineering,
and Mechanical Engineering, Napier College, Edinburgh

Abstract

Results of coefficient of performance measurement tests and theoretical
predictions for vertical and horizontal-axis wind turbines have
frequently been presented in the literature in recent times. The
detailed experimental instrumentation, computer or analogue data logging
and subsequent software processing details are usually omitted from what
has frequently been an expensive and time consuming exercise.

This paper describes a test instrumentation system which illustrates how
the theoretical predictions can to some extent be tested using a
relatively low cost, technically straight-forward, microcomputer system
incorporating standard off-the-shelf equipment and using an easy to
understand software package developed at, and available from, Napier
College.

Nomenclature

C_p = wind turbine power coefficient

ρ = density of air

V = average wind velocity at the wind turbine, m/s

A = area of wind intercepted by the wind turbine, m^2

I = total wind turbine enertia, kgm^2

ω = mechanical speed, rad/s

α = angular acceleration, rad/s/s

T = mechanical torque, Nm

r = cross-arm radius, m

λ = tip speed ratio

σ = solidity

P = power, W

θ = overall angle of rotation of blade

C_t = aerofoil thrust coefficient

W = resultant air velocity relative to a blade element

Suffices

d - downstream

u - upstream

Introduction

The main aim of constructing a model vertical-axis wind turbine was to
measure the coefficient of performance and blade loadings for a wide
range of possible machine configurations and, therefore, provide a basis
for comparison with theoretical predictions.

The principle factors which influence the performance of vertical-axis
wind turbines are generally well understood, though there has been a
tendency for published studies simply to quote the results obtained with
no detailed mention of the method of recording and analysing this data.
This paper presents a detailed account of a low cost microcomputer data
logging and software processing package which was assembled using
standard off-the-shelf hardware.

Wind Turbine Details

The wind turbine model tested was a small straight-bladed vertical-axis type designed such that the basic configuration and dimensions could be easily altered, eg blade span and profile, number of blades, blade pitch setting, radius to blades. The dimensions of the turbine for the purpose of the wind tunnel tests were: blades NACAOO15, 0·150 m chord, span 1 m; diameter 1·23 m. The blades used were extruded aluminium, and the cross arms were of light alloy construction fitted with a fibre-glass fairing.

Wind Tunnel Tests

General Description: The entire wind rotor model assembly was mounted in the working section of an open jet 1·52 metre diameter wind tunnel located at the University of Strathclyde. A single anemometer, type BPG152 MK1 was mounted in the wind tunnel - this anemometer was calibrated insitu from a Pitot-static tube. To safeguard against mechanical failure or overspeed, a safety interlock was fitted between the turbine brake and the wind tunnel control system. All control equipment and instrumentation was remotely mounted in a control area of the laboratory.

Experimental Procedure: The object of the tests was to determine from the acceleration test technique (1) the performance characteristic of the model when tested for a variety of configurations, ie 3-blades, 4-blades and pitch offset. These tests were conducted such that for each test configuration the wind tunnel was operated at a constant wind velocity and the turbine was allowed to accelerate from low rotational speed to a steady speed state. A range of wind velocity settings were then chosen between 5 m/sec and 12 m/sec.

Equipment and Instrumentation

The data collection system adopted was based on the IEEE 488 bus standard 'Digital Interface for Programmable Instrumentation', also known as the G.P.I.B., 'General Purpose Interface Bus'. Using this bus system, it is much easier to interconnect measuring instruments for computer controlled experiments or automatic test systems. Relieved of the major problem of system construction by the ready-made cables, connectors,   and the standardised logic and protocol of a G.P.I.B. which is fully implemented in most desktop microcomputers, means that the required assembly can be put together quickly and just as quickly dismantled for re-deployment of the instruments.

Using the IEEE bus up to fifteen devices can be linked together in one system, all using the same standard stackable connector. Each device attached to the bus can perform in at least one of three different modes known as talker, listener or controller.

When acting as a talker, the device puts data on the bus, for example a digital voltmeter (D.V.M.) sending its reading. As a listener, the device takes data from the bus, for example a signal generator setting up a required frequency. In the controller mode the device controls all communication activity on the bus.

Because of the high speed of the interface and the fact that is is eight bit parallel, this gives rise to two limitations when using an IEEE interface. Firstly, the maximum distance between two devices is five metres and secondly, the maximum distance from one end of the cable to the other is twenty metres.

A data collection system was assembled, Figure 2, in which two digital voltmeters reading analogue input signals, placed this data on the bus via a G.P.I.B. unit. This unit in which the standard protocol for talker and listener functions is used and each of the data, hand shake and management signal lines are correctly terminated transmits the two readings merged together as one character string to the controller. Most controllers have ample string handling functions in which separation of the two readings is simple. Following a request for data by the bus controller, it would be undesirable for the G.P.I.B. unit to transmit this data whilst it is changing, therefore a necessary facility of the unit must be to delay transmission until the external equipment indicates that data is valid. In this way a simultaneous recording of any two transducer outputs can be made. When run with controller software programmes written in BASIC, this equipment was capable of transmitting six readings per second - this data recording speed being higher than that necessary to measure the performance characteristic for this particular wind turbine model made it necessary to introduce a software delay into the programme. In recording strain gauge outputs as a function of rotor position, an assembly language version of that part of the basic programme to address the G.P.I.B. will give adequate recording speed.

Considering the variety of I/O, especially the IEEE 488 and the parallel user port facilities available on the CBM Commodore microcomputer, a 32K, '4032' series unit was selected as the controller together with a 4040 dual floppy disk drive unit.

The transducer signals required to obtain a Cp/λ characteristic using the acceleration method are rotor speed, wind speed and elapsed time. Rotor speed can be determined by means of a self energising magnetic probe, the output of which is produced by cutting the lines of magnetic force with a rotating sixty-toothed wheel. This output can then be connected to a digital counter which will indicate the current R.P.M. value. In order that this reading be recorded by the computer it is necessary that the counter should have b.c.d. (binary coded decimal) and hold reading facilities. This type of unit is particularly suitable for outdoor work where the absence of a low voltage regulated power supply to the pickup head is desirable. Such a unit as this is likely to cost in the region of £150 - a low cost alternative (< £8) used in these tests utilised an opto electronic device combining an infra-red light emitting diode, chopped by a toothed wheel, and photo transistor, the frequency output of which, in the absence of a b.c.d. counter, was converted to an analogue voltage via a frequency to voltage converter I.C., Figure 3. This analogue voltage, together with the output voltage of the anemometer, was recorded simultaneously via the D.V.M. and G.P.I.B. by the computer. On accepting these two inputs via the bus the computer software then referenced the controllers internal clock time and thus three simultaneous values were recorded.

Theoretical Aspects

Streamtube theory (2)(3)(4)(5) has been used as a basis from which to predict the aerodynamic performance of the wind turbine. This theory accounts for the difference in induced velocities at the upstream and downstream positions of the blades and takes into consideration expansion of the stream tube through the turbine. The theory uses repeated application of Bernoulli's equation to a streamtube, application of the force momentum relationship, consideration of the forces acting on the blade element, hence by iteration the velocities upstream and downstream may be calculated. The power contributed per streamtube, due to the torque on the upstream and downstream blade element may then be obtained from:

$$\delta P = \frac{\omega r^2 \rho \sigma \delta \theta}{2\pi (V_u + V_d)} \left\{ V_d W_u^2 C_{tu} + V_u W_d^2 C_{td} \right\}$$

The overall power may be determined by numerical integration of this equation. The increment $\delta \theta$ is set by choosing the number of circumferential stations between $\theta = -90^0$ to $\theta = +90^0$. Increments of 10^0 has been found to give sufficient accuracy.

The power coefficient may then be obtained from:

$$C_p = \frac{POWER}{\frac{1}{2}\rho A (V_\infty)^3}$$

Results

The wind turbine was allowed to accelerate from a low rotational speed, unloaded except by its own inertia, to a steady speed condition. Measurement of wind velocity and angular acceleration as a function of time enables the complete C_p vs λ characteristic to be obtained.

Using the computer software programs the power coefficient was determined from:

$$C_p = \frac{POWER\ OUTPUT}{\frac{1}{2}\sigma A (V_\infty)^3} = \frac{T\omega}{\frac{1}{2}\sigma A (V_\infty)^3} = \frac{(I\alpha)\omega}{\frac{1}{2}\sigma A (V_\infty)^3} ,$$

and the velocity ratio from, $\lambda = \frac{r.\omega}{V}$

A standard laboratory test enabled the inertia value to be determined. By curve fitting to the data obtained from this test the rotor torque can be corrected for bearing friction and aerodynamic drag losses.

All data recording and subsequent analysis was carried out using microcomputer programmes written in Commodore BASIC. Sequential files containing raw data items, intermediate data (eg transducer voltages converted to actual quantities) and final results (eg Cp, torque, power) were created on floppy disk. This information could at any stage be transferred to a printer or X-Y plotter for display. In order that a more refined analysis could be carried out using a larger computer, the files created by this portable microcomputer system can easily be transferred to a host mini computer (PRIME 650) via an RS232 serial link, thus giving access to better graphics and statistical packages for curve fitting and display.

Blockage Correction: Given the dimensions of the wind turbine model tested and the diameter of the wind tunnel available it was apparent that the wind velocity measured during the tests would not be a true value of V_∞ and also the very high blockage factor meant that no attempt could be made to use any meaningful wind tunnel corrections. Indeed at high wind turbine rotational speed the actual wind velocity decreased due to air being expelled from the wind tunnel circuit.

Results of tests using 3 blades are shown in Figure 4. The comparison between the experimental results and the theoretical predictions is not good, however the repeatability of experimental results is very good.

Summary

The principle objective of the work carried out to date has been to design and test a low cost data recording and analysis system suitable for laboratory and outdoor site use in measuring wind turbine performance parameters. This system, which is fully described in the text, together with the equipment schedule listed in appendix A, is easily capable of further expansion, eg monitor a greater number of transducer inputs, link-up to a host mini computer. All software programmes were written in Commodore '4000' series BASIC and were designed in a generalised way to cater for a wide variety of test situations. This software package can be made available to anyone wishing to assemble a similar instrumentation system such as listed in appendix A. The computer programmes will record the raw data values, process any number of test runs within the storage capacity of the 35 track floppy disk unit and output coefficient of performance, power and torque information to an X-Y plotter, floppy disk or printer.

In testing the wind turbine model in the controlled environment of the wind tunnel it was readily apparent that the blockage factor was far too high to provide accurate wind tunnel test results. However, the data obtained gave an indication of the wind turbine performance which could then be used to assess the relative effects of modifications to the turbine configuration. An attempt to repeat the measurements at an outdoor site is currently in progress and it is hoped to publish this information in the near future.

References

1. Sharpe, D J. A Theoretical and Experimental Study of the Darrieus Vertical-Axis Wind Turbine. School of Mechanical, Aeronautical and Production Engineering, Kingston Polytechnic, 1977.

2. Read, S and Sharpe, D J. An Extended Multiple Streamtube Theory for Vertical-Axis Wind Turbines. 2nd BWEA Wind Energy Workshop, Cranfield, 1980.

3. Strickland, J H. A Performance Prediction Model for the Darrieus Turbine. International Symposium on Wind Energy Systems, Cambridge, 1976.

4. Bannister, W S. A Theoretical Analysis of Small Vertical-Axis Wind Turbines. International Symposium on "Applications of Fluid Mechanics and Heat Transfer to Energy and Environmental Problems", University of Patras, Greece (July, 1981).

5. Stacey, G and Musgrove, P. Developments on Vertical-Axis Wind Turbine Streamtube Theories. 4th BWEA Wind Energy Workshop, Cranfield, 1982.

Acknowledgements

The authors gratefully acknowledge the many contributions made by colleagues on the Wind Energy Group at Napier College and to other workers in the field of wind energy research. Thanks are also due to the Science Research Council for their support via Grant Ref. GR/B52717. The work is being carried out at Napier College, Edinburgh and is being supported by Lothian Regional Council.

Appendix A

Schedule of Instrumentation:

Commodore-4032 microcomputer, 5¼ inch dual floppy disk drive, model 4022 printer.

Farnell Instruments - IEEE 488 General purpose interface (omnibus OB2), 4½ digit autoranging multimeter DM141, SWIB switching unit.

J J Instruments - PD4 plotter and software ROM.

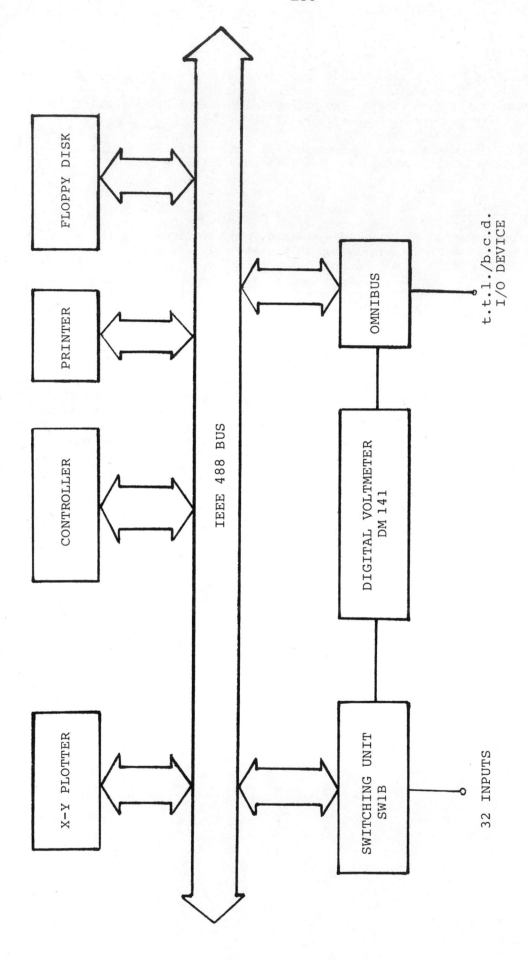

FIG.2 — IEEE BUS INSTRUMENTATION LAYOUT

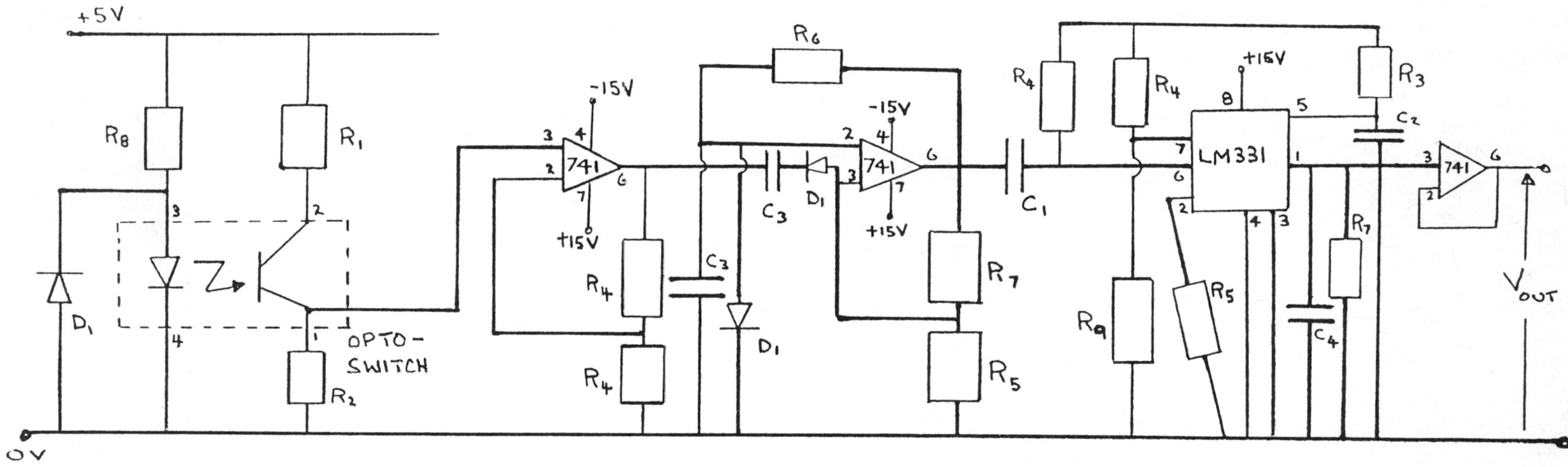

RESISTORS

R_1 = 1 k	R_6 = 56 k
R_2 = 4.7 k	R_7 = 100 k
R_3 = 6.8 k	R_8 = 470 k
R_4 = 10 k	R_9 = 680 k
R_5 = 12 k	

CAPACITORS

C_1 = 4.70 pF
C_2 = 0.01 μF
C_3 = 0.1 μF
C_4 = 4.7 μF

DIODE, D_1 = IN414B

OPTO-SWITCH RS 306061

FREQUENCY-TO-VOLTAGE I.C.-
NATIONAL SEMICONDUCTOR LM331N

OPERATIONAL AMPLIFIERS -
RS 305311

FIG.3 - SHAFT SPEED ENCODER

WBC-10

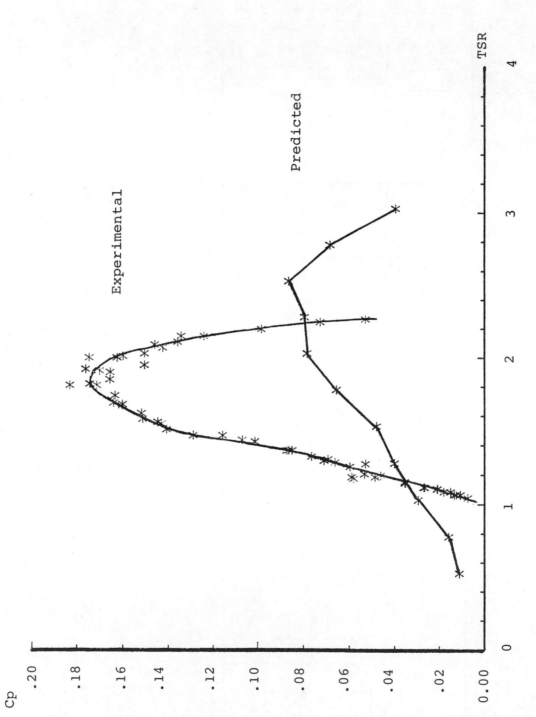

FIG.4 - Cp/λ FOR 3-BLADE CONFIGURATION

INTERPRETATION AND ANALYSIS OF FIELD TEST DATA

M. B. ANDERSON
SIR ROBERT McALPINE & SONS LTD

ABSTRACT

Three sources of error associated with field test data are discussed and a number of methods for minimising them are suggested. Two statistical techniques for deriving the performance characteristics from field test data are examined and are validated using data derived from measurements made on a six metre diameter horizontal axis wind turbine.

INTRODUCTION

The paper is divided into three separate but related sections. The first is concerned with the siting of anemometer with respect to a turbine. In order that the power vs. wind speed relation or the performance characteristics can be determined it is necessary to make an accurate measurement of the wind velocity at the turbine. Three sources of error in this measurement are discussed and a number of techniques for minimising them are suggested. In the final two sections these techniques are used firstly, to determine the power output vs. wind speed relationship and secondly, to determine the performance characteristics of a six metre diameter turbine.

WIND VELOCITY MEASUREMENT

Three sources of error arise when using an anemometer to determine the wind velocity:

a) <u>Physical separation</u> Due to natural fluctuations in the wind the instantaneous velocity at the two locations may not be the same even though, under some circumstances, the mean velocity and standard deviation are the same.

b) <u>Spatial averaging</u> A single anemometer measures the wind velocity at a point in space whereas a turbine averages the velocity over its swept area.

c) <u>Flow modification</u> The physical presence of the turbine will modify the air flow, causing in some places a decrease in the wind speed and in others an increase.

The error introduced due to physical separation of the anemometer and wind turbine depends on whether the separation is longitudinal, vertical or lateral. The mean velocity measured at two locations separated either longitudinally or laterally will in general be the same, whereas for vertical separations this will not be true due to the presence of wind shear. It may be possible to correct for wind shear if the functional relationship between wind speed and height is known, however this is not recommended.

A measure of the correlation of the longitudinal velocity fluctuations at two locations is the cross-correlation coefficient

$$R_{\shortparallel}(\underline{\xi}, T) = \frac{\overline{v(t)\, v'(t+T)}}{\sigma_{\shortparallel}\, \sigma_{\shortparallel}'} \tag{1}$$

where v, v' and $\sigma_{\shortparallel}$, $\sigma_{\shortparallel}'$ are the velocities and standard deviations respectively, measured at two points separated both spatially ($\underline{\xi}$) and temporally (T). If the two locations are coincident then the correlation coefficient for zero time lag (T = 0) will be unity, but in general $R_{\shortparallel}/\sigma_{\shortparallel}\sigma_{\shortparallel}'$ for non-zero values of ξ and T it will always be less than unity and the peak will be displaced from the origin (T = 0). The peak value of the correlation coefficient for longitudinal separations will be greater than those corresponding to vertical and lateral separations due to high frequency components (small eddies) persisting in the direction of the mean flow.

The variance of the difference in the velocities at the two locations can be derived as

$$\sigma_{\shortparallel}^{2} = \overline{\left[v(t) - v'(t+T) \right]^{2}}$$

$$= \sigma_{\shortparallel}^{2} + \sigma_{\shortparallel}'^{2} - 2 R_{\shortparallel}(\underline{\xi}, T) \tag{2}$$

For lateral and longitudinal separations it can be assumed that $\sigma_{\shortparallel}^{2} = \sigma_{\shortparallel}'^{2} = \sigma_{v}^{2}$, therefore we obtain

$$\frac{\sigma_{\shortparallel}^{2}}{\sigma_{v}^{2}} = 2 \left[1 - \frac{R_{\shortparallel}(\underline{\xi}, T)}{\sigma_{v}^{2}} \right] \tag{3}$$

The maximum value of the cross-correlation coefficient is 0.92, giving from equation (3), $\sigma_{\shortparallel}/\sigma_{v}$ = 0.4. The corresponding value for a lateral separation is 0.75. For a turbulence intensity of 15% (rural, Ref. 1), the r.m.s. difference of the velocities measured at the longitudinally and laterally separated anemometers are 6% and 11% of the mean velocity respectively.

It is important when determining the power coefficient vs. tip speed ratio characteristics that an accurate measurement of the instantaneous wind velocity at the turbine is available, i.e. a high correlation coefficient. The correlation coefficient can be increased by averaging the instantaneous values. This has the effect of reducing the high frequency uncorrelated components of the wind velocity fluctuations. For both longitudinal and lateral separations a significant increase in the coefficient can be achieved with relatively short averaging periods. A certain amount of averaging of the wind velocity fluctuations is acceptable as this reduces the second source of error; namely that a single anemometer measures the wind velocity fluctuations at a point whereas a turbine tends to average the velocity over its swept area. It can be shown (2,3) that the averaging time necessary is approximately proportional to the diameter of the turbine and inversely proportional to the wind speed. A 1.5 second averaging gives approximately the effective wind speed seen by a six metre diameter rotor in a mean wind speed of 5 m/s.

The third source of possible error is caused by the turbine modifying the flow in its immediate vicinity. It is possible to estimate this error by modelling the turbine as either a source or doublet distribution; their strength being related to the rotor thrust. Another method, developed by Anderson (2), is to determine the velocity field due to a semi-infinite number of expanding coaxial vortex rings. This method, like the previous ones, is not valid for regions within the wake of the turbine as it depends on the assumption that the air can be treated as inviscid. This is a valid approximation for regions upwind of the turbine, where the velocity gradients are small. The measurement of the wind velocity within the wake of a turbine is open to large errors and is obviously an undesirable location for performance measurements.

Figure 1 shows an isovent contour map, from the method of ring vortices, for a turbine operating at a tip speed ratio of ten with a thrust coefficient 0.85. The incident wind velocity profile was assumed to be uniform. The contour map shows that the centre-line velocity deficit 1 rotor-diameter up-stream of the rotor is approximately 4.5% of the free-stream. Moving half a rotor diameter off-axis reduces this velocity deficit to 4.0%.

From measurements made on a six metre diameter turbine operating at tip speed ratio of approximately ten it was possible to estimate the velocity deficit up-stream of the rotor at two locations. The first 1.8R (R = radius) up-stream and 0.8R off-axis, and the second 3R up-stream and on-axis. Shown in Figure 2 is a plot of sixty second averages of the wind velocity measured at the first location against the free-stream velocity measured at a location 5R (R = blade radius) off-axis in the plane of the rotor. From an orthogonal regression analysis the gradient of the line was evaluated to be 0.96 ± 0.01 giving a velocity deficit of approximately 4%. The scatter about the regression line indicates that the correlation coefficient was less than unity (0.91) and was caused by a) fluctuations in wind direction, b) variations in the thrust coefficient and c) the lateral separation of the two anemometers. For the second location a similar analysis revealed that the gradient was 0.98 ± 0.01 giving a velocity deficit of 2%. The correlation coefficient was measured to be 0.87. For both locations the experiment was repeated with the turbine stationary in order to check for any systematic errors. It was found for both locations that these errors were small compared with those associated with the determination of the velocity deficit itself. Close agreement was found to exist between the experimental results and the theory, Figure 1.

At first sight it would appear that siting an anemometer up-stream of the turbine and correcting its reading according to a reference anemometer would give an accurate measurement of the incident wind velocity. However, this assumes that a) the thrust coefficient of the rotor and b) the wind direction do not vary during the period of measurements. It was, therefore, necessary to examine these assumptions before this method could be adopted. From Figure 1 it can be seen that for an anemometer situated 2R up-stream and initially on axis, a 20 degree change in the wind direction will cause the velocity deficit to change by less than 0.5%. This indicates that fluctuations in wind direction are unlikely to pose a problem in this respect. However, they will reduce the correlation between the measured wind velocity and that experienced by the tubing due to a lateral separation being introduced. Anderson (2) has shown that, for most practical purposes, variations in the thrust coefficient, hence velocity deficit, will be small in comparison with errors arising from other sources.

CONCLUDING REMARKS RE: WIND VELOCITY MEASUREMENT

From this discussion on the siting of an anemometer with respect to the turbine a number of conclusions can be drawn:

a)   To obtain meaningful results the anemometer should not be placed in the wake of a turbine unless measurements are explicitly required for this region.

b)   For an estimate of the instantaneous incident velocity field the anemometer should be situated up-stream of the rotor and the relevant correction factors applied. These will include i) 'time of flight' to maximise the correlation coefficient, ii) filtering to model the spatial averaging of the rotor blades and iii) the measurement of the free stream velocity to take into account interference effects.

c)   If instantaneous values of the wind speed are not required the measurements should be filtered to increase the correlation especially when there is a lateral separation between the anemometer and the wind turbine. Additionally, the anemometer should be situated in a position such as to minimise interference effects, i.e. greater than two diameters away.

POWER VS. WIND SPEED

In this section the aerodynamic power output vs. wind speed relationship of a six metre diameter turbine is determined from experimental results using the method of 'bins' developed by Akins (4). This is a straightforward technique in which the sampled power output is assigned to a particular bin which is determined by the instantaneous wind speed. A counter associated with that bin is also incremented. At the end of a particular set of measurements the average power within each bin is evaluated.

Shown in Figure 3 are the results of using this form of analysis on a six metre diameter rotor when controlled to operate approximately at a tip speed ratio of ten. The aerodynamic power, rotational speed and wind speed were sampled and averaged every second and assigned to velocity bins of width 0.2 m/s. The wind velocity was measured at a location 1.3R up-stream and was corrected according to a reference anemometer before being binned. The total length of the data record was 3600 samples (1 hour); the mean wind speed and aerodynamic power output over the period were 5.6 m/s and 1.22kW respectively, and the turbulence intensity was 23%. To investigate the effects of reducing the correlation between the measured wind velocity and that experienced by the turbine the analysis was repeated with an anemometer situated 4R up-stream and 5R off-axis. The results are also shown plotted in Figure 3. For both sets of data typical errors are shown indicating the spread of power measurements within each bin. The scatter can be attributed to the correlation coefficient being less than unity and the turbine being unable to respond to fast fluctuations in both wind direction and velocity. It can be seen that a reduction in the correlation coefficient causes the scatter within each bin to increase, the power at low wind speeds to be overestimated, and the power for higher wind speeds to be underestimated. This dependence of the binned data on the correlation has been noted by other authors e.g. Akins (5); Lundsager, Fransen and Christensen (6). It is possible to gain some insight into this dependence if a number of assumptions are made regarding the statistical nature of the velocity fluctuations. Firstly, it will be assumed that the wind velocity V experienced by the turbine is normally distributed about $\bar{V}$ (the mean wind speed for a particular set of measurements) with a standard deviation σ_V .

Secondly, the measured wind velocity V_m at a location other than that coincident with the turbine is normally distributed about V with a standard deviation σ_{mv}, given by equation (3), i.e. the root mean square of the difference in the wind velocity fluctuations as measured and that experienced by the turbine. Therefore, the probability of a power measurement P_{Ai} due to an actual wind velocity, V occurring at the turbine being associated with a velocity, V_m is

$$P_R(V_M) = \int_0^\infty P_R(V_M - V) \, P_R(V - \bar{V}) \, dV \tag{4}$$

where $P_R(V_m - V)$ and $P_R(V - \bar{V})$ are Guassian probability distributions with means and standard deviations of V, $\bar{V}$ and σ_{mv}, σ_v respectively.

To determine the average power associated with a particular velocity, V_m it will be assumed that there is a linear relationship between the aerodynamic power output P_{Ai} and V, such that

$$P_{Ai} = kV \tag{5}$$

The average power associated with a particular bin, V_m of infinitesimal width is therefore

$$P(V_M) = \frac{\int_0^\infty kV \, P_R(V_M - V) \, P_R(V - \bar{V}) \, dV}{P_R(V_M)} \tag{6}$$

The lower limit of the integral can be extended to $-$infinity when $\sigma_v \ll \bar{V}$, thus under these conditions

$$P(V_M) = kV_M \left(\frac{\sigma_v^2}{\sigma_v^2 + \sigma_{mv}^2} \right) + k\bar{V} \left(\frac{\sigma_{mv}^2}{\sigma_v^2 + \sigma_{mv}^2} \right) \tag{7}$$

From this equation it can be seen that when $\sigma_{mv}^2 \neq 0$ the slope of the measured power curve is $< k$ and that a permanent off-set will exist at $V_m = 0$. As σ_{mv}^2 tends to zero and σ_v^2 remains constant the measured power curve will become coincident with the theoretical relationship, equation (5). With reference to Figure 3 this can be seen to be the case, i.e. reducing σ_{mv}^2 (increasing the correlation) increases the slope of the measured data. As σ_v^2 becomes large (uniform distribution) equation (7) will tend to equation (5). This may be achieved by taking measurements at different mean wind speeds and combining the results.

By equating equations (5) and (7) it can be shown that the point of intersection occurs at $V_m = \bar{V}$. This obviously will not necessarily be true for any non-linear relationship beween the power output and the wind speed.

The question at this point can be raised as to the effect of averaging the data for short periods of time (increasing the correlation) on the power output vs. wind speed relationship. For a linear relationship it will not introduce any additional effects besides increasing the slope and reducing the scatter within each bin. For a non-linear relationship, the power output associated with a given bin will be biased upwards by an amount dependent on a) the averaging time, b) the turbulence intensity and c) the magnitude of the non-linear terms in the power output vs. wind speed relationship.

Assuming the velocity fluctuations are normally distributed during the period of averaging, then for a cubic relationship the power output will be factored by $(1+3a^2)$; where 'a' is the turbulence intensity. For averaging times of less than 10 seconds it is unlikely that this factor will be significant, however averaging will increase the correlation and hence the slope of the measured power output vs. wind speed characteristics.

POWER COEFFICIENT VS. TIP SPEED RATIO

A number of methods for determining the power coefficient vs. tip speed ratio characteristics have been suggested (Akins 4; Sharpe 7; Taylor et al. 8). The method suggested by Akins (1978) is a simple extension of the previously discussed method of 'bins'. It has been successfully applied (Worstell 9) to determine the power coefficient vs. tip speed ratio of a two-bladed Darrieus vertical axis wind turbine operating at a constant rotational speed. The advantages of constraining the turbine to operate at a constant rotational speed are two-fold. The first is that the shaft power need not be corrected to take into account angular acceleration and the second is that a range of tip speed ratios will automatically be scanned due to the natural fluctuations in the wind speed.

Another method of binning has been suggested by Taylor et al. (8). In their approach the performance parameters were binned according to the apparent instantaneous tip speed ratio. As will be shown, this method is open to large errors when the turbine is controlled accurately to operate at a constant tip speed ratio.

An alternative power measurement technique used by Sharpe (7), based on rotor acceleration, was not tried because it has been shown (10) that this method is very susceptible to large errors due to fluctuations in the wind velocity which are not correlated between anemometer and turbine.

METHOD OF BINS

a) Velocity

This method, as proposed by Akins (4), is only strictly applicable to turbines which are constrained to operate at a constant rotational speed. For variable speed, even assuming perfect correlation between the measured wind speed and that experienced by the turbine, it is apparent that a unique relationship between bin velocity and the tip speed ratio will not exist for those fluctuations to which the turbine can only partially respond.

A detailed discussion of the application of this method to turbines operating at a constant rotational speed will not be given here as it has previously been reported by several authors (3,4,8) and has been touched upon in the previous section. However, several comments pertaining to the interpretation of results derived by this method will be discussed.

For a given rotational speed a unique relationship between bin velocity and tip speed ratio will exist; the higher the bin velocity the lower the tip speed ratio. It is, therefore, apparent that the statistical analysis carried out in the previous section is also applicable in interpreting the performance characteristics. Thus incoherent velocity fluctuations between the anemometer and turbine locations will cause the power coefficient at low tip speed ratios to be underestimated and at high tip speed ratios to be overestimated. It is possible to reduce this effect by combining data sets from measurements made at different mean wind speeds; this approach has been successfully adopted by Sandia Laboratories.

b)   Tip Speed Ratio

The method developed by Taylor et al. (1981) relies on averaging all the instantaneous power coefficient values corresponding to the measured tip speed ratio lying within a specified bin.

The instantaneous power coefficient, C_{pi} and the tip speed ratio, λ_i are defined as

$$C_{pi} = \frac{P_{Ai}}{\frac{1}{2} \rho \pi R^2 V_i^3} \tag{8}$$

and

$$\lambda_i = \frac{\Omega_i R}{V_i} \tag{9}$$

where P_{Ai} is the instantaneous aerodynamic power, Ω_i is the instantaneous angular velocity and V_i the instantaneous wind velocity after being corrected for proximity effects, 'time of flight' and spatial averaging. The results of using this method of analysis on measurements made on a 6 metre diameter turbine are shown in Figure 4. Also shown for comparison is the theoretical prediction. The data is the same as that used in Figure 3 (3600 samples each averaged over 1 second). It is clear that the experimentally derived power coefficients do not agree with the theory. At tip speed ratios less than eleven it is underestimated and above eleven overestimated. This can be explained if a number of assumptions are made regarding the statistical nature of the instantaneously measured tip speed ratio. Firstly, it will be assumed that the actual tip speed ratio of the turbine, λ is normally distributed about $\bar{\lambda}$ (the mean tip speed ratio for a particular set of measurements) with a standard deviation of σ_z. Secondly, the measured tip speed ratio, λ_m (which will not in general be the same as λ due to the measured velocity being different from that experienced by turbine) is normally distributed about λ with a standard deviation, $\sigma_{M\lambda}$ given approximately by

$$\sigma_{M\lambda} = \frac{\bar{V}}{\bar{\lambda}} \sigma_{MV} \tag{10}$$

where σ_{mv} if given by equation (3).

Therefore the probability of the instantaneous power coefficient, C_{pi} at an actual tip speed ratio of λ being associated with a measured tip speed ratio of λ_m is

$$P_R(\lambda_m) = \int_0^\infty P_R(\lambda_m - \lambda) P_R(\lambda - \bar{\lambda}) \, d\lambda \qquad (11)$$

where $P_R(\lambda_m - \lambda)$ and $P_R(\lambda - \bar{\lambda})$ are Guassian probability distributions with means and standard deviations of λ, $\bar{\lambda}$ and $\sigma_{m\lambda}$, σ_λ respectively.

The instantaneous power coefficient, C_{pi} can be related to the actual power coefficient through the relationship

$$C_{pi}(\lambda_m) = C_p(\lambda)(\lambda_m/\lambda)^3 \qquad (12)$$

Thus the average power coefficient of a bin λ_m of infinitestimal width from equations (11) and (12) is

$$C_{PM}(\lambda_m) = \frac{\int_0^\infty C_p(\lambda)(\lambda_m/\lambda)^3 P_R(\lambda_m \lambda) P_R(\lambda - \bar{\lambda}) \, d\lambda}{P_R(\lambda_m)} \qquad (13)$$

This integral has been numerically evaluated for a range of $\sigma_{m\lambda}$ and σ_λ using the theoretically predicted $C_p(\lambda)$ characteristics of a six metre diameter turbine. The results are presented in Figure 5 as a fraction of the theoretical power coefficient. From this figure it can be seen that the method of bins, based on the tip speed ratio, will overestimate the power coefficient at high tip speed ratio and underestimate at low tip speed ratio, as experimentally observed. Increasing either the range over which the turbine operates or the correlation coefficient (reducing $\sigma_{m\lambda}$) reduces the error. This effect was observed experimentally by low pass filtering the data with a moving average prior to binning; (Figure 4). This had the effect of increasing the correlation between the measured and actual tip speed ratios.

The variance of the probability distribution ($\sigma_{m\lambda}^2 + \sigma_\lambda^2$) of the measured tip speed ratio can be determined from the number of samples within each bin. From two anemometers, in an equivalent spatial configuration to that of the measuring anemometer and the turbine, the standard deviation ($\sigma_{m\lambda}$) can be calculated from equation (10). Assuming that the theoretical performance characteristics given in Figure 4 (note the magnitude of the assumed performance characteristics are unimportant, only the shape is required). The corrected experimental performance characteristics are shown in Figure 6. Reasonable agreement is now seen to exist between the theoretical prediction and the experimental data.

Figure 7 shows the effect of combining several data sets (i.e. increasing (σ_λ^2), each for a difference range of tip speed ratios, for a 6 metre diameter turbine. It can be observed that close agreement exists between the predicted and experimental data. The discrepancies may be accounted for by inaccuracies in the reproduction of the blade aerofoil which would undoubtedly lead to the profile drag being underestimated. From a numerical analysis it was found that the discrepancy in the maximum power coefficient was consistent with the manufacturing tolerance of the blades.

CONCLUSIONS

It has been shown that the major source of error associated with field test data arise from the lack of correlation between the wind velocity measured by an anemometer and that experienced by the turbine. A number of methods for minimising this error have been suggested and are used to determine the performance of a horizontal-axis wind turbine.

The various statistical techniques for determining the performance from field test data have been reviewed and extended to include variable speed operation. When the results are interpreted correctly they are found to be in close agreement with the theoretically predicted performance.

Care, however, must be taken when interpreting the results at the limits of the range of measurements, because as shown they are open to large systematic errors. The statistical techniques used in this paper are of a general nature and could easily be extended to cover the measurement and interpretation of other parameters besides power.

ACKNOWLEDGEMENTS

It is my pleasure to thank M. Ryle, D. M. A. Wilson and P. F. Scott for their contributions to this work.

REFERENCES

1. Counihan, J., 1975. 'Adiabatic atmospheric boundary layers : A review and analysis of data from the period 1880-1972', Atmos. Environ., Vol. 9, pp. 871-905.

2. Anderson, M.B., 1981. 'An experimental and theoretical study of horizontal-axis wind turbines', Ph.D. Thesis, University of Cambridge.

3. Bossanyi, E.A., 1981. 'Wind turbines in a turbulent wind : Energy output and the frequency of shut-downs', Wind Engineering, Vol. 5, No. 1, pp.12-28

4. Akins, R.E., 1978. 'Performance evaluation of wind energy conversion systems using the method of bins', Sandia Labs., Rep. SAND 77-1375.

5. Akins, R.E., 1979. 'Wind characteristics for field testing of wind energy conversion systems', Sandia Labs. Rep. SAND 78-1563.

6. Lundsager, P., Fransen, S. and Christensen, C.J., 1980. 'Analysis of data from the Gedser wind turbine 1977-1979', Riso National Labs., Rep. RISO-172242.

7. Sharpe, D.J., 1977. 'A theoretical and experimental study of Darrieus vertical axis wind turbine', Kingston Polytechnic Research Report.

8. Taylor, G.J., Turner, J.D. and Fordham, E.J., 1981. 'Preliminary performance measurements on the 5-metre horizontal axis wind turbine at University College, Swansea', IEA, Wakes and Clusters Technical Meeting, CEGB.

9. Worstell, M.H., 1979. 'Aerodynamic performance of the 17-metre Darrieus wind trubine', Sandia Labs., Rep. SAND 78-1737.

10. Musgrove, P.J., and Mays, I.D., 1978. 'Development of the variable geometry vertical axis wind turbine'. Proc. of the Second International Symposium on Wind Energy Systems, Amsterdam.

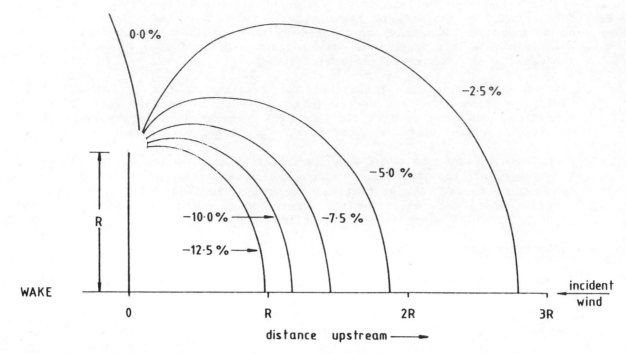

Figure 1   Regions of possible reduction in the incident wind velocity due to
semi-infinite number of expanding coaxial ring vortices representing
a wind turbine with a thrust coefficient of 0.85 operating at tip
speed ratio of ten.   Only half the rotor is shown.

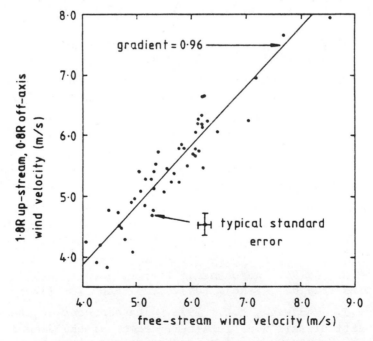

Figure 2   Plot of sixty second averages of the wind velocity measured at a
location 1.8R (R=radius) up-stream and 0.8R off-axis against the
free stream velocity of a six metre diameter turbine.

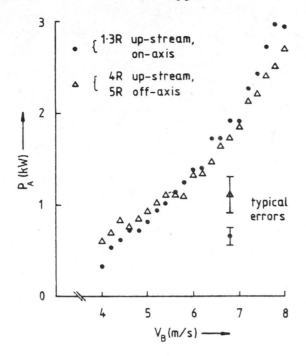

Figure 3   Aerodynamic power output for a six metre diameter turbine as a
function of the binned wind velocity.

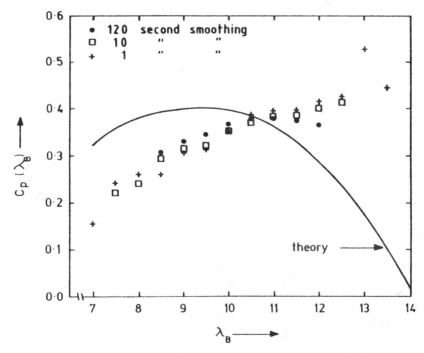

Figure 4   Comparison of theory (solid line) with experimental points (symbols)
derived by the method of 'bins' based on tip speed ratio for a six
metre diameter turbine.   The effect of smoothing the data prior to
binning is also shown.

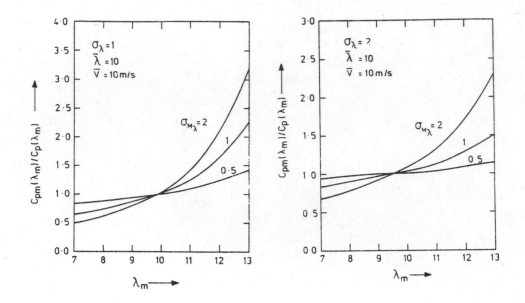

Figure 5   The effect of correlation between the measured wind speed and that experienced by a turbine on the measured performance characteristics.

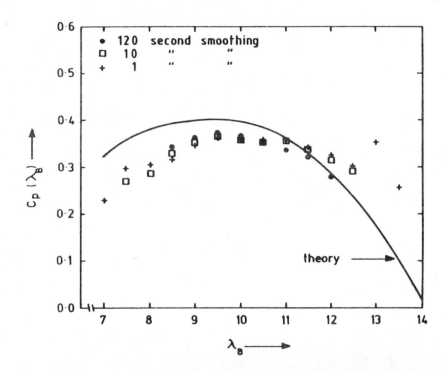

Figure 6   Comparison of theory (solid line) with corrected experimental points (symbols) derived by the method of 'bins' based on tip speed ratio for a six metre diameter turbine.   The effect of smoothing the data prior to binning is also shown.

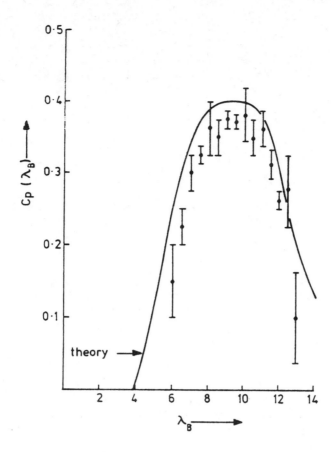

Figure 7   Comparison of theory (solid line) with experimental points (symbols) derived by the method of 'bins' based on tip speed ratio for a six metre diameter turbine.

CONSTRUCTION, COMMISSIONING AND OPERATION
OF THE 300 kW WIND TURBINE AT CARMARTHEN BAY

T.Young and D. McLeish
James Howden and Company Limited, Glasgow

D. Rees
Central Electricity Generating Board, Swansea

Abstract

Large scale wind power in the U.K. is gradually moving towards reality. A few medium sized wind turbines are now in operation, or in process of construction. This points the way to clusters of similar machines and even larger machines.

One of the most significant events was the announcement by the C.E.G.B. in 1980 of its intention to pursue a policy leading to the commercial use of wind power. The major practical outcome of this so far has been the building of the 300 kW wind turbine at Carmarthen Bay in South Wales. This is being used to gain experience in all aspects of wind turbine potential.

Introduction

The C.E.G.B. have publicly declared a three pronged investigation into the economical use of wind power for the production of electricity. The first stage was the production of a specification for the supply of a proven machine. This was the subject of competitive tender and was won by James Howden & Co. in conjuction with W.T.G. Energy Systems.

The MP5-200 is a three bladed, upwind machine with a synchronous generator. It is constructed with a lattice type tower. The rated output is 200 kW at 27 m.p.h. with 300 kW being obtained at 33 m.p.h. Hub height is 24.4 m. The rotor diameter is 24.4 m with a synchronous speed of 30 r.p.m. A gearbox is used to increase the speed to 1000 r.p.m. at the generator shaft. The connection to the C.E.G.B. grid is shown in Fig. 1.

An aerogenerator is by definition a generation station and as such must comply, when operated by the C.E.G.B., with the C.E.G.B. Safety Rules. The rules state:-

3.1 (III) in the case of low and medium voltate apparatus, the primary means of achieving safety is by isolation from the system.

(IV) when work is to be carried out on mechanical plant, the primary means of achieving safety is by isolation from the system followed by draining, venting and purging as appropriate.

In order to comply with these Safety Rules the aerogenerator was fitted with a locking pin on the low speed shaft and on the yaw system. All circuit breakers and isolators can be locked in the 'off' position and removable links were installed in the supply circuit to the hydraulic pump motor. This ensured that the system fully complied with the Safety Rules and the Permit to Work scheme.

Planning

The construction of the aerogenerator required very careful planning in terms of both materials and manpower.

It was necessary to ensure that the appropriate personnel were available on site when required. Also, the supply of materials to the site had to be co-ordinated with progress so that items of equipment were available when needed and expensive items, like cranes, fully utilised. The contract was awarded in August 1981 and work proceeded closely to programme; foundation work took place in June 1982, the machine was erected during July and August and officially inaugurated in November 1982.

The aerogenerator has a ladder which gives access to the nacelle. This ladder was to be supplied with a railock device which would allow the individual to climb the ladder freely but should he slip or fall it would immediately lock thus preventing injury. After consultation with local maintenance staff a hooped ladder was specified. Padlockable hatches were installed at the bottom of the ladder and at the entrance of the nacelle.

As a maintenace aid a hoist was installed on one leg of the tower. The hoist was also to be used to inspect the blades during the scheduled maintenance periods. It operated by pulling itself up a steel rope and was guided on rails. The Health and Safety Executive classified the hoist as an elevator because of the rails. This in turn would have required interlocking doors at the top and bottom and the whole rail assembly to be totally enclosed. This would have greatly affected the aerodynamic flow around the tower and for this reason the hoist is no longer installed on the aerogenerator.

Construction

The aerogenerator was constructed on the site of an old PFA lagoon. Although the load bearing capability of the PFA is satisfactory, the stiffness is comparatively low and thus a raft type foundation was required to give the correct dynamic characteristics of the complete system. The foundations were constructed with reinforced concrete which was poured in a continous operation. The holding down bolts for the tower were installed during this period.

The tower had previously been assembled at the manufacturers works and to ensure a trouble free erection on site each part of the tower had been marked for identification. A relatively small crane (20 tonne) was used during the erection of the tower.

The bedplate was shipped to site and the drive train components assembled on it at site. The drive train shafts were lined up twice, namely, on the ground and after the rotor was installed. This was found to be necessary due to deflections of the support structure due to the rotor weight. The machinery bedplate is attached to the top of the tower though a slewing ring and an intermediate piece called the tower cap. This sub-assembly, with the geared yaw motors and yaw brakes attached, was built up on the ground. Before being craned to the top of the tower, the hydraulic system and operation of the yaw assembly was tested with the bedplate sitting on railway sleepers. After the bedplate assembly was attached to the top of the tower, the sheet steel nacelle cover was fitted. The machine cabin control pannel (MCCP) was given a functional test while the bedplate was still on the ground. The MCCP is normally used during maintenance periods.

The three blades were shipped in a cradle to prevent damage. The blades were installed into the hub individually and supported by a tripod at the tip end. When the three blades had been installed in the hub the top of the hub was lowered to clamp the blade spars and the hub lightly torqued up. The next stage was to install the tips into each of the blades. A temporary hydraulic supply was then used to check the operation of the tips and finally to bring them into line with the main blades. At this point the pitch angle of each complete blade was set by the use of a jig applied to the end of the tip sections. When this has been completed the hub was torqued to its final value.

The complete rotor had to be lifted and fitted to the low speed shaft. The method devised was to use a large (80 tonne) crane to support two of the blade spars near their root ends, while the smaller crane took the weight of the third spar. In this fashion, with both cranes lifting together, the rotor was raised well clear of the ground while still in a horizontal position. Gradually the large crane would take more of the total weight until the rotor was in a vertical position and the smaller crane was unhooked. The rotor was then lined up with the nacelle and the hub secured to the shaft.

The computer control panel and the generator switchgear panel were installed in the Control Room at the base of the tower. It was originally intended that the switchgear panel would be energised immediately to provide site service however it was decided to use a diesel generator and commission the switchgear at the same time as the computer panel.

Commissioning

The process of commissioning proceeded during the entire site programme and included static and funtional checks. The mechanical checks included operational checks on the yaw system, brakes and tips. The general commissioning philosophy was to test the operation of each individual item, then check the operation of the various sub-assemblies, and finally to check the complete system. In this sequential manner the aerogenerator was put through an exhaustive commissioning schedule.

The generator switchgear panel and computer control panel were subjected to the normal C.E.G.B. commissioning procedure which included injection testing of the relays, transducer calibration and pre-synchronising checks.

During the latter stages of commissioning the aerogenerator was run up to just below synchronous speed on several occasions to check the software routines. Indeed the first and subsequent occasion on which the generator was synchronised to the grid was done manually. Each stage of the operating procedure was undertaken first manually then subsequently on automatic operation.

The last stage to give the aerogenerator a 30 hour run in the automatic mode. During this period it was allowed to operate in an automatic 'unattended' mode although personel were present during this period.

Having passed this test the aerogenerator was now ready for full operation.

Operation

The machine was first synchronised with the U.K. grid on Monday 1st November, 1982. This was accomplished with manual control of the rotor speed via the tip flaps and observation of the built in synchroscope. This point signalled the begining of the first 100 hours of on-line operation, before the first scheduled stop for maintenance purposes. The first 30 hours were used for on-line acceptance tests. It was found that in certain low wind conditions the machine would repeatedly start and stop. This was exacerbated by torsional-oscillation-induced power swings, which could shut the turbine down due to a reverse power trip. To reduce this effect, a small hysteresis was built in by increasing the low wind speed start up to 11 m.p.h., while leaving the low wind speed cut-out at 9 m.p.h. Similarly the high wind speed limits are 35 m.p.h. at cut-out and 30 m.p.h. at cut-in.

Fig. 2 shows the anticipated power curve for the machine and some data recorded during the first few months. The data is compiled by using the "method of bins". All the hourly readings of instantaneous wind speed and power are grouped into bands of wind speed. The average wind speeds and corresponding average power from each band are plotted on Fig. 2. Some of the scatter is accounted for by the power swings, and the fact that in some bands not enough reading were avilable to give a good average.

In normal operation the machine is controlled by the supervisory program in the microprocessor. The wind conditions are sensed by a combined speed and direction indicator mounted at the rear of the nacelle. If the mean wind speed is between 11 m.p.h. and 30 m.p.h. for 5 minutes the rotor is yawed into the wind, the shaft disc brakes are released and the tips are brought to the run position. Rotor acceleration is controlled by the computer via the tips at about 0.5 r.p.m. per second up to 28 r.p.m. Beyond this, the speed is controlled prior to synchronization. At 30 r.p.m. the contactor is closed and the turbine is on-line. When conditions dictate the computer initiates an automatic shut-down by gradually reducing the power under tip control, below 20 kW the contactor is opened and the disc brakes are applied to stop and park the rotor.

A summary of the operation of the machine during the first 100 hours in shown in Fig. 3. It must be borne in mind that all the operational experience has been gained during this period, which is too short to form a true picture of the machine's capabilities. In addition, the turbine has been operated on a "9 to 5" basis so overall performance has been limited.

No information has as yet been processed from the data logging system; all data has come from the much more limited printouts of the microprocessor teletype, and the various meters on the control panel.

Shortly after re-starting from the 100 hour maintenance period, the rotating hydraulic coupling which feeds oil through the low speed gearbox shaft to the tip actuator seized and unscrewed itself from the shaft. This brought about an immediate failsafe emergency shutdown and repairs to this area have been recently completed.

Conclusion

The building of the Carmarthen Bay wind turbine has been extremely useful in providing valuable experience for those involved. Up to now the machine has not been in operation for long enough to be able to form an accurate impression of it's long term capabilities. However, it is confidently expected that this machine will provide a great amount of important data for many years to come.

Fig. 1 Grid Connection Diagram

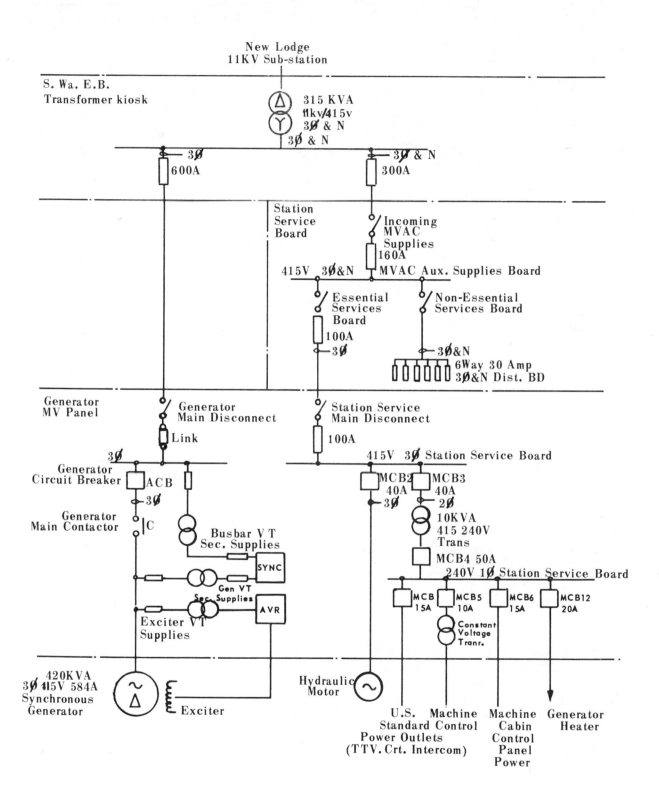

CARMARTHEN BAY WIND GENERATOR
OPERATIONAL DIAGRAM

Fig. 2 Comparison between expected and recorded power output

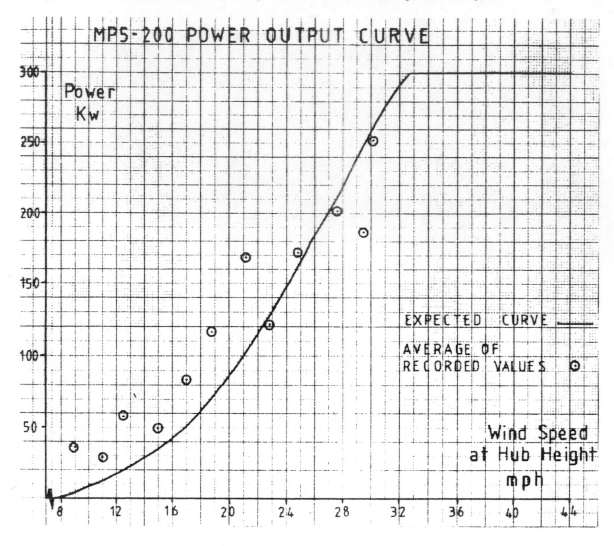

Fig. 3 Summary of operation from 1.11.82 to 16.1.83

ITEM		RESULT
Total hours run on on-line	hr	96
Total energy generated	kwhr	11920
Station essential services	kwhr	1390
Net energy generated	kwhr	10530
Mean power	kw	110

DISTURBANCES IN ELECTRICITY SUPPLY NETWORKS CAUSED BY A WIND ENERGY SYSTEM EQUIPPED WITH A STATIC CONVERTER

C.J. Looijesteijn
Netherlands Energy Research Foundation ECN,
P.O. Box 1, 1755 ZG   PETTEN, The Netherlands

Abstract

Within the frame work of the Netherlands Wind Energy Research Programm an experimental 25 m diameter wind turbine has been built at the ECN institute located on the north-west coast of The Netherlands. The turbine is not intended as a prototype, but as an instrument for data collection and measurements. Special attention has been given to the electrical power conversion system. The wind turbine is equipped with a DC-current machine and a static DC to AC converter. This is not a commonly used system for wind energy conversion. The main reasons for choosing this type of system are the following:
- a variable rotation speed of the rotor shaft can be obtained
- the system can be operated with a constant and high power factor for varying wind speeds
- torque fluctuations caused by wind fluctuations can be reduced.

A disadvantage of this system however are the costs, and the harmonic distortion of the supply voltage of the elctricity grid.
This paper describes the method and results of measurements made on the 400 kW converter, and includes the evaluation of the expected disturbances of the supply voltage.

Introduction

Since the introduction of alternating current in electricity supply networks, the utilities have to deliver a sinusoidal voltage waveform, which has to be as pure as possible. Distortion of the sinusoidal waveform is known as "grid pollution".
Grid pollution introduces several problems such as:
- increasing losses in transformers, induction and synchronous motors etc.
- failure of electric and electronic devices
- overload of capacitors
- decrease of the accuracy of kWh-registration instruments.

The limitation of disturbances in electricity supply networks is established in the GENELEC-Standard (1). This European Standard was accepted by the CENELEC-members in 1975. The CENELEC-standard however is not applicable to harmonic disturbances. For that reason the Netherlands National Committee of IEC and CENELEC has made an additional recommendation (2). This recommendation is very important for power conversion systems of wind turbines.
In the application of large scale wind energy systems, it is usual to couple the turbines to the electricity grid.
Some methods of grid coupling are the following (3):
- using an induction generator directly, coupled to the public grid
- coupling of an DC-generator to a DC- to AC-converter; the converter being coupled to the grid
- by applying a synchronous generator connected to a rectifier, while the rectifier is connected to a DC- to AC-converter; the converter being connected to the grid
- the application of a capacitor-excited induction generator connected to a rectifier and a DC- to AC-converter.

All converters mentioned above are static converters.
The most important components of those converters are semi conductors (thyristors) which are installed in the power lines of the converter. Such applications however, cause disturbances which can exceed the permitted levels which are established in the earlier mentioned recommendations.

In this paper the results of measurements made on the 400 kW converter of the 25 m HAT will be discussed. Therefore a short description of the applied converter will be given.

Description of the installation

General description of the turbine

At the location of the Netherlands Energy Research Foundation (ECN), Petten, The Netherlands, a 25 m horizontal axis wind turbine (HAT) has been built and was completed in July, 1981 (see figure 1). The turbine is located in the dunes (10 to 18 m high) at a distance of about 500 m from the sea shore.
The main characteristics of the turbine are summarized in table 1 and de-scribed earlier (4), (5). A special feature of this wind turbine is its electrical system (DC-generator followed by DC-AC converter), which allows a wide interval of rotor shaft speeds. For this reason we will describe the electrical system in more detail.
The turbine has not been built as a commercial portotype, but as a measuring device in order to obtain more knowledge about the behaviour of large WECS.

Electric power conversion system

The wind turbine is equipped with a DC-machine and a static DC to AC conver-sion system. This system consists of a line commutated converter coupled to the 10 kV public grid via a transformer.
The main reasons for choosing this type of electrical system were:
a. The rotor speed is allowed to vary over a wide interval, so that a con-stant tip-speed ratio control strategy can be realised and can be com-pared with a constant rpm control strategy.
b. Wind fluctuations will cause smaller torque fluctuations, using a control strategy with a variable rotor speed, so that the dimensions of the gear box can be chosen smaller.
c. The system can be operated with a constant and high value of the power factor (cos Φ) over a wide interval of the load, so that less reactive currents will be introduced in the public grid, compared with wind tur-bines equipped with induction generators.
d. In the future, a modified version of this electrical system, consisting of a synchronous AC-generator, combined with a rectifier could be used in wind turbine clusters. A fairly large number of wind turbines can then be coupled to one single DC-AC-converter. A turbine equipped with such a type of power conversion system has been built and will be put into ope-ration in 1983 (7).
e. The DC-machine can be operated as a generator as well as a motor, so that the turbine behaviour can also be studied without wind in a wide interval of rotor speeds. Moreover, the turbine can be put to work easily at a lower wind speed by making a motor start.

The generator and the control circuits

The generator is a 4-pole DC shunt machine with an independent field. The independent field-current supply allows a constant voltage control. The principal variable of the control circuit (see fig. 2) is the rotor speed, which is measured by a small tacho generator. In the range 0 to 900 rpm (0-45 rpm of the rotor shaft) the output voltage of the generator is proportional to the rotor speed. From 900 rpm up to 1600 rpm (45-80 rpm on the rotor shaft) the field controller stabilizes the output voltage at 720 V, a voltage level at which the DC-AC converter reaches a rather good efficiency and power factor.

The DC- to AC-converter

Fig. 3 shows the principle scheme of a 3-phase fully controlled current bridge. The power section consists of an inverter bridge with six thyristors (T1 to T6) supplying a DC-voltage (E) to the 3-phase current line. The numbers 1 to 6 show the sequence by which the thyristors are fired or triggered. Each phase (R, S, T) has its thyristor control set which generates line synchronous control pulses in a periodical sequence. These control pulses trigger the thyristor at a moment determined by the line voltage period (line commutation). The converter of the 25 m HAT consists of two, 3-phase, fully controlled line commutated current bridges, which are connected in series at the DC side (see fig. 4). Each current bridge consists of an assembly of thyristors. At the AC-side the current bridges are connected to a 6-phase current line, created by a high voltage transformer. At the common primary side three coils are connected in star. The secondary side of the transformer consists of two separated parts, of which the coils are connected in star and delta, respectively. The choice of a 6-phase current bridge has been based on the demand to reduce the harmonics at the AC-side (public grid). Fig. 5 and 6 show the sequence of the conductance of each thyristor as a function of time. The thyristors are ciclically connected to the 6-phase of the AC-grid, allowing current and thus energy to flow into the grid. The figures show that once the circuit is in continuous operation no special requirements are needed for the commutation. At a certain value of the line voltage the conductive state of the thyristor ends.

Harmonic Analyses

From the Fourier analysis it is a well known fact that every periodic phenomenon can be reduced to a sinusoidal fundamental wave and sinusoidal quantities of which the frequencies are an integer of the fundamental frequency. By using the Fourier analysis it can be predicted which harmonic components of the current will be introduced in the output lines of the converter. This proces applied on the block wave form of the current will lead to the result that only the odd harmonics and the sines terms of the Fourier series will be left. Which higher harmonic order components will be left depends on the configuration of the converter.
For example: The first higher harmonics of the current introduced by using a 3-phase fully controlled bridge are the 5th and the 7th. The first higher harmonic currents injected into the output lines of a 6-phase fully controlled bridge (thus 12 thyristors) are the 11th and 13th, which have a lower energy content as the 5th and the 7th.

Table 2 gives an overview of the maximum of each harmonic current which will be introduced by using several types of line commutated converters. The calculations made for this table are based on an ideal DC-current at the DC-side of the converter and an infinitely short communication time. In this paper this percentage is called: "The theoretical maximum".
Note: If the impuls rate (numbers of thyristors) of the converter increases, the first higher harmonic component which occurs has a higher number, and a lower percentage of the fundamental frequency.

Measurement Procedure

General

In this paper the point of common coupling (P.C.C.) with other consumers, is defined as the point in the supply network, that is electrically nearest to the consumer for whom the converter is proposed, at which other consumer loads are or may be connected.
The following measurement procedure has been used to determine the harmonic contents of voltage and current wave forms produced by the converter.

Converter connected to the grid, turbine not in operation

The harmonic contents of the voltage waveform have been measured at the P.C.C. The harmonic contents of the current wave form, which are equal to those of the no-load current of the transformer, has been measured at the same point as above. This measurement gives an estimation of the harmonic background of voltage distortion which already exists in the public grid.

Converter connected to the grid, turbine in operation

The harmonic contents of the voltage and the current waveform have been measured at the same point of common coupling as mentioned above. Amplitude and relative phase angle of each harmonic component has been recorded together with the measured power output of the converter.

At the primary side of the high voltage transformer, current and voltage are measured with measuring transformers. The output terminals of these measuring devices are connected to the spectrum analyzer and a transient recorder. The current measurement is carried out at shunts which have been placed in series with each line of the measuring transformer.
The output-terminals of the earlier mentioned measuring transformers are also connected to 4 transducers: Voltage, current, power and phase angle. Each of these transducers consists of a converter with a current output (4-20 mA). Their current output is led to the computer interface. At the moment that the spectrum analyzer starts to record the amplitude and relative phase angle of current and voltage, a data-collection programme which runs on the computer, starts to collect the most important parameters defining the state in which the measurement was realized.
This specially developed data-collection programme has a minimum sample frequency of 2.5 Hz and a measuring time which is equal to the measuring period of the spectrum analyzer.
When the measurements are completed, the obtained data are processed by an evaluation programme which combines the data of the spectrum analyzer with the data of the data collection programme calculates the results and stores them in a file. An example of such a file is given in table 3.
The evaluation programme includes an option to make a subtraction of each harmonic recorded above from that recorded earlier (background). This subtraction will produce the contribution of the converter to the harmonic voltage distorsion at the P.C.C.
From a fundamental point of view however, this method of subtraction is not fully correct, because the substraction is not a vectorial subtraction. For that reason the indications "gross contribution" and "net contribution" are introduced. The contribution are the measured harmonic components of the current and voltage wave form without the subtraction of the earlier measured background. The net contribution is the measured harmonic contents with a subtraction of the background for each harmonic.
The remaining measurements are carried out in accordance with the recommended practices for wind turbine testing as described in (6).

Results

General remarks

A total of 22 measurements were made at several percentages of full load. The minimum average power during the measurement was 32 kW (10%) and the maximum average power was 267 kW (90%). During the measurements the power fluctuations had an average standard deviation of approximately 30%. This means that a wide range of the opertional power has been covered by the measurements.

As a result of errors in the synchronisation of the ignition of the thyristors, it is possible to introduce a certain a-symmetry in the line current of the transformer. In this paper the possible causes of such an error will not be discussed.

An a-symmetric ignition will also lead to the introduction of even harmonics at the AC-side of the converter.

In a stationary situation the symmetry of the thyristor ignition is measured at a fire angle of 150° el. In that situation the errors in the symmetric ignition were erroneous.

An evaluation of all the measurements which have been carried out, shows that even harmonics of the current are less than 1% of the fundamental frequency.

Harmonic currents at the AC-side of the converter

An evaluation of all the measurements which have been made on the harmonic currents shows that they do not have a higher net percentage of the fundamental frequency in relation to the calculated fraction given in table 2. The calculated average percentage of the harmonic currents shows a certain a-symmetry in the measured percentages compared to the theoretical maximum percentages (given in table 2). The percentages of the 11th and the 23th are relatively higher than the 13th and the 25th in relation to the theoretical maximum percentages. A comparison of these results is given in table 4.

Some possible causes for this marginal a-symmetric behaviour are:
- a small deviation in the windings of the coils of the high voltage transformer;
- a certain reduction of the 13th and the 25th harmonic as a consequence of the non ideal DC-voltage at the DC-side of the converter, on which the calculations of the maximum percentages are based.

Also a comparison of the influence of the control strategy at which the turbine was operating during the measurements has been carried out.

The results of this comparison show that the control strategy has no influence on the relative contribution of the harmonic currents.

Harmonic voltages at the AC-side of the converter

The injection of the harmonic currents as described earlier will lead to a distortion of the basically pure sinusoidal voltage wave form which is supplied by the grid. The extent in which the distortion takes place is dependent on the grid impedance and the amount of the harmonic currents.

Fig. 10 shows the results of measurements of voltage distortion The figure shows the voltage distortion as a function of the power. The percentage of each harmonic voltage component is given as a fraction of the fundamental frequency. From fig. 10 it can be seen that the extent in which the distortion of the voltage takes place is approximately proportional to the power. The permitted levels of the maximum voltage distortion of the 11th and the 13th harmonic of the voltage are respectively 0,4% and 0,3% of the fundamental frequency (see table 5). The voltage distortion exceeds the permitted levels of the earlier mentioned recommendations at an average power of about 200 kW.

Remarks

The above mentioned conclusion however, requires some additional comments:
As a result of the over dimensioning of the high voltage transformer
(500 kVA instead of the necessary 300 kVA) the harmonic voltages will give a
higher distortion at the primary side of the transformer (public grid). In
the case of a well dimensioned transformer of 300 kVA the harmonic voltage
will be reduced and be kept more at the secondary side of the transformer.
As the harmonic voltage distortion is approximately proportional to the
power, it can be limited to an acceptable level. Therefore it is necessary
to retain a stable power output. From this point of view a control strategy
which converts power fluctuations into fluctuations of the rotor speed is
very important.

Conclusions

- Calculations based on an ideal DC-current of the generator at the DC-side
  of the converter and an infinitely short commutation time, predict a so
  called "theoretical maximum percentage" of each harmonic current which
  results from the application of a certain type of converter. Measurements
  carried out on the converter of the 25 m HAT at Petten show that the measured harmonic currents do not contain a higher percentage as calculated.
- There is certain a-symmetry in the measured percentages of the harmonic
  currents compared to the theoretical maximum percentages of these harmonic
  currents.
  The percentages of 11th and the 23th are relatively higher than the 13th
  and the 25th compared to the theoretical maximum percentages.
- Some possible causes for this slight a-symmetric behaviour are:
  - a small deviation in the windings of the coils of the high voltage
    transformer;
  - a certain reduction of the 13th and the 25th harmonic as a consequence
    of the non ideal DC-voltage at the DC-side of the converter, on which
    the calculations of the maximum percentages are based.

- From the point of view of the disturbances introduced in the electricity
  grid the harmonic currents are not as important as the harmonic distortion
  of the voltage.
- As a result of a over-dimensioning of the high voltage transformer
  (500 kVA instead of the necessary 300 kVA) the harmonic voltages exceed
  the permitted levels according to the recommendation of the Netherlands
  National Committee of IEC and CENELEC at an average power output of
  200 kW.
- The harmonic voltages are approximately proportional with the power.
  Temporary overshut of the rated power will easily lead to an exceeding of
  the permitted levels of harmonic voltages according to the recommendation.
  From this point of view a control strategy which converts power fluctuations into fluctuations of the rotor speed is recommended.

References

(1) EN 50.006 The limitation of disturbances in electricity supply networks caused by domestic and similar appliances equipped with electronic devices.

(2) CHC 75/10 Aanbevelingen voor toelaatbare harmonische stromen bij niet huishoudelijke toestellen. Vereniging van Directeuren van Electriciteitsbedrijven in Nederland.

(3) De Zeeuw, W.J., General overview of various systems. Proceedings First International Workshop on Power conversion systems, Petten, The Netherlands, January 1983, will be published).

(4) Schellens, F.J.G., The 25 m experimental horizontal axis wind turbine. Proceedings Third International Symposium on Wind Energy Systems, Copenhagen, Denmark, August 1980.

(5) Dekker, J.W.M., Lekkerkerk, F., Looijesteijn, C.J., Valter, G.P. Operating Experience Control and Measurements made on a 25 mtr Horizontal Axis Windturbine.
Proceedings Fourth International Symposium on Wind Energy Systems, Sweden, September 1982.

(6) Recommended Practices for Wind Turbine Testing.
Volume 7, Quality of Power Edition 1982, edited by Lionel J. Balland and Roy H. Swansborough (will be published).

(7) Technical specification of the NEWECS 25.
FDO(B)-82-181, published by FDO, P.O. Box 379, Amsterdam.

Table 1. Main characteristics of the 25 m horizontal axis wind turbine at
         ECN, Petten, The Netherlands.

Geometric data:

Rotor

Type	2 bladed, upwind, with pitch angle regulation over 90° interval
Diameter	25 m
Hub height	22 m
Rotor speed	from 40 to 80 rpm
Blade weight	total weight 567 kg, epoxy plastic part: 281 kg
Cone angle	5°
Tilt angle	5°
Tip speed ratio	optimal value, $\lambda = 8$
Tower	14 meter
Base	8 meter

Materials

Blade	glass fibre and carbon fibre epoxy plastic, with a steel root disk
Tower	steel
Base	concrete

Gearbox

two stage gearbox

Ratio	1 : 20
	first stage planetary ratio 1 : 5.6
	second conventional ratio 1 : 3.571

Generator

DC-shunt with independent excitation

Rated power	400 kW
Number of poles	4
Voltage	720 Volt

DC-AC converter

600 kW, 6 phase thyristor mutator

connection to the grid   500 kW tranformer, 330 to 10.000 V

Yawing system

Electric, yawing velocity $0.2° \text{ s}^{-1}$ over an interval of 720°.

Control method

Manual, semi-automatic, or automatic, via a PDP 11/34 process computer all three methods under supervision of a programmable logic microcomputer (PLC).

Wind speed regime

$V_{in} = 6 \text{ ms}^{-1}$, $V_{rated} = 13 \text{ ms}^{-1}$, $V_{out} = 17 \text{ ms}^{-1}$.
Wind speed measured at three different heights at 4 masts around the turbine.

Table 3

MEASUREMENT NUMBER;	33
IDENTIFICATION NUMBER;	30606001
DATE;	10-AUG-82
STARTTIME;	09:12:34
STOPTIME;	09:13:46
AVERAGE WINDSPEED AT START OF MEASUREMENT;	10.5
AVERAGE WINDSPEED AT END OF MEASUREMENT;	10.6
STANDARD DEVIATION;	0.33
AVERAGE VOLTAGE;	10296.
STANDARD DEVIATION OF VOLTAGE;	22.7
AVERAGE CURRENT;	9.7
AVERAGE POWERFACTOR;	-.7417
AVERAGE POWER;	128.
STANDARD DEVIATION OF POWER;	35.5
AVERAGE ROTATION SPEED;	1179.

NUMBER OF HARMONIC COMPONENT	PHASE ANGLE	PERCENTAGE OF FIRST HARMONIC COMPONENT OF:		
		-CURRENT	-VOLTAGE	-POWER
1	45	100.00	100.00	100.00
3	11	0.91	0.32	0.00
5	15	3.49	0.34	0.02
7	-115	1.24	0.39	0.00
9	-69	1.27	0.04	0.00
11	1	9.41	0.55	0.07
13	11	6.61	0.16	0.01
15	0	0.00	0.00	0.00
17	0	0.00	0.00	0.00
19	0	0.00	0.00	0.00
21	0	0.00	0.00	0.00
23	20	4.20	0.19	0.01
25	31	3.09	0.13	0.00
27	0	0.00	0.00	0.00
29	0	0.00	0.00	0.00
31	0	0.00	0.00	0.00
33	0	0.00	0.00	0.00
35	72	2.67	0.05	0.00
37	65	1.92	0.03	0.00

Table 2.

Maximum percentage of the harmonic currents by line commutated converters.
The percentages are given as a part of the fundamental frequency.

Harmonic order	Frequency (Hz) (in case of 50 Hz grid)	Number of thyristors			
		3	6	12	24
1	50	100	100	100	100
2	100	50	-	-	-
3	150	-	-	-	-
4	200	25,0	-	-	-
5	250	20,0	20,0	-	-
6	300	-	-	-	-
7	350	14,3	14,3	-	-
8	400	12,5	-	-	-
9	450	-	-	-	-
10	500	10,0	-	-	-
11	550	9,09	9,09	9,09	-
12	600	-	-	-	-
13	650	7,69	7,69	7,69	-
14	700	7,14	-	-	-
15	750	-	-	-	-
16	800	6,24	-	-	-
17	850	5,83	5,83	-	-
18	900	-	-	-	-
19	950	5,26	5,26	-	-
20	1000	5,00	-	-	-
21	1050	-	-	-	-
22	1100	4,54	-	-	-
23	1150	4,35	4,35	4,35	4,35
24	1200	-	-	-	-
25	1250	4,00	4,00	4,00	4,00

Harmonic order : The ratio between the harmonic frequency and the fundamental frequency.

Frequency : The frequency of the harmonic in case of using a 50 Hz grid.

Number of thyristor: This is an indication of which type of converter is used.

Table 5.

Permitted level of voltage distortion.

As a basis for the recommendation of the permitted level of voltage distortion the dutch National Electrotechnical Committee, have used the CENELEC standard.

The following tabel gives the maximum acceptable levels of voltage distortion as a function of the harmonic number.

Harmonic number	Permitted voltage level in PPC in percentages of the fundamental frequency
2	0.2
3	0.85
4	0.2
5	0.65
6	0.2
7	0.6
8	0.2
9	0.4
10	0.2
11	0.4
12	0.2
13	0.3
14	0.2
15	0.25
16	0.2
17	0.25
18	0.2
etc.	etc.

Table 4.

Comparison of the measured harmonic currents and the calculated theoretical maximum percentages of the harmonic current.

The theoretical maximum percentage of the fundamental frequency is established at 100%.

In factor In/Int is the quotient of the measured percentage of the harmonic current and the theoretical maximum percentage.

Control strategy	Constant rotor speed (1)		Constant tip speed ratio (3)	
	In/Int % (gross)	In/Int % (net)	In/Int % (gross)	In/Int % (net)
Harmonic order number				
11	102	100	102	100
13	82	81	84	83
23	94	90	74	69
25	80	77	85	82
35	87	84	91	87
37	74	70	74	70

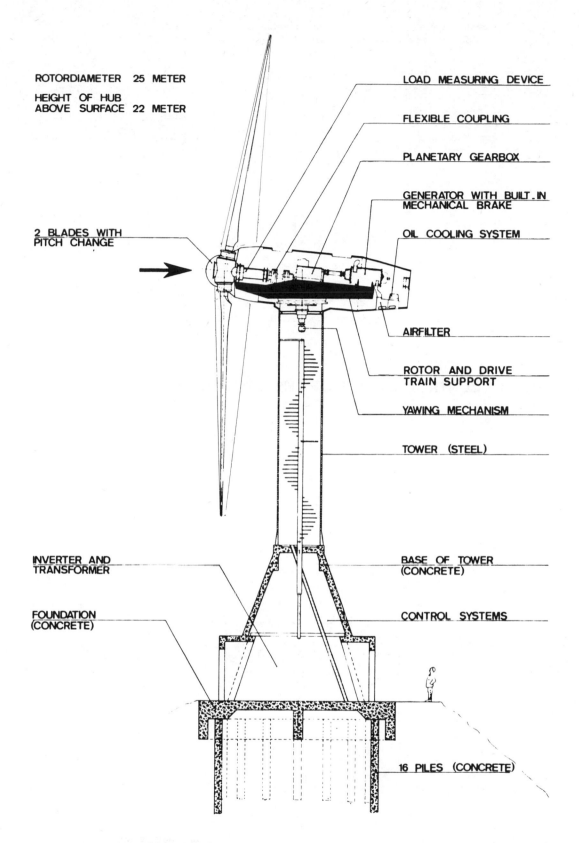

ROTORDIAMETER 25 METER

HEIGHT OF HUB
ABOVE SURFACE 22 METER

2 BLADES WITH
PITCH CHANGE

LOAD MEASURING DEVICE

FLEXIBLE COUPLING

PLANETARY GEARBOX

GENERATOR WITH BUILT.IN
MECHANICAL BRAKE

OIL COOLING SYSTEM

AIRFILTER

ROTOR AND DRIVE
TRAIN SUPPORT

YAWING MECHANISM

TOWER (STEEL)

INVERTER AND
TRANSFORMER

BASE OF TOWER
(CONCRETE)

FOUNDATION
(CONCRETE)

CONTROL SYSTEMS

16 PILES (CONCRETE)

FIG.1 EXPERIMENTAL WINDTURBINE WITH HORIZONTAL AXIS

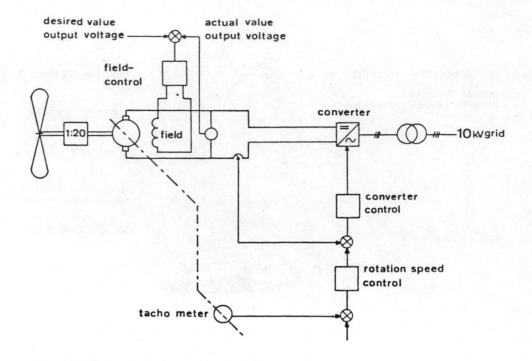

FIG.2   CONTROL CIRCUIT OF THE CONVERTER

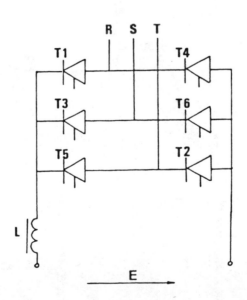

FIG.3   3-PHASE LINE COMMUTATED CONVERTER

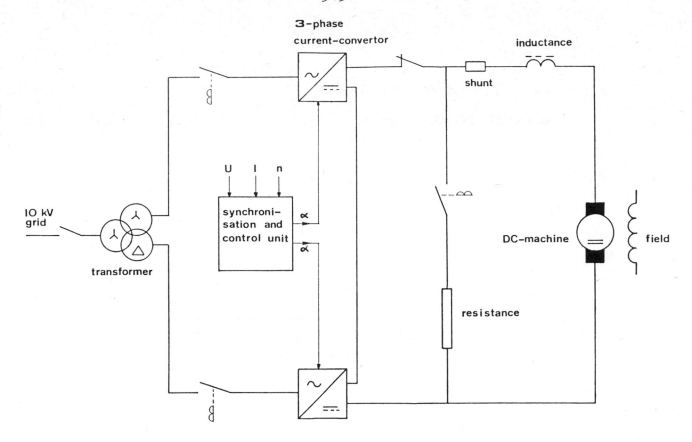

FIG.4 SCHEME OF ELECTRIC POWER CONVERSION SYSTEM

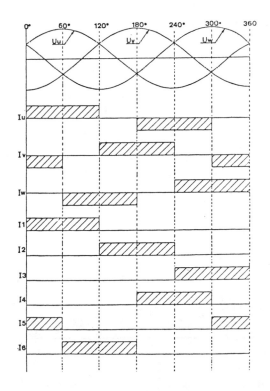

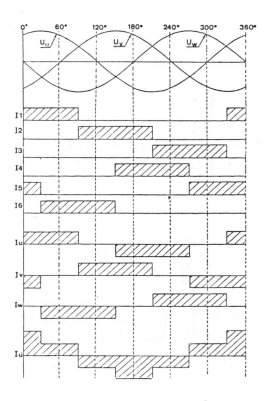

FIG.5 Response in time of the six thyristor-currents as well as the line currents of a 3-phase current bridge fed via a star-star connected transformer

FIG.6 Response in time of the six thyristor-currents as well as the line currents of a 3-phase current bridge fed via a star-delta connected transformer

FIG.7 SPECTRUM OF THE CURRENT

AVERAGE POWER; 128 kW

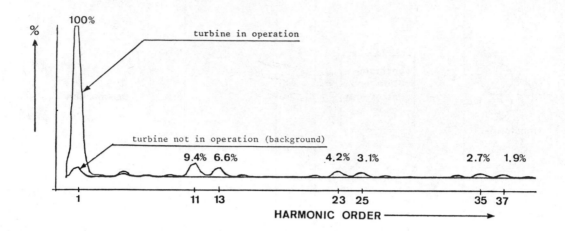

FIG.8 SPECTRUM OF THE VOLTAGE

AVERAGE POWER; 128 kW

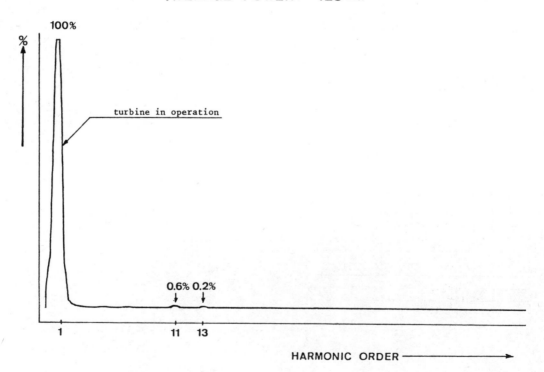

FIG.9 HARMONIC CURRENTS AS A FUNCTION OF POWER

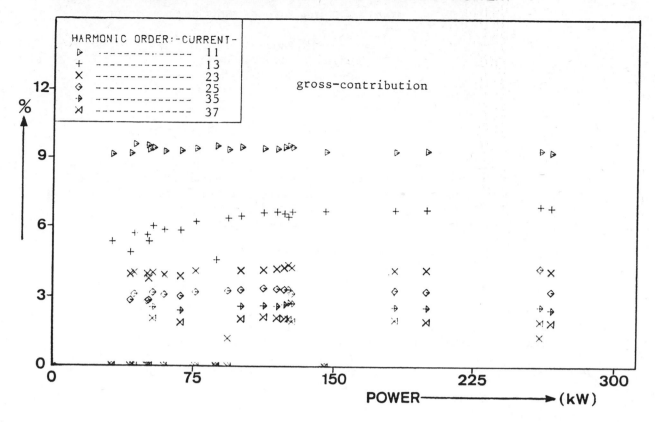

FIG.10 HARMONIC VOLTAGES AS A FUNCTION OF POWER

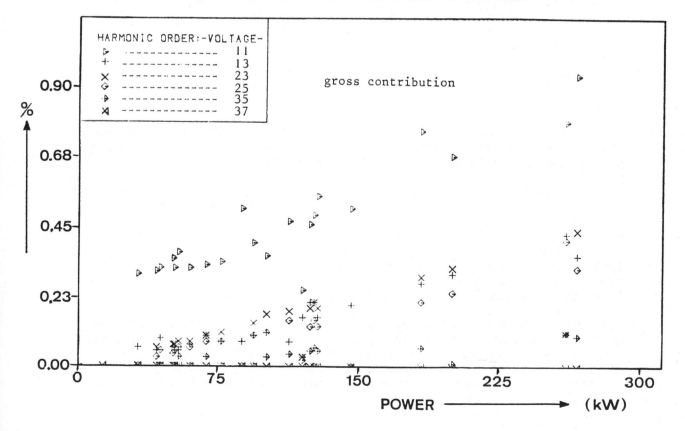

WIND POWER FOR SHIP PROPULSION

C.T.Nance
Medina Yacht Co. Ltd.
Cowes, Isle of Wight

Abstract

The dramatic rise in the price of oil in the 1970's revived interest in
the possible use of wind power to assist the propulsion of ships. Modern
materials, and advances in aerodynamics, have both made available improved
versions of traditional rigs and also offered prospects of the use of
newer concepts such as aerofoils, wind turbines, and kites.

To identify, from so wide a range of possibilities, the form and proportion of rig which best meets the requirements of ships of many types
and sizes is a problem which requires considerable study.

The Department of Energy has an interest to encourage the fuel saving
which would result from the introduction of wind-assistance systems for
ships and, from its Wind Energy programme, it already has a wide appreciation of the technical disciplines involved. The paper suggests that the
Department be encouraged to set up a Wind Propulsion Steering Committee.

Introduction

The revival of interest in the use of wind power to propel ships stems
from the dramatic rise in the price of oil in late 1973. But whereas for
thousands of years man has used the wind through the medium of soft sails,
he now has at his disposal, in addition to much improved materials for
sails and modern aerodynamically efficient sailing rigs, many alternative
means of converting wind energy into propulsive thrust.

It is now almost eight years since publication of a study commissioned by
the US Maritime Administration concluded that, at the then fuel price of
$ 11.25 a barrel, the economic position of the sailing ship was close to
equal footing with powered ships. The sensitivity analyses in that report
showed that, if their estimates were correct, on half of the routes they
studied the sailing ship would probably become more economical than oil-
powered ships at a fuel price of $ 13.25 a barrel.

It seems highly likely that, if that same conclusion had been reached about
some revolutionary new fuel, in unlimited supply, and guaranteed not to
increase in price in real terms, substantial resources would have been
expended in promoting it; research teams in many parts of the world would
be trying to improve upon it; Governments would be spending large sums
encouraging its use; and shipping lines would be impatiently withholding
orders for new ships while pressing for clarification as to the best type
of engine in which to use the new fuel for their specific operations.

Yet because those words were said about wind, which is free, and because
no one can benefit by manufacturing it, or marketing it, or improving it,
there is no one to lobby politicians or industrialists to finance the
necessary research and development. This paper examines briefly the
principal ways in which the power of the wind can be used to assist ship

propulsion, and considers whether, if the traditional sources of support
for R & D are not available, the time has come for those now concerned
with the development of wind energy for power generation to become
involved in this additional field.

General Observations

There are several very different ways in which wind energy might be used
to drive ships, and to make the best choice between them involves many
technologies, and operational and economic analyses in substantial depth.
Because the supply of wind energy varies from one route to another, and
on many routes from season to season or even day to day, and because in
addition not only is it variable in direction, but different systems are
direction-sensitive to differing degrees, the best choice of system is
likely to vary from one route and trade to another. Further, the optimum
balance between fuel power and wind power will depend not only on these
parameters, but on the particular economics of the ship and cargo
concerned. In many cases we must contemplate blending together energy
from fossil fuel and thrust or energy from the wind in variable proportions; a mode of operation normally referred to as motor-sailing. Wind
energy must not therefore be considered as a rival to oil, or to coal, or
to whatever other fuel may be used in future, but as a source of energy
available to complement whatever fuel is used.

There is another factor pointing to this probability.. It is now generally
appreciated that it was not so much the fact that fuel was cheap (which
in relation to the then wage levels it was not) or their safety record,
which ousted the sailing ship from the seas, as much as the revolution in
maritime transportation which fuel-power permitted by way of accurate
voyage and cargo planning. It follows that forms of wind propulsion or
wind assistance must be sought which are not only more efficient in
performance and little if any more demanding in crew size than the equivalent all-fuel ship, but which are also capable of operating with the
regularity, and within the range of port to port speeds, that would be
acceptable for fuel-powered ships.

Possible Wind Propulsion/Assistance Systems

Wind, being directional, has momentum as well as energy; and it follows
that the options open for converting it into propulsive force are much
wider than those contemplated within the traditional 'wind energy industry'
as we know it today. Indeed at first sight, it might be thought that to
use it in the form of direct thrust, as with sails down wind, or of direct
lift, as with sails to windward, would be the only effective forms of use,
since to convert its energy to rotary motion, and then to convert that
rotary motion back into thrust through a water-screw, would seem to
involve losses so great that the system could not be competitive.

But there are factors which mitigate this situation: a ship so propelled
can travel in any direction, including head to wind (which for a ship with
sails is an almost ninety degree no-go sector - almost one quarter of the
whole), and this confers the benefit of being able accurately to follow
shipping lanes, separation zones, etc, irrespective of the direction of,
or change of direction in, the wind; further such a ship can claw itself
off a lee shore, or even out of a harbour, into the teeth of a gale.
Indeed it seems likely that, if a wind turbine system could be developed

which would also operate as an auto-gyro developing direct thrust when
the wind was in favourable directions, a system fully competitive with
direct-thrust-only systems might conceivably result.

The range of options can best be grouped under six categories (see Fig 1):
Soft sail square rigs; Soft sail fore-and-aft rigs (of both of which
there are many variants); Rigid and semi-rigid aerofoils; Magnus effect
and similar devices; Wind turbines; and Airborne sails (kites).

Square Rig

Justification for the introduction of a modernised form of traditional
square rig, as proposed by Windrose Ships Ltd of UK, rests on the fact
that it has been proved in practice, its capabilities and limitations are
well understood, and, after making allowance for improved performance and
reduced crew strengths resulting from modern developments in materials and
machinery, and accurate weather forecasting etc, voyage predictions and
economic assessments can be made very accurately.

Alternative proposals, such as the Dynaship (a range of design studies
extending to 100,000 dwt based on the ideas of Wilhelm Prölss developed
during the 1960's and 70's by the Schiffbau Institut of Hamburg University) are extensively modified from the classic square rig to give
aerodynamic cleanness, using cantilever masts and dispensing with fore
and aft stay-sails.

The square-rig, in whatever form, is best suited to long voyages,
especially those where the course can be varied to follow the great
belts of strong and favourable winds, such as the classic grain trade
routes between Europe and Australia. But for many cargoes, routes of
this length are not nowadays sailed by small or even medium sized ships,
so for these cargoes there might well prove to be a minimum size below
which such ships may be uneconomic. It is not known whether any square
rigged ships are yet on order, although a design study for a ship of
30,000 dwt was recently funded by the EEC to be undertaken in Belgium.

Fore and Aft Rig

Study of historical routes shows that wind direction has a far more marked
effect upon the short routes than upon the long ones. Whereas sailing
ship routes of 2,000 miles or more rarely showed a difference betwen
voyage lengths in opposite directions of more than 50%, many of the
shorter routes were almost impossible. To take perhaps the worst example,
at certain times of the year the voyage length between Madras and Calcutta
was 780 miles in the one direction, and 2,730 miles, i.e. $3\frac{1}{2}$ times as far,
in the other.

Thus, in general, the shorter the route the closer-winded must be the
ships. And since many of the first wind-powered or wind-assisted ships
are likely to be of small or medium size, and since small ships are likely
to prove competitive only on short routes, it follows that many of the
first generation requirements will be for rigs whose windward performance
is good.

Because the leading edge of fore and aft sails can be accurately
constrained in shape, and because their greater aspect ratio reduces

their induced drag, fore and aft rig is in principle closer winded than square rig. This fact makes it particularly suitable for motor-sailing.

Recent cooperation between the Institut fur Schiffbau mentioned above, funded by the German Government, and the Indonesian Government, has led to the design of a range of fore and aft rigged ships from 900 to 2,500 tons, the first of which is understood to be now building in the Indonesian Government Shipyard.

A parallel American development by Windship Company of Massachusetts comprises a fore and aft sail of some 3,000 sq ft installed upon a single rotating mast in the bow of a Greek-owned tanker of 3,000 tons. Trading over a period of 18 months in direct comparison with her all-fuel-powered sister ships, confirmed fuel savings of 24% and a 5% speed increase are reported to have been made. This is a remarkable achievement.

Aerofoils (Wing Sails)

Theory suggests the aerofoil as one of the most efficient of all possible rigs. Such sails have been devised for racing catamarans in many different forms, and in 1973 a catamaran so rigged took the world speed record for her sail area class, and remained undefeated for almost six years. Aerofoils have gained considerable credibility by the choice of folding aerofoils for the first wind-assisted oil tanker, the Japanese Shin Aitoku Maru of 1,600 dwt. A further vessel has now been built, and two 2,100 dwt cargo ships are on order and due for delivery in the spring of this year.

An alternative solution to folding the aerofoil is provided by a rig developed by Aerosystems of Beaulieu, pioneered on the yacht Gallant, which uses an elliptical wing, capable of being varied in area similarly to the Chinese junk by lowering the sail one segment at a time, thus approaching the efficiency of the rigid aerofoil while maintaining the convenience of a soft-sail rig.

The most recent and exhaustive study yet made, funded by the US Govt (1), favoured the use of aerofoils and concluded that "motor sailers with wing (aerofoil) sails possess an economic advantage up to at least 40,000 tons cdwt". A trial wing of 300 sq feet has already been manufactured by Windship Company, designed to feather successfully in winds of up to 150 knots, and static tests have to date shown that it can do so satisfactorily without flutter in winds of up to 60 knots maintained for several hours.

A British initiative has recently been taken by Wingsail Systems Ltd of Hamble to propose the installation of two tri-plane wingsail units on a 2,000 dwt coastal tanker. These units, controlled in part by fins rather than wholly by computer instructions, is claimed to have a higher efficiency than the Japanese or American systems previously mentioned, because of its ability to react to rapidly changing directions of relative wind.

Magnus Effect Devices

A rotating object positioned in an airstream experiences a force at right angles to the relative direction of motion. Anton Flettner made practical use of this effect for ship propulsion when in the 1920's he developed a system by which a vessel could be propelled in a wind by the thrust developed by two or more vertical cylinders, rotated by the use of a small proportion of the normally installed power of an equivalent vessel. He showed that these cylinders need only have a surface area of about a tenth of the sail area which would be required to produce a similar performance. Two ships were installed with this device: the Buchau, a converted topsail schooner, later renamed the Baden-Baden, and the Barbara, specially constructed to employ this system as auxiliary to her main diesel motors. The ships are reported to have operated successfully, their voyages including crossing the Atlantic.

Such a system would have the advantages that it could be developed, designed and manufactured within industries already in being, and can be maintained ashore and afloat, and operated at sea, by trades already current; its manpower requirement should be no greater than that required for a powered ship; and if such a system could be satisfactorily engineered, it might offer the possibility of being installed on existing ships, or ships now in manufacture or undergoing design.

Wind Turbines

The operational advantages of the wind turbine have already been mentioned. As an engineering system, it also possesses some of the advantages mentioned above for the Flettner rotor. Furthermore the extensive investment now going on throughout the world into the development of aerogenerators for electrical power production should much reduce the effort involved in the development of wind turbines for marine propulsion.

Towards the end of 1978, the Department of Industry approved a project to study the feasibility of propelling by wind turbine a ship of about 4,000 tons. The theoretical element of this study was completed in September 1979, and suggested (2) that if the wind turbine could be so designed that it could develop direct thrust, as suggested previously in this paper, the resulting ship could be expected, on suitable routes, to be economic by the time she could be developed, built and put into service. Although this deduction is purely theoretical, and awaits physical confirmation, it is relevant that a New Zealander, J. Bates, who has installed a wind turbine in his 30 foot yacht, has reported a performance at least equal to his normal rig. This and other indications have encouraged another New Zealander, R.A. Denney, resident in England, to decide to seek sponsorship for an entry for the 1984 "Observer" single handed transatlantic race in a wind-turbine driven trimaran: a most economical way for any company interested in entering this field to assess its practicality on a much smaller outlay than a normal development programme.

Airborne Sails (Kites)

Until recently the kite, a solution which offers potential benefits, seems to have been almost entirely ignored. The reason for this is not clear: several people, among them scientists such as Benjamin Franklin, and military men such as Colonel Cody and Admiral Cochrane, have experimented in the past with kites for propulsion with some degree of success.

On small sailing boats, using a multiple-tier sporting type of kite, a performance can be achieved to windward about as good as with a square sail rig. Off the wind, having almost no heeling moment, performance is spectacular, and the world speed record for a sail area of 300 sq ft was broken in 1982 by the kite-drawn catamaran "Jacobs Ladder", which is now seeking sponsorship for an assault on the world's absolute speed record.

There are of course substantial problems to be overcome, notably the development of efficient launching, control, and recovery systems. But the economics of a rig which may well prove to be cheaper than all others, lighter in weight than all others, and to have a zero windage drag when not in use, are powerful incentives to contemplation.

Rig Selection

The factors affecting rig selection have already been mentioned. As one outcome of the wind turbine study referred to above, a generalised computer programme was prepared (3) which enables the performance of any type of rig to be studied on the basis of the fuel-engine power required to maintain a given ship speed at different wind speeds and angles. (Fig 2). Although these curves are notional, and must not be interpreted in any absolute sense, since the basis for comparison varies from circumstance to circumstance, the dramatic differences in performance of the rigs at different wind angles is at once apparent.

Before converting such theoretical power savings into an economic assessment, many practical factors must be taken into account, as the effect of these varies not only from rig to rig but from one wind to fuel-engine-power ratio to another. The influence of each rig on capital and running costs, including maintenance, reconditionings, manpower, insurance, and so on, and also any additional costs which the rig may cause to the ship itself (for example, if with a new ship, a different and more costly hull shape would be needed) must also be assessed. An overall economic comparison between various wind-assisted options, and between the best such option and an equally modern all-fuel ship, can then be made.

The Task of Research and Development

While the factors mentioned in the preceding paragraph are primarily the concern of naval architects, ship operators, and marine economists, and thus fall rightly within the purview of the Maritime Technology Committee of the Department of Industry, this committee can only wisely exercise its responsibilities to support project development if it has available to it basic data (such as the curves presented in Fig 2, but of course drawn from a much sounder base than the very primitive data now available will allow). It will also need to be given outline guidance as to the physical characteristics of the different rigs, and whether the fundamental problems associated with each (for example, with kites, the relationship between stability, performance, and manoeuvrability; and the launching, control, and retrieval problems) are within sight of satisfactory solution.

Given such information from an independent and authoritative source, the MTC would then be in a position to appraise proposals, whether from Industry, from shipowners, shipbuilders, or rig-system manufacturers, and hence to ensure that Government encouragement and, where appropriate, support, is given to those projects most likely to result in economical and successful ships.

It is clear that, to obtain the necessary basic data, and to solve the practical but basic problems such as those just mentioned, some independent (and to guarantee independence, preferably government-related) source must be identified to sponsor a coherent programme; first of studies to identify the essential minimum of data and trials needed, and then of wind tunnel tests and small-scale prototypes of systems and equipments which are foreseen as typical.

Clearly this source must have knowledge of the basic disciplines involved, and access to such facilities as wind tunnels. As the defence implications of wind assistance for ships is, in the quantitative sense, likely to be very small, it cannot be expected (as with civil aviation) that development of the basic technologies might be sponsored by Defence. As the technologies involved, and the facilities required, are to a great extent in some cases identical, and in others similar, to their equivalent in the wind energy field, the deduction seems inevitable that those who now sponsor the wind energy programme and exercise a persuasive influence over its development should play some equivalent part for wind propulsion if this country is to match effectively the developments elsewhere, some of which have been mentioned above.

Further, while the energy saved from wind-assistance for ships could not approach that from power generation, it would be by no means inconsiderable. (World-wide, over 90 million barrels of crude, some $3 billion, a year, has been estimated (4)). To encourage the U.K. proportion of such savings, would surely be a legitimate responsibility of the Department of Energy.

Conclusion

Now that U.K. has in hand a programme of work leading to full scale prototypes for the use of wind energy to generate electrical power, the Department of Energy may wish to consider whether the time has come for it to play some part in the field of wind power for shipping. The setting up of a Wind Propulsion Steering Committee would seem to be a logical first step.

References

1. "Wind Propulsion for Ships of the American Merchant Marine", U.S.Dept of Commerce, Maritime Administration, March 1981.

2. Nance C.T., "Role of the Engineer in the Windship Revolution", D of I Symposium on Commercial Sail, June 1979, and;
   Rainey R.C.T., "The Wind Turbine Ship", RINA Symposium on ditto, Nov.80

3. AWASH (Atkins Wind Assisted Ship) programme, under D of I contract, 79

4. "American Wind Power", Marine Engineers Review, Jan 1983

Diagrams

Fig 1. Six major Categories of Windship

Fig 2. Fuel-power Requirements v. True Wind Angle

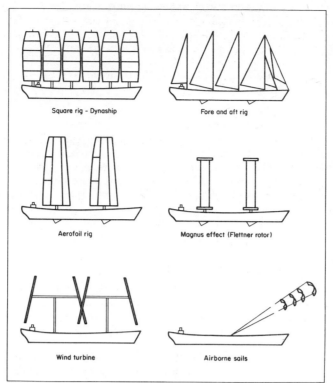

Fig.1. Six major Categories of Windship

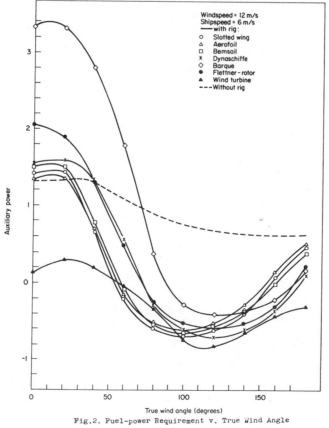

Fig.2. Fuel-power Requirement v. True Wind Angle

AERODYNAMICS AND STRUCTURAL COMPLIANCE

J C Dixon and R H Swift

Faculty of Technology, The Open University, UK.

Abstract

The tower diameter necessary for a wind turbine depends upon stiffness requirements (natural frequency), and on strength requirements. The applied aerodynamic forces, for survival conditions, applied to the tower, nacelle and rotor, and the consequent tower root bending moment, are considered, and also the forces during normal operation. Possibilities for reduction of survival moments, through changes of parking strategies, and of operating thrust through change of operating strategy, are investigated. Some implications for tower design are revealed, especially for the case of offshore application where wave loads result in a high substructure cost.

Introduction

Wind turbine tower design involves consideration of both stiffness (natural frequency) requirements and strength requirements. The latter may be divided into operating conditions and survival conditions. For offshore machines it is also necessary to consider the implications of wave loading, which gives a considerable incentive to minimise the diameter below the design wave crest level.

The design survival windspeed is typically in the range 50 to 60 m/s for a 3 s gust although it can be markedly higher in hurricane zones (eg US). Contributions to the survival tower root moment come from the rotor, the nacelle and the tower, in varying proportions. It is convenient to work here in terms of the specific drag area $C_D A_F/S$, ie the component drag area (drag coefficient times frontal area, $C_D A_F$), referred to the hub-height for the correct moment, non-dimensionalised by the rotor swept area S. The thrust coefficient, ie the effective rotor drag coefficient in normal operation, is of order 1.0, and the ratio of survival to rated speed is about 5, so a total survival specific drag area of roughly $1/5^2 = 0.04$ corresponds to the same force as the operating condition, implying that it is undesirable to greatly exceed this value. A study of existing designs suggests that specific drag area contributions are typically 0.01 to 0.05 for the tower, 0.005 to 0.06 for the nacelle, and 0.002 to 0.06 for the rotor.

Survival forces - tower and nacelle

Considering the simplified case of a tower of height H, constant diameter D_U (upper tower) in a uniform windspeed V, the aerodynamic force and moment are

$$F = 0.5\rho V^2 D_U H C_D \quad ; \quad M = F H/2$$

The effective force at the tower top for the correct moment is simply F/2. Hence the tower specific drag area is simply

$$A_{St} = D_U H C_D/2S$$

The effect of an exponential wind speed-height relationship $V = V_1(h/h_1)^\infty$ may be incorporated without difficulty, resulting in an additional factor $1/(1+2\infty)$ on the force, and $1/(1+\infty)$ on the moment; about 0.81 and 0.89 respectively for $\infty = 0.12$. At a representative survival speed and tower diameter, the Reynolds number is about 10^7, well supercritical, so $C_D \approx 0.6$. Also, typically, H = 0.7D, giving

$$A_{St} \approx 0.27 D_U/D$$

Figure 1 illustrates D_U/D data for some shell type towers. The best modern design practice evidently realises values $D_U/D \approx 0.03$ and $A_{St} \approx 0.008$.

The nacelle contribution depends upon the parking orientation. Side-on to the air-flow, it presents a short and generally rectangular outline, for which 3-dimensional flow effects are significant. The cross sections found vary from sharp square or rectangular to circular. Considering the nacelle to be subject to a uniform windspeed,

$$A_{Sn} = A_{Fn} C_D / S$$

where A_{Fn}, nacelle frontal area, implies frontal to the appropriate flow direction. Figure 2 shows representative specific nacelle areas (A_{Fn}/S), for which the variation is considerable. The appropriate drag coefficient is not entirely easy to assess because of the three dimensional nature of the flow, and the likelihood of interference effects with the tower. Rounding of the section edges has a considerable effect (Figure 3), with a marked reduction in C_D for $r/h = 0.2$ for a 2D flow, and even smaller values in 3D flow (Hoerner, Ref 1). The data shown is for Reynolds number $Re \approx 10^5$, but is qualitalively confirmed at higher values. Thus it appears that in 2D flow, the drag coefficient for the nacelle section will be from 2.0 down to 0.5, possibly even lower, according to the height/width ratio and the degree of attention to radiusing, or better streamlining.

The nacelle height/length ratio is typically about 0.25; Figure 4 shows the effect of h/l on drag (through 3D flow), although, of course, this is without the interference of tower or rotor. The cylinder result is for $Re \approx 10^5$, but the plate result gives some confirmation that R_e is not very crucial for this effect. Typically

$$C_D/C_{D\infty} = 0.6 + 0.4\ e^{-20\ h/l} \qquad \text{(plate)}$$

$$C_D/C_{D\infty} = 0.6 + 0.4\ e^{-14\ h/l} \qquad \text{(cylinder)}$$

At $h/l \approx 0.25$, the drag is reduced in the factor 0.6. Thus the overall nacelle C_D is likely to be 1.2 down to 0.3.

With an actual specific nacelle side area range of 0.004 to 0.06, and drag coefficients 0.3 to 1.2, but noting that in existing designs these are negatively correlated, the range of specific drag areas is 0.005 to 0.018 It appears feasible to achieve a specific side area of 0.004 with a drag coefficient of 0.3, ie $A_{Sn} \approx 0.001$. Actually, other nacelle styles, eg KaMeWa, may do better, but contrarily, there may be compromises due to other accoutrements, including, in the extreme, a helicopter landing pad on the nacelle.

When aligned into the wind, the nacelle will have a markedly lower drag. The specific area can be as low as 0.001, with a drag coefficient of 0.2 or less (Ref 1).

Survival forces - rotor

The contribution of rotor blade drag depends primarily upon rotor solidity, pitch control, and parking strategy. Typical modern large-scale design practice uses a total solidity σ, the blade area over swept area, of close to 4%, with a tip solidity σ_T, the sum of tip chords over circumference, of about 1.6%. The parked area presented to the flow generally depends on the degree of pitch control. Typically about 25% of the blade radius is controllable (Mod-2 30%, 5A 20%), giving control of about 40% of the swept area, and about 15% of the blade area. The relationship between these figures depends on the chord distribution, but is typically as shown in Figure 5. The full range of this curve is relevant because proposals range from virtually full pitch control (Hamilton Standard WTS-4) to zero (ie pure stall control, possible alternative to Boeing Mod 5B, Ref 2). Presented fully to the flow in an operating pitch position, but stationary, the rotor is approximately equivalent to a strip of high aspect ratio - typically 30 overall - ie h/l = 0.03, giving a drag coefficient, from Figure 4, of about 1.7. It is interesting to note that if an adequate flow path can be provided at the blade roots (eg through the teeter hinge area) then the blades will be effectively independent, with $h/l \approx 0.06$, giving a significantly lower $C_D \approx 1.4$.

If the blades are fully pitchable, and in the feathered position, then the average incidence angle will be typically one quarter of the pitch twist, ie about 4 deg. This is likely to shift the laminar flow wing sections outside their drag bucket, but even allowing for some surface roughness the effective overall $C_D \approx 0.01$ (Ref 3), two orders of magnitude below 1.7. The specific drag area of the rotor, parked across the wind is, therefore, $A_{Sr} \approx \sigma C_D (1 - F_{PA})$, and varies with different degrees of blade area under pitch control, according to Figure 6, from 0.004 to 0.07. However, alternative parking strategies are available, which are an improvement for blades of large non-pitchable area.

The most obvious possibility is to park with the rotor horizontal and with the wind along the rotor. This will reduce the rotor drag to a very small value, although at the small expense of accepting the larger lateral nacelle drag.

The minimum practical total specific drag area is probably achieved by using feathered full pitch control, with a reasonably streamlined nacelle and slim tower (WTS-4), allowing

$$A_S = A_{St} + A_{Sn} + A_{Sr} \quad \approx 0.008 + 0.001 + 0.004 \quad \approx 0.013$$

Interference effects are likely to increase this, but it does appear realistic to remain below 0.02. For a typical partial span control type (Mod-2), parked with the rotor cross-wind,

$$A_S = 0.008 + 0.002 + 0.055 \quad \approx 0.065$$

several times the minimum achievable, whilst, with the rotor along the wind,

$$A_S \approx 0.008 + 0.005 + 0.002 \quad \approx 0.015$$

It is of course necessary to consider the implications of various control system failures on survival. We do not have space to expand on this topic here. We will, however, make two observations. Firstly, it is proper to use a reduced return wind for the failed state according to its probability (Tornkvist, Ref 4). Secondly, the conjunction of pitch control, yaw control and mainshaft rotation are not all necessary to achieve good parking. However, it does seem that the provision of various control systems is a more economic safeguard against control failure than to increase strength. Proper optimisation of these factors is of course complex, and is not practically feasible in the absence of an extensive cost framework.

Should there be sufficient economic incentive to shift the designs in the low parking drag direction, it seems likely that survival design could be based on a specific drag area of as little as 0.015, in contrast to a more typical 0.060 found today. As a matter of practical interest, for a typical 100 m rotor at 70 m, in a 60 m/s wind, the actual moments are 18 and 72 MN m respectively.

Operating forces

The permissible thrust during normal operation must be somewhat lower than the permissible survival thrust, both because of the greater safety factor appropriate for a normal load state as opposed to an extreme one, and also because of need for a reduced safe working stress in the face of high-cycle fatigue loads. This ratio is likely to be in the range 1.2 to 2.0 for horizontal axis machines, depending on the success of the tower compliance and control systems in reducing the fatigue loads.

In normal operation the effective specific drag area of the tower will be reduced from that for the survival case, by about 50%, because it is subject to a reduced local airspeed. The corresponding value $A_S \approx 0.14 \, D_U/D$ gives $A_S \approx 0.004$ for a typical compliant design and 0.010 for a rigid design, and can be considered practically negligible compared with the rotor thrust coefficient.

The operating rotor thrust coefficient does not appear to have been very widely discussed in the literature despite its obvious design importance. Simple actuator-disc momentum and energy analysis gives the well known results

$$C_T = 4a(1-a) \quad ; \quad C_P = 4a(1-a)^2$$

with a maximum thrust coefficient $C_T = 8/9$, corresponding to $C_P = 16/27$, and a parabolic variation of C_T with varying interference factor a, Figure 7. The reducing value of C_T is evidently unrealistic, (a > 0.5 is unphysical in the simple model anyway), the final return to zero corresponding to D'Alembert's zero drag paradox. Rather, one might anticipate a terminal C_T value of 1.2 corresponding to the real flow around a solid disc. In fact, a rather higher mean value appears amongst the rather wide scatter of Figure 7, which incudes data from various sources as presented and added to by Anderson et al (Ref 5), plus data from Viterna and Janetske (Ref 6). The simple theoretical relationship seems adequate for a < 0.4. Adopting a linear tangent line as illustrated results in the C_T v C_P curve of Figure 8. The notable observation from this figure is that a C_P of 0.5 corresponds to a C_T of 0.6, and that beyond this point any increase of C_P has a severe cost in terms of C_T. This suggests that where a low thrust is desirable, very marked savings might be possible for a quite moderate reduction in C_p.

Unfortunately, this information is of only limited value to the designer. In the absence of specific tests, the axial interference factor is subject to significant uncertainty, because it depends upon the coefficient of power extracted from the windstream, rather than the more easily measured electrical power. The differences are the generator, gearbox and shaft losses, and the rotor losses through viscosity and turbulence including stalling. This last may be considerable, especially, of course, for a stall-power-controlled rotor.

As an example of a partially pitch-controlled rotor, Figure 9 shows a thrust-speed curve calculated for Mod-2 from the C_P curves given by Lowe and Engle (Ref 7). The principal assumptions underlying this calculation are that the lift-drag ratio remains approximately constant, and that C_P and C_T should be related through a reduced rotor area calculated according to Prandtl's correction factor (Ref 8). The former is probably acceptable over a limited but adequate range from the peak C_P to the rated C_P. The claimed peak C_P can then only be achieved if the mean effective section lift/drag ratio exceeds about 55, which value we have therefore applied throughout. The peak thrust is found to be 320 kN ($C_{TR} = 0.53$) at the rated speed. At this speed C_P is already reduced significantly from its peak value, with a correspondingly greater reduction in C_T (from Figure 8). More important than the absolute value is the general shape; specifically a thrust reduction beyond the rated speed.

An interesting contrast arises for the stall-controlled rotor, an example of which has been described in detail by Rasmussen (Ref 9). The axial thrust rises approximately linearly with windspeed, Figure 10. This implies a total power extraction rising as the square of the speed, with power regulation through very great stalling losses. This obviously represents a serious problem where minimum tower diameter is sought.

Scope for reduced operating forces and moments

If the allowable operating thrust is to be reduced below the original maximum (ie rated) thrust, then a change of operating strategy is required, in which both the thrust and power are reduced over a speed range, Figures 11 and 12. In fact, a limited application of this strategy has some appeal in any case, because of the difficulty of obtaining fast and accurate control response at the rating point. Since the power loss is initiated at a corner of the characteristic, the energy yield loss must initially be of second order, not first. From curves such as 11 and 12, plus an adopted wind speed distribution, the energy yield density against mean thrust may be calculated (eg as Figure 13 for an unrestricted Mod-2) and thence may be found the energy yield loss under conditions of restricted working thrust, Figure 14.

The most satisfactory non-dimensionalisation of a restricted thrust F_L appears to be to define a thrust limit coefficient C_{TL} against the characteristic wind speed and swept area:

$$F_L = C_{TL} 0.5 \rho V_{ch}^2 S$$

and, with a Weibull wind model, to non-dimensionalise the rated power against the swept area and the characteristic wind speed to give a characteristic power coefficient C_{ch} (we have previously used a 'specific power coefficient' which is a non-dimensionalisation of the rated power against mean windspeed - the preferable coefficient in any particular case depends on the method of wind characterisation):

$$P_R = C_{ch} \, 0.5 \rho V_{ch}^3 S$$

$$= C_{ch} \, V_{ch} \, F_L / C_{TL}$$

$$F_L = (P_R/V_{ch})(C_{TL}/C_{ch})$$

$$\text{Also} \quad F_R = C_{TR} \, 0.5 \rho V_R^2 S$$

$$P_R = C_{PR} \, 0.5 \rho V_R^2 S$$

$$F_R = P_R \, C_{TR} \, V_R / C_{PR}$$

$$\text{Thus} \quad F_L/F_R = (C_{TL}/C_{TR})(C_{PR}/C_{ch})(V_R/V_{ch})$$

It is of interest to explore the allowable power as a function of rotor diameter for a given limit tower moment, where the hub-height is proportional to rotor diameter D. At a given speed, a larger D increase the power until the limit moment is reached. Then a larger D, because of the greater moment arm, results in a smaller allowable thrust and power, Figure 15.

Implications for tower design

It is well known that for a given stiffness or strength requirement, the largest diameter (commensurate with an acceptable wall thickness) uses the least material. It is less widely realised that for a given bending stiffness, a smaller diameter is actually stronger, since $Z = I/R_o$.

Douglas (Ref 10) gives information on the tower design for Mod-2, including wall thickness requirements. For example, at the bottom of the constant diameter section (48 m below the hub) the fatigue operating requirement is 18 mm, operating plus zone-3 seismic requirement is 22 mm, and the extreme wind requirement is also 22 mm (outer diameter 3.05 m). From this we conclude that, in view of the possibilities of drag reduction considered above, the design is actually operationally limited, not survival limited. For the UK, without the seismic requirement, the fatigue requirement is the true limitation, and this could be realised by some changes to the parking strategy or pitch control fraction, and that further reductions in tower bending moment would be feasible with changed operating strategies. To change to a very compliant design, the stiffness must be reduced by a factor of 4, but with the same strength. This would require a much smaller diameter (since $R_o = I/Z$) and more material, and so is probably uneconomic.

The optimisation conditions for offshore application are rather different; the wave loadings govern the design and cost of the substructure (Ref 11), encouraging use of a small diameter tower. Since the substructure cost is a substantial part of the total (eg 30%, Ref 12), the optimum diameter is likely to be noticeably smaller than for land designs (eg $D_T/D \approx 0.02$ instead of 0.03), with advantage being taken of the moment reduction possibilities described here. Also, or alternatively, the tower could be waisted to minimise wave loads; there is no practical difficulty in providing a yield moment of 50 MN m with a mild steel section of outer diameter only 2 m (73 mm wall). Figure 16 gives an indication of the substructure cost (25 m depth), expressed as p/kWh, as a function of D_T/D. The compliant type is significantly better than the rigid type, although it seems unlikely that the very compliant type will show any marked overall advantage. One important advantage of changing from a rigid to a compliant tower is to reduce the problem of the seabed compliance.

We have considered here only the horizontal axis machine. Conditions for the vertical axis type are rather different. If this has two opposed blades, the axial thrust has an amplitude approximately equal to its mean value. Thus at least three, possibly four, blades will be required, or the L-format adopted, if a slim tower is to be used. Alternatively, a completely different approach must be adopted, using several relatively widely spaced legs, and rigid dynamics. In the absence of adequate comparative data on the various types of vertical axis turbines, we have not attempted any detailed analysis of this type of machine.

Conclusions

The tower design of the modern compliant-tower teetered-rotor turbine, such as Mod-2, is essentially operationally limited, since there is considerable scope for survival drag reduction. Operational loads are governed by the power loading, and the operating control strategy. For offshore machines, it is important to minimise wave loads, so minimisation of tower diameter is especially important, and will probably lead to increased attention to the possibilities suggested here.

References

1   Hoerner S F; Fluid-Dynamic Drag, Hoerner Fluid Dynamics, New Jersey, US, 1965. (pp 3.12, 3.13, 3.16).

2   Douglas R R; Conceptual design of the Boeing Mod-5B, WW5, US, 1981.

3   Abbott I H and von Doenhoff A E; Theory of Wing Sections, Dover, 1959.

4   Tornkvist G; Load cases for WECS, 3rd ISWES, Lyngby, Denmark, BHRA, 1980.

5   Anderson M B, Milborrow D J and Ross J N; Performance and wake measurements on a 3 m diameter horizontal axis wind turbine, 4th ISWES, Stockholm, 1982.

6   Viterna L A and Janetske D C; Theoretical and experimental power from large horizontal axis wind turbines, WW5, US, 1981.

7   Lowe J E and Engle W W; The Mod-2 wind turbine, EPRI ER 1110 SR, US DOE Conf 790352, USA, 1979.

8   Prandtl L; in Durand W F, Aerodynamic Theory, (Division L, Airplane propellers, H Glauert, p 265), Dover 1963.

9   Rasmussen F and Pedersen T F; Measurements and calculations of forces on the blades of a stall regulated HAWT, 4th ISWES, Stockholm, BHRA, 1982.

10 Douglas R R; The Boeing Mod-2, NASA N80-16457.

11 Dixon J C and Swift R H; Structure design for compliant tower offshore wind turbines, BWEA 4th ACWES, Cranfield, 1982.

12 Dixon J C; Large offshore wind turbines: system design and economics, BWEA 3rd ACWES, Cranfield, 1981.

Nomenclature

Symbol	Units	Description
a	–	axial interference factor
A_B	m^2	blade area
A_D	m^2	drag area ($C_D A_F$)
A_F	m^2	frontal area
A_S	–	specific drag area (A_D/S)
C_{ch}	–	characteristic power coefficient ($P_R/0.5\rho V_{ch}^3 S$)
C_D	–	drag coefficient
$C_{D\infty}$	–	drag coefficient for 2D flow
C_P	–	power coefficient
C_T	–	thrust coefficient
C_{TL}	–	limited thrust coefficient
D	m	rotor diameter
D_B	m	substructure base diameter
D_T	m	tower diameter (low)
D_U	m	tower diameter (upper)
F	N	force, thrust
F_R	N	thrust at rated speed
F_L	N	limited thrust
F_{PR}	–	pitch control fraction (by radius)
F_{PA}	–	" " " (by blade area)
F_{PS}	–	" " " (by swept area)
h	m	component occupied height, altitude
H	m	hub height, altitude
I	m^4	section second moment of area
l	m	length
M	N m	moment at tower root
P_R	W	rated power
r	m	corner radius
R	m	section radius
S	m^2	rotor swept area
V	m/s	wind speed
V_{ch}	m/s	Weibull characteristic windspeed
V_R	m/s	rated wind speed
Z	m^3	section modulus
$\propto$	–	wind speed-height exponent
ρ	kg/m^3	air density
σ	–	rotor solidity (A_B/S)
σ_T	–	tip solidity

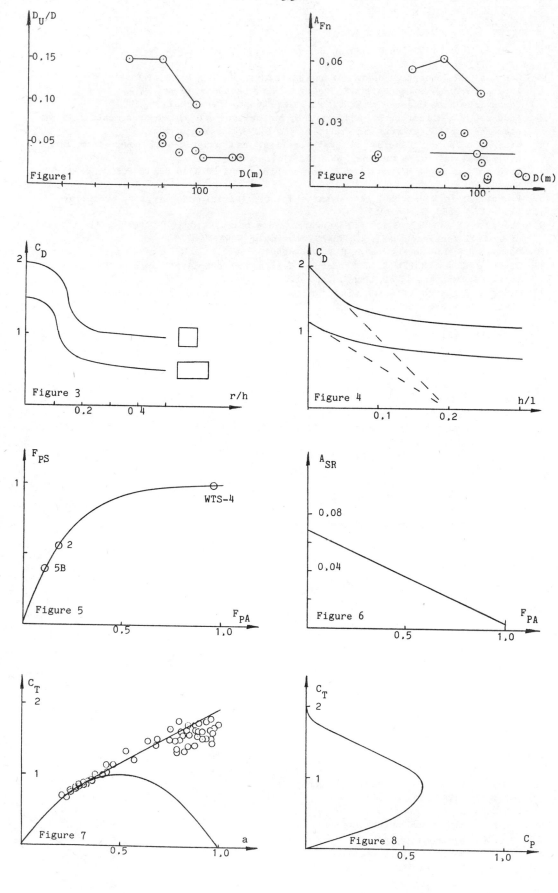

Figure 1

Figure 2

Figure 3

Figure 4

Figure 5

Figure 6

Figure 7

Figure 8

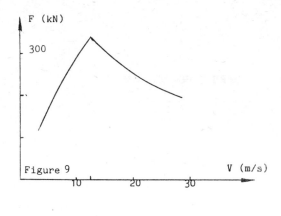

Figure 9

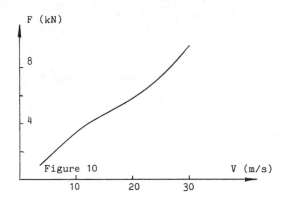

Figure 10

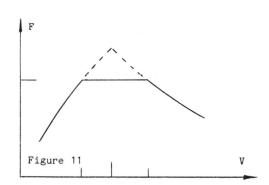

Figure 11

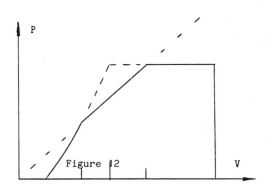

Figure 12

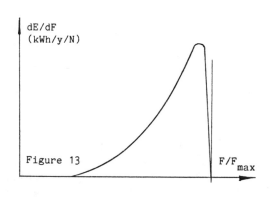

Figure 13

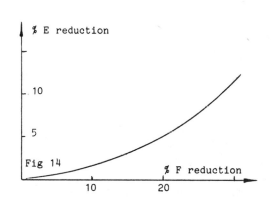

Fig 14

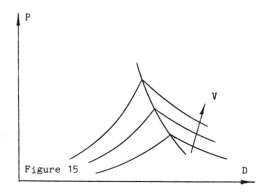

Figure 15

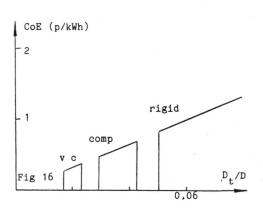

Fig 16

OFFSHORE WIND-TURBINE SUB-STRUCTURE DESIGN

R H Swift and J C Dixon

Faculty of Technology, The Open University, UK.

Abstract

Design options for support structures for large offshore wind turbines are considered, with special attention to influence of the principal design parameters on the cost. A 90 m diameter teetered-rotor compliant-tower turbine is considered, mounted on a concrete gravity support structure. A range of possible shapes of the support structure are examined, with a view to optimising the shape with regard to construction costs. Wave forces are calculated using stream function theory. A computer program has been developed to perform appropriate design calculations, and to find total material quantites and corresponding costs. Various results are shown, giving some indication of a possible optimum substructure shape.

Introduction

There have been a number of substantial studies of the prospects for large offshore wind turbines to generate electrical energy in large quantities (Refs 1-6). These investigations agree that the principal disadvantages of offshore siting are the cost of the extra supporting substructure and the extra grid connection requirements. The advantages are prospectively higher wind speeds, a very large total resource, and less environmental problems.

In this paper we give some preliminary results of an investigation of the relative costs of various kinds of substructure. The most economical installations are generally expected to be in relatively shallow coastal waters, typically 10 to 40 m deep; we have here considered a representative depth of 25 m, so the conclusions are probably valid in general terms over the depth range of interest.

Attention is focussed on concrete gravity support structures of a kind suitable for mass production, since a typical development would require several hundred units. The structures would be ballast filled, although costs for this are not included here. Concrete possesses advantages over steel as an offshore construction material because maintenance is reduced. The gravity base has some advantage over piled foundations because the on-site installation time is likely to be much lower. However, the foundation mechanics of gravity bases in shallow waters are not as well understood as their deep water counterparts, and more work is required in this area.

Structural configuration and material costs

A computer program was developed to investigate the influence of structural configuration on relative costs. Referring to Figure 1, the input data for a given structure consisted of the outer diameters, and levels at transition heights (Z1, D1, Z2, D2, Z3, D3), and the base height, Z3. On the basis of this information, the required wall thicknesses for the structure were obtained, along with the diameter D4 for the base. Quantities of concrete, shuttering, reinforcement and prestressing strand were obtained, in order to estimate costs. All structures were assumed to be ballasted. Only circular bases were considered in this study, although there is some precedent for square bases (Ref 7), which may have some advantage. For a given design moment and vertical force, the width of a square base is less than the diameter of the equivalent circular one, which may be an important consideration if existing construction facilities are to be used for the largest possible turbines. On the other hand, it is possible that a square base would increase scour problems.

The costs of the construction materials were taken from Ref 8, but to alleviate difficulties caused by inflation, all costs were expressed as multiples of the cost of concrete, on the expectation that the comparative costs would be relatively insensitive to inflation. The actual total cost of the structures can be estimated by multiplying the unit costs by the current concrete cost per cubic metre (about 60 £/m^3 in 1983).

Loading estimates for design

In the initial costing exercise, all structures were designed for typical 50 year survival conditions, on the assumption that the structure would support a turbine similar to the Boeing Mod-2, mounted on a steel shell tower. The adoption of a compliant upper tower enables the diameter of the support structure to be minimised in the region of the most severe wave loading. Wind loading on the turbine was evaluated on the assumption that the turbine was parked, but that control system failures resulted in the worst possible configuration. In a more detailed design, the loading cases resulting from multiple control failures should be associated with a lower probabilty return wind, and possibly with a relaxation of the stress limitations for the concrete. The design would then be optimised by economic analysis based on failure risk, but this optimisation requires more detailed costs than were available to us. A survival wind of 41.5 m/s at 10 m height was adopted, with a velocity height exponent of 0.1.

Although the wind loading is of some direct significance, it is the wave loading that represents the major proportion of the total survival design moment for the support structure. The main role of the wind force is in designing the tower, which then influences the wave loads. It is the usual practice to select a design wave height from the wave statistics available for the site, and to use this to evaluate the forces and moments on the structure. It is also possible to evaluate a design force or moment directly from the site statistics, without reference to the intermediate step of evaluating a design wave height. We have described a method for doing this

(Ref 9), in which the geometric form of the structure is shown to have a significant effect on the form of the long-term distribution of wave forces and moments. This method is, however, more complex, so for the present purpose of initial screening of designs, we have adopted the conventional design wave approach, with a 17 m high wave of period sufficiently short to break in the design depth of 25 m.

Wave forces were evaluated using Morison's equation (Ref 10), which idealises the force on a structural element as the phased sum of a drag and a fluid acceleration related component. Semi-empirical drag and added mass coefficients are required to use the equation, and although a considerable body of literature exists concerning possible values for these coefficients (eg Ref 11, 12) there is considerable scatter in the results, which are in any case mostly for flow conditions and scales much smaller than full scale. For the flow regimes anticipated in this study, however, values of 0.7 and 2.0 for the drag and added mass coefficients respectively appear to be reasonable, and have accordingly been adopted.

In addition to the in-line wave forces, it is known that transverse forces arise from vortex shedding, although present evidence suggests that these forces will be much smaller than the inline force for the Reynolds numbers under consideration (Ref 13). A transverse force coefficient based on results of Sarpkaya was used in this study, although it is accepted that this area requires further investigation (eg Ref 14).

The estimation of water particle velocities and accelerations requires the use of a suitable wave theory. Stream function theory (Dean, Ref 16) is generally regarded as being the most suitable wave theory over a wide range of depths, and we have developed a computer program to calculate the wave kinematics according to this theory. For the present study, a 15th order function was found to be suitable.

Wave impact forces were not considered in this study, since the wave used for design was a spilling breaker, and unlikely to be capable of imparting a substantial impact force.

Design of support structure

The support structure was designed in prestressed concrete, based on a characteristic strength of 50 N/mm^2. An allowance was made for gain in strength until commissioning (Ref 16) assuming an elapsed time of 3 months. Prestressing strand of type 12K15 was adopted as being representative of the type commonly used in offshore construction (Ref 17). It was required that the tensile strength of the concrete should not be exceeded, i.e. no cracking would be allowed. In addition to stresses arising from in-line and transverse wave forces, further stresses will arise due to local deformations under the action of exterior hydrostatic and hydrodynamic pressures, and allowance was made for this in the design. It was found that the bases could be designed at their critical sections as reinforced concrete members, although some light prestressing would be desirable to resist cracking.

The foundation was designed for no tension in the base in plane bending, and using a factor of safety against bearing capacity failure of 1.75, according Brinch Hansen's method (Ref 18), which is suitable for the design of eccentrically loaded foundations. It was assumed that the foundation consisted of marine clay with a cohesive strength of $75 \, kN/m^2$ (Ref 19). When these two criteria had been satisfied, it was found that the factor of safety against sliding was adequate, on the assumption that steel skirts would be provided to utilise the shear strength of the foundation. Sand foundations were not specifically considered in this part of the study. The capacity of the base to spread load evenly was also analysed, because the simple analysis of plane bending assumes that the base is perfectly stiff. In practice, this would not be so, and tensile stresses may arise in the base even though the assumption of plane bending suggested otherwise. This situation was checked approximately by analysing a narrow strip across the diameter of the base as a beam on an elastic foundation.

The design implications of scour around structures located on a sandy seabed were not considered in this study, although it is worthy of note that laboratory studies of scour around model piles subjected to a combination of wave and current action (Ref 20) demonstrated that increasing the size of the structure merely tended to worsen the situation. The effects of scour around a shallow water offshore structure have been documented by Bishop (Ref 21), where it was again reported that increasing the base size in order to combat foundation failure associated with scour was of no benefit. Two main approaches may be adopted to combat the effects of scour; either dumping graded riprap or installing flexible skirts around the perimeter of the base, to encourage the deposition of material by locally reducing water particle velocities (Ref 22). Of the two approaches, the latter seems cheaper and easier to maintain, although it would appear to require model testing to validate a specific design.

Results of costing analysis

Figure 2 summarises the results for the caisson-type structure in which it is shown that the value of the diameter D3 is not very critical to the total cost, provided that the depth of the caisson is chosen appropriately. Figure 3 shows that for some extra cost, the diameter of the caisson could be minimised, thus showing a possible actual saving in cost if existing dry dock facilities could be used through the adoption of a minimum diameter caisson.

Figures 4, 5 and 6 show unit costs for frustrum type support structures, using base slab thicknesses of 1.5, 2.0 and 2.5 m. The cost is not very sensitive to variations in the thickness of the base slab, but the 1.5 m thick slab produces marginally the cheapest solution. Reducing the level of Z2, the change in section, appears to produce continuous savings in cost, the cost of the extra wall thickness for the frustrum and tower being offset by savings in base quantities arising from a reduction in wave forces and moments. However, reference to Figure 2 suggests that the total costs will rise again as Z2 approaches Z3, the level of the top of the base. A ring beam would be required at level Z2 to resist the horizontal component of the prestressing force arising from the change in direction of the tendons. As Z2 diminishes, the sensitivity of the cost to the diameter D2 at the change in section diminishes.

The optimum design amongst those specifically investigated is with $Z3 = 1.5\,m$, $Z2 = 15\,m$, $D2 = 5\,m$, $D3 = 15\,m$, with a cost of approximately 4500 units, corresponding to a real cost of around 270 k£ at present day concrete prices (about 60 £/m^3). The materials requirements of this particular structure are: 1119 m^3 (2726 t) of concrete, 110 t of reinforcing steel, 46.8 t of prestressing steel, plus 906 m^3 of ballast, with 658 m^2 of plane shuttering and 1550 m^2 of curved shuttering. The total cost includes labour and site related costs, and corresponds to a gross value of 243 £/m^3 of concrete, or 101 £/t of concrete. There would be some increase of the total cost with further detailing of the design, which we anticipate would be of the order of 15%. The total value quoted would correspond to approximately 125 £/kW for a Mod-2 turbine, although the size of the turbine could be increased and the cost of the structure per kW correspondingly decreased, since the design is determined much more by wave than by wind loading. Figure 7 shows the variation in base diameter with diameter of frustrum bottom, for three sets of Z2 and D2. The diameter of the caisson becomes less sensitive to D2 as Z2 is decreased. Also, the diameter of the base may be reduced by lowering Z2 and hence reducing wave forces and moments. It was also found that base diameter was comparatively insensitive to variations in base depth.

Design implications of seabed stiffness

The effect of varying seabed stiffness on the first modal frequency for example rigid and compliant designs was investigated, and the results are shown in Figure 8. In order to realise a rigid design, it is necessary to found the structure on a seabed with an elastic modulus of at least 2 GN/m^2, which would correspond to sandstone, whereas the compliant design is suitable for a wide range of seabed materials.

Summary

A study has been presented of the relative costs of various configurations of support structures for offshore wind turbines. It was found that the solution utilising the frustrum of a cone was considerably cheaper than the hollow caisson type. Cost savings could be achieved by keeping the top of the frustrum at a low level to minimise eccentricities due to wave loading. An approximate rate of 125 £/kW was obtained based on a support structure for a Mod-2 type wind turbine, although larger turbines could easily be accommodated without incurring a large cost penalty, since the design was determined primarily by wave and not wind loading. It should therefore be possible to realise costs significantly lower by considering larger wind turbines.

References

1. Assessment of offshore siting of WTGs, Taywood Ltd for UK DeN, 1979 (unpublished).

2. Simpson P B and Lindley D, Offshore siting of WTGs in UK waters, BWEA 2nd W E C, Cranfield Multiscience, 1980.

3. Kilar et al, Design study and economic assessment of WECS, ERDA E(49-18)2330, 1979, USA

4. Hardell R and Ljungstrom O, Offshore based WTS for Sweden, BHRA 2nd International Symposium on WES, BHRA 1978.

5. Wind power at sea, National Swedish Board for Energy Source Development, 1979 (unpublished).

6. RSV and Hydronamic, Offshore siting of WECS, 1980 (unpublished).

7. Antonakis C J, A problem of designing and building for a structure at sea, Proc. ICE, Pt. 1, 1972.

8. Woolley M V and Tricklebank A H, Mass production of concrete wave energy devices, Journ. I Struct E, Vol. 59A, No. 2, 1981

9. Swift R H and Dixon J C, Design wave and design force estimation for shallow water offshore structures, Paper No. OTC 4614, Proc. Offshore Tech. Conf., Houston 1983.

10. Morison J R et al, The force exerted by surface waves on piles, Pet. Trans. AIME, Vol. 189, pp 149-154, 1950.

11. Hogben N et al, Estimation of fluid loading on offshore structures, Proc. ICE, pt. 2, 1977.

12. BSRA, A critical evaluation of the data on wave force coefficients, Rep. no. W-278.

13. Sarpkaya T, In line and transverse forces on cylinders in oscillatory flow at high Reynolds numbers, Paper no. OTC 2533, Proc. Offshore Tech. Conf., Houston 1976.

14. Shigemura T, Wave forces on an inclined circular cylindrical pile, Proc. Conf. Coastal Eng., Vol. 2, Ch. 108, 1980.

15. Dean R G, Stream function representation of non-linear ocean waves, Journ. Geophys. Res., Vol. 70, pp 4561-4572, 1965.

16. Derrington J A, Construction of MacAlpine/Sea Tank gravity platforms at Ardyne Point, Argyll, Proc. Conf. Design and construction of offshore structures, ICE, 1976.

17. Long J E, Experience in prestressing and grouting concrete offshore structures, Proc. Conf. Design and construction of offshore structures, ICE, 1976.

18. Brinch Hansen J, A revised and extended formula for bearing capacity, Danish Geotechnical Institute Bulletin no. 28, 1970.

19. Tomlinson M J, Foundation design and construction, Pitman, 1978.

20. Machemehl J L and Abad G, Scour around marine foundations, Paper no. OTC 2313, Proc. Offshore Tech. Conf., 1975.

21. Bishop J R, Experience with scour at Christchurch Bay Tower, National Maritime Institute, Rep. no. NMI R120, 1981.

22. Brebbia C A and Walker S, Dynamic analysis of offshore structures, Newnes-Butterworths, 1979.

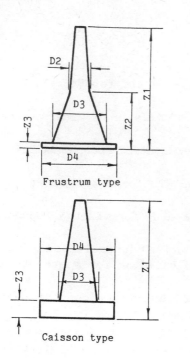

Fig. 1

Structure types and dimension definitions

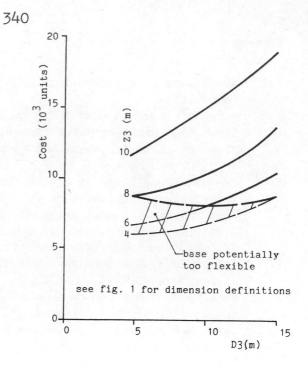

Fig. 2

Unit costs of caisson type structure

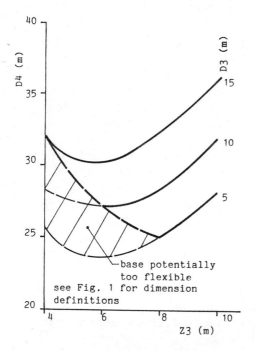

Fig. 3

Variation in base and tower dimensions
for caisson type structure

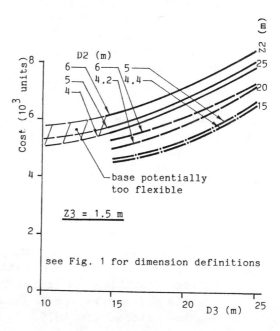

Fig. 4

Unit costs of frustrum type structure

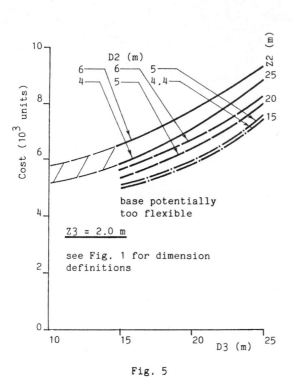

Fig. 5

Unit costs of frustrum type structure

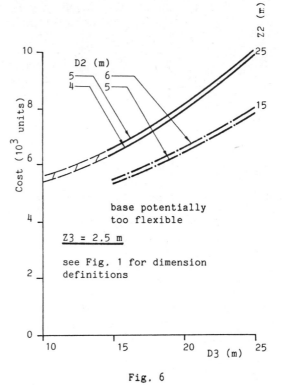

Fig. 6

Unit costs of frustrum type structure

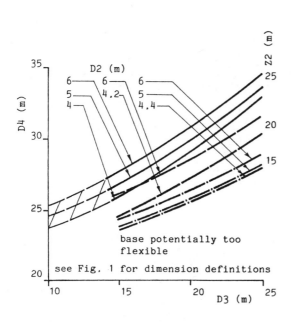

Fig. 7

Variations in main dimensions for
frustrum type structure

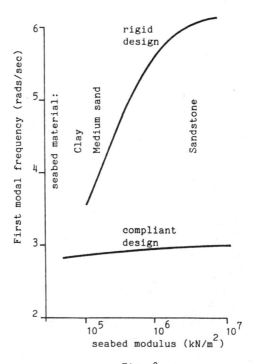

Fig. 8

Effect of seabed modulus upon
first modal frequency

EXPERIMENTAL AND THEORETICAL PERFORMANCE ANALYSIS
OF THE TORNADO CONCENTRATOR

F. HAERS and E. DICK
Department of Machinery
State University of Ghent
9000 Gent, Belgium

Abstract

The tornado type wind energy concentrator was introduced by J.T. Yen [1].
It consists of a large tower which collects the wind to form a vortex.
The low pressure core in the vortex sucks the air through the turbine in
the bottom of the tower.

In this paper the results of wind tunnel tests on logarithmic spirally
shaped towers are presented. The influence of several tower parameters
is discussed. Optimal values are indicated. Experimental model laws are
derived. In combination with theoretical flow models, a performance prediction in function of Reynoldsnumber for large scale systems can be formulated.

Flow field in the tornado concentrator

The wind tunnel models are logarithmic spirally shaped (fig. 1).

$$r = r_o e^{\alpha\theta} \quad (r_o = 100 \text{ mm}; \ \alpha = 0.1; \ 0 \leq \theta \leq 2\pi) \qquad (1)$$

Results for models of type (1) with closed bottom walls have been reported by Windrich et al. [2]. The present study concentrates on the influence of the turbine flow in the tower. The pressure drop across the turbine is simulated by a screen. The influence of tower aspect ratio (H/d_o),
turbine diameter (d_1/d_o) and turbine load (screen porosity a) is analysed.

Figure 1 shows total and static pressure, tangential and axial velocity
versus radius for several locations (z) above the tower bottom wall. A
large screen porosity (a) has been selected to obtain a large difference
with the axisymmetrical curves, obtained in [2], for towers with closed
bottom walls.

The incoming turbine flow perturbs the central part of the tangential velocity curve for z = 15 mm. This shows that the mixing of the turbine-
and the tower flows has not yet been completed. For higher values of z,
nearly symmetrical profiles are obtained, indicating a completion of the
mixing process. Near the tower top axial symmetry is lost again because
of the interaction with the top stream above the tower.

The static pressure curves reveal an adverse pressure gradient along the
vortex axis for the ascending flow. This explains the change in profile
shape for the axial velocity with increasing height above the tower
bottom.

The total pressure above the bottom opening is higher than the total pressure in the immediate surroundings of the opening. This is due to the incoming turbine flow with high total pressure, due to the high screen porosity and the corresponding low pressure drop. For higher values of z, where the mixing of tower- and turbine flows has been completed, the profile shape is axisymmetrical again.

By increasing the turbine diameter, less symmetrical profiles are obtained and the maximum axial velocity decreases.

By changing the tower aspect ratio (H/d_O) pressure- and velocity profiles of unchanged shape are obtained, with different total pressure drop in the vortex core and different maximum tangential velocity. For low tower aspect ratios, air is sucked in the vortex core at the tower top too. This return flow vanishes for aspect ratios higher than 2.5. The occurence of return flow can be understood by comparing the axial flow with an axisymmetrical boundary layer with adverse pressure gradient. The pressure gradient along the vortex axis is due to the diffusion of the axial flow. For large pressure gradients, separation occurs. This happens for low aspect ratios ($H/d_O < 2.5$). For $H/d_O > 2.5$ the flow can accept a higher diffusion since no return flow is observed.
The maximum diffusion coefficient :

$$\left((p_{s,min,top} - p_{s,min,bottom})/\frac{1}{2} \rho v_\infty^2\right)/(H/d_O) \qquad (2)$$

is reached when near the tower top just one point of zero axial velocity occurs (i.e. onset of return flow). This happens for $H/d_O = 2.5$. This value of tower aspect ratio is found to be optimum, independent of other system parameters.

As turbine load is decreased (screen porosity in increased), the profiles of pressure and velocity become less symmetrical. Hence a high turbine load is required to obtain good axisymmetrical working conditions for the turbine.

Power coefficient C_p

The power coefficient is defined as the ratio of the useful power to the power available in an area equal to tower frontal area (= $H.d_O$). The useful power is calculated as :

$$P = \int_{screen} \Delta p_{t,screen} . w_1(r,\theta).2\pi r \, dr \qquad (3)$$

Figure 2 shows the power coefficient in function of the system parameters. This figure demonstrates the importance of the turbine load. The maximum power coefficient is reached for tower aspect ratio : $H/d_O = 2.5$, turbine diameter : $d_1/d_O = .7$, and screen porosity : a = .2 (or turbine load : $\overline{\Delta p_t}/(1/2 \, \rho w_1^2) = 16$).

Experimental model laws

Figure 3 shows the influence of screen porosity and tower aspect ratio on the turbine flow, characterized by the mean axial velocity through the bottom opening (w_1). It is found that the flow rate curves are similar for different H/d_O. The turbine mass flow shows a maximum for $H/d_O = 2.5$, corresponding to optimum diffusion coefficient.

Since the turbine load is simulated by a screen, the pressure drop is proportional to the square of the velocity and optimum turbine flow rate corresponds to optimum extractable power.

Changing the turbine diameter with a fixed H/d_o results again in similar curves.

Figure 4 shows the axial force coefficient ($C_{D,ax}$) versus flow rate (w_1/v_∞). The axial force coefficient is a mean total pressure drop across the turbine (screen), defined as :

$$C_{D,ax} = C_{P,T}/(w_1/v_\infty) = \overline{\Delta p}_{t,screen}/(1/2\ \rho v_\infty^2)$$

with $C_{P,T}$ the power coefficient refering to turbine disk area. This figure reveals that the maximum mean axial velocity attains only about 60 % of the free stream velocity. This should be increased by a suitable designed turbine inlet.

Theoretical flow model

Figure 5 shows the pressure- and tangential velocity profiles for a = .23 (optimum porosity) at height z = 15 mm. The core region of the vortex can be considered to have solid body rotation and the outer region of the vortex can be modelled by a free vortex flow (v.r = K). The pressure distribution is well represented by simple radial equilibrium in the outer part of the vortex :

$$\frac{1}{\rho}\frac{dp}{dr} = \frac{v^2}{r} = \frac{K^2}{r^3}\tag{4}$$

In the core region the pressure can be considered to be constant. By integration of (4), this pressure is given by :

$$\frac{1}{\rho}(p_{s,\infty} - p_c) = \frac{v_\infty^2}{2}\left[\left(\frac{R}{r_c}\right)^2 - 1\right]\tag{5}$$

Figure 6 shows the different flows in the system. The frontal incoming air which is collected in the core is called the primary mass flow ($\dot{m}_p$). The turbine flow is called the secondary mass flow ($\dot{m}_s$) and the flow entering the free vortex region is called the tertiary mass flow ($\dot{m}_t$). Section (2) is a section in which complete mixing of primary and secondary flows has just been completed.

The energy exchange between the flows and the turbine can be calculated as the difference in energy flux between a position far upstream of the tower and section (2) :

$$P = \rho w_2 \pi r_{c2}^2 \left(\frac{p_{s,\infty}}{\rho} + \frac{1}{2}v_\infty^2\right) - \int_0^{r_{c2}} \left(\frac{p_{c2}}{\rho} + \frac{1}{2}v_2^2 + \frac{1}{2}w_2^2\right)\rho w_2 2\pi r\ dr\tag{6}$$

The integral can be restricted to the core region since the free vortex flow in the outer region has a stagnation enthalpy which is equal to the free stream value.

Combination of (5) and (6) gives :

$$P = \rho \pi r_c^2 w_2 \left(\frac{1}{4}v_m^2 - \frac{1}{2}w_2^2\right)\tag{7}$$

in which v_m is the maximum tangential velocity at r_{C2}.

When no power is extracted by the turbine, it follows from (7) that :

$$w_2 = w_{2o} = v_m/\sqrt{2} \tag{8}$$

(7) gives :

$$C_P = \frac{P}{H.d_o \frac{1}{2} \rho v_\infty^3} = \frac{\pi}{8\sqrt{2}} \left(\frac{v_m}{v_\infty}\right) \left(\frac{d_o}{H}\right) \left(\frac{w_2}{w_{2o}}\right) \left[1 - \left(\frac{w_2}{w_{2o}}\right)^2\right] \tag{9}$$

Maximum power coefficient is reached for : $w_2/w_{2o} = 1/\sqrt{3}$ (10)

For $H/d_o = 2.5$, (9) gives : $C_{P,max} = .04275 \ (v_m/v_\infty)$ (11)

For $H/d_o = 2.5$, $d_1/d_o = .5$ and $a = .23$ (optimum geometry) : $v_m/v_\infty = 1.32$ and (11) gives : $C_{P,max} = .06237$. The theoretically predicted value is in good agreement with the experimental value $C_P = .087$, but is underpredicting it. This is due to two reasons. Firstly, the value of v_m/v_∞ to be inserted into (11) is not the experimental value, but a theoretical peak value (fig. 5). Secondly, in calculating the power by (3) local $\Delta p_{t,screen}$ values and w_1 values are used. Hence the calculated power is higher than the value that would be obtained using mean values of $\Delta p_{t,screen}$ and w_1. Based on mean values, the optimum value of C_P is 4.6 %. This value should be compared with the theoretical result from (11).

Prediction of the power coefficient for large scale systems

Formula (11) shows that the power coefficient of a system is proportional to the maximum tangential velocity that can be maintained in the tower.

Since in absence of viscosity the tangential velocity profile would be a free vortex profile, the maximum tangential velocity is viscosity dependent. The experimental velocity profile can be represented by :

$$v = \frac{k_1}{r}\left[1 - \exp(- Re.r^2/2R^2)\right] \tag{12}$$

Hence :

$$v_{max} = .4513 \ k_1 \sqrt{Re}/R \tag{13}$$

For the static pressure distribution a consistent formulation is used :

$$\frac{p_{t,\infty} - p_s}{\rho} = \frac{k_2^2}{2r^2}\left[1 - \exp(- Re.r^2/\alpha R^2)\right] \tag{14}$$

Both equations are connected through the condition of radial equilibrium, in the outer region (4). When radial equilibrium is also used in the core region, by integration from $r = 0$ to infinity, it is found that $\alpha \cong \sqrt{2}$.

Three similar logarithmic spirally shaped towers ($d_o = .258$ m, .326 m and .55 m) with a tower aspect ratio 2.5, were tested for several windspeeds. Each of these towers were tested in a wind tunnel of suitable size. In order to obtain similar testing conditions, the frontal inlet opening of the tower was connected directly to the exit section of the wind tunnel. The tests were effected on towers with a closed bottom wall. This is allowable since the viscous effects are identical for towers with and without opening. The velocity and pressure distributions were measured for all cases just above the tower bottom wall ($z = 15$ mm).

The parameters k_1, k_2, Re and α are obtained by a least square fit for all studied configurations. The results show that k_1 and k_2 are equal within a few percent for all cases. The value of α is different from the theoretical value $\sqrt{2}$. This shows that the assumption of radial equilibrium in the vortex core is not valid. This is illustrated in figure 5. Figure 7 shows the obtained values of Re for all studied cases as a function of the test Reynoldsnumber :

$$Re_{test} = d_o \cdot v_{ref}/\nu \tag{15}$$

Since the velocity has no influence on the value of the fitted Reynoldsnumber except for very low values of the test Reynoldsnumber, the fitted Reynoldsnumber can be written as a function of d_o only :

$$Re_{fit} = a \cdot (d_o)^b \tag{16}$$

A unique set of values for a and b cannot be extracted from the test results. The extreme possible sets are : a = 14, b = .4 and a = 12, b = .25.

By (11) and (12) it is seen that the maximum power is proportional to $\sqrt{Re_{fit}}$. Hence, taking into account formula (16), for large towers (e.g. d_o = 50 m) the available power can increase by a factor 2 to 3.

In figure 4 it is seen that in unloaded conditions the w_{1o} is much smaller that v_∞. Without losses this value would be :

$$w_1^2/2 = v_\infty^2/2 + (p_{s,\infty} - p_c)/\rho \cong 1.5\ v_\infty^2 \tag{17}$$

or $w_1 \cong 1.7\ v_\infty$.

This shows that by a careful design of turbine inlet and by decreasing the losses due to the mixing of primary and secondary flows, the mass flow to the turbine can be approximately doubled.

Based on these results, an estimate of the maximum obtainable value of C_P for a large scale system can be made :

$$C_P = .087 * 2 * (2\ to\ 3) = .35\ to\ .52 \tag{18}$$

Conclusion

By a combination of theoretical and experimental methods, an estimate of the maximum attainable power coefficient, based on tower frontal area, was made for the Tornado concentrator. The predicted value is somewhat higher than the one obtained for a classical horizontal axis wind turbine. This shows the possibilities of this innovative wind energy system.

References

1. J.T. YEN. Tornado type wind energy system. 10th Intersoc. Energy Convers. Eng. Conf., Univ. of Del., Newark, 1975.

2. J. WINDRICH, B. HENZE, J. FRICKE. Proceedings of the 4th International Colloquium on Wind Energy, Brighton, U.K., August 1981.

Acknowledgements

This research is supported by the Belgian Institute for the Encouragement of Scientific Research in Industry and Agriculture (I.W.O.N.L.-I.R.S.I.A.)

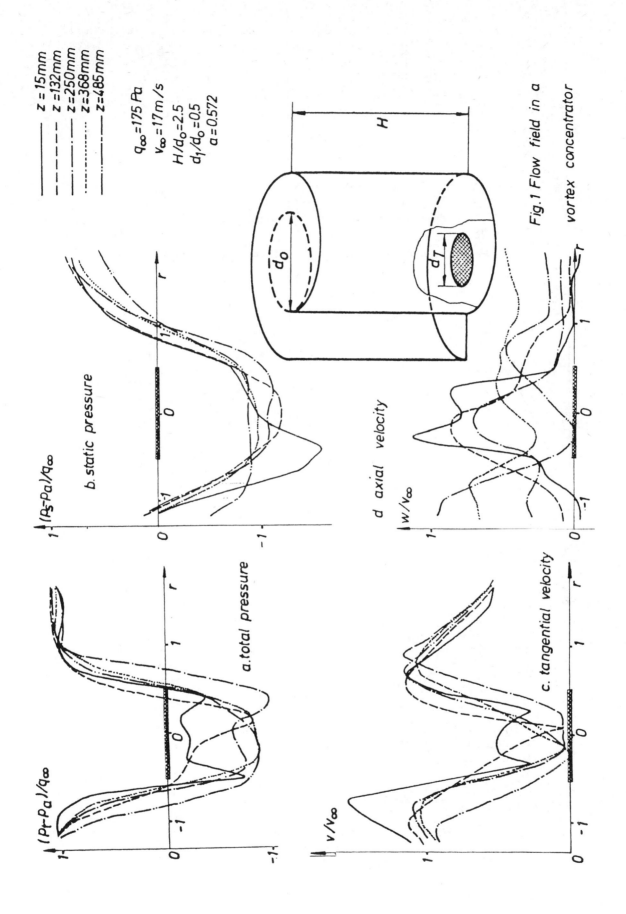

Fig.1 Flow field in a vortex concentrator

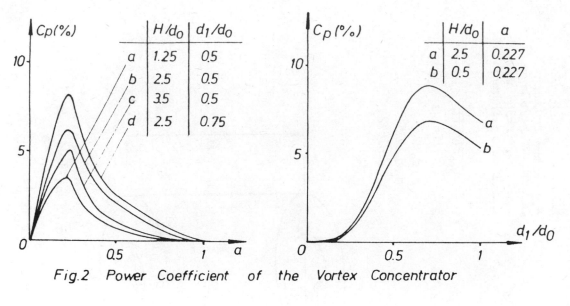

Fig.2 Power Coefficient of the Vortex Concentrator

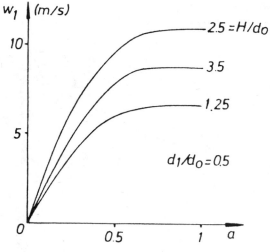

Fig.3 Secondary Flow Rate

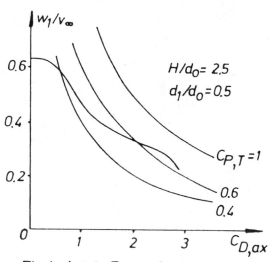

Fig.4 Axial Force Coefficient

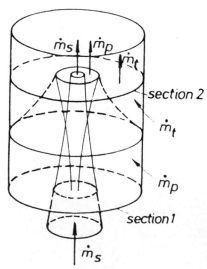

Fig.6 Important Mass Flows

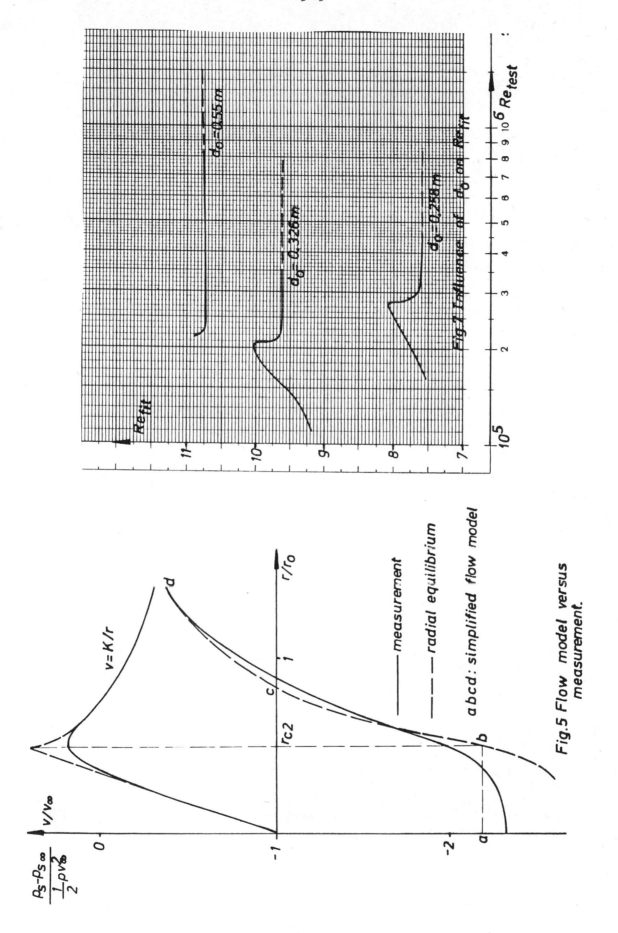

Fig.7 Influence of d_0 on Re_{krit} :

Fig.5 Flow model versus measurement.

OPTIMISING THE DESIGNS OF FLEXIBLE-SAIL WIND-TURBINES

P.D. Fleming and S.D. Probert
Applied Energy Group,
School of Mechanical Engineering,
Cranfield Institute of Technology,
BEDFORD MK43 0AL

Abstract

The advantages of sail-type wind-turbines are outlined and methods of augmenting their power outputs described. Data are presented for power augmented horizontal-axis and vertical-axis wind-turbines.

Sail-type wind-turbines

These are lighter in weight than the rigid, multi-vane, horizontal-axis and the metal-vaned, vertical-axis types of wind-turbine. They are cheaper to construct and can be readily fabricated by semi-skilled labour using a variety of indigenous skills and materials in Third World countries.

In addition, flexible sail wind-turbines possess the following major attributes;-

1. self-starting capabilities: a relatively high torque is produced at low rotational speeds;

2. useful power is harnessed even from very low speed winds; and

3. the sails provide a degree of self-regulation — when used with elastic sail attachments, they "spill the wind" at high speeds, so only needing to be furled during infrequent, excessively-high winds.

Flexible sail wind-turbines are inherently slow running (with optimal tip-to-wind speed ratios of approximately unity). They have large solidities and high starting torques, so readily overcoming the relatively large torques required to start the majority of irrigation pumps.

Small wind-driven water pumps have been used sucessfully in many rural areas of the world. They continue to operate to this day in those regions where they have not been superceeded by low-cost electric or diesel pumps (1). As there are many ($\sim 10^6$) such wind-driven pumps still in use around the world, this implies that the economically-viable potential market for them is very large — may be as high as one hundred million systems globaly (2).

HORIZONTAL-AXIS WIND-TURBINE

This 0.64m diameter wind-rotor was based upon a standard cycle wheel, so enabling the tip-fins to be rigidly mounted by attaching them to the rim. Rotating or stationary hub-fairings could also be easily positioned on the wind-turbine.

The shroud

The shrouded wind-turbine — a rotor surrounded by what is essentially a cylindrical circular wing that directs an aerodynamic "lift" force radially inwards — is capable of at least trebling the power output of a wind-turbine (3). As a reaction to the "lift" force directed towards the shroud centre-line, a cross-wind force acts upon the air, so deflecting the flow outwards, causing its streamtube to widen downwind of the rotor. Previous research (4) showed that the shroud has a considerable power augmentaion effect on sail-type wind-turbines. However it is relatively expensive, so inhibiting its future widespread incorporation in inexpensive wind-turbine systems. Thus cheaper concepts for augmenting wind-turbine power ouputs have been investigated.

Tip-fins

The theory predicting the maximum power coefficient for wind-turbines (i.e. the so-called Betz limit), is only valid for ideal rotors applying a steady axial force to the air stream. Real wind-turbines, that produce radial, as well as axial forces are not accounted for by the Betz analysis and they may have much greater power outputs than this one-dimensional theory suggests.

Van Holten (5), from data obtained with a simulated propeller-type wind-turbine, showed that the diffusion ratio is dependent upon the mean radial force so exerted by the wind per unit length of the circumference of the rotor plane. This suggests that to achieve diffusion of the flow, a full shroud may not be required. Replacing the shroud by tip-fins (which are also referred to as tip-vanes) concentrates the cross-wind forces on the aerodynamic surfaces near the circular periphery of the wind-turbine. This arrangement, shown in Fig. 1, causes a deflection of the airstream in a manner analogous to that occurring in shrouded wind-turbines.

The power enhancement is greater for a shrouded rotor than for a rotor with tip-fin augmentors. This occurs because the drag forces on the shroud do not adversely affect the power output of the wind-turbine. However, with the tip-fin augmented wind-turbine, induced and viscous drag forces ensure that the power augmentation capability is less than that achievable with the shrouded wind-turbine.

By utilising relatively inexpensive aerodynamically-shaped tip-fins, the induced drag can be reduced substantialy. As the cross-wind forces do no work on the fluid, no energy is lost to establish a a force field, and so no torque is required to drive the tip-fin. This implies that they need not experience induced drag. But the drag is also dependent upon tip-fin shape, as well as upon the "tilt" angle. The tested tip-fins are shown in Fig. 2.

Centre-bodies

The centre-body, shown in Fig. 3, consists of a hub-fairing and an after-body. The hub-fairing generally rotates (with the rotor) whereas the after-body is usually stationary and encloses the power take-off equipment. The centre-body's intended function is to interfere favourably with the wind passing through the rotor, thereby increasing the speed of air impingement (and hence the couple imposed) upon the sails, which are located predominantly near the outer rim of the rotor.

If a streamlined body is introduced into a uniform air flow, the average local velocity of the stream is increased as the air passes over the body. Thus placing a steamlined hub-fairing symmetrically about the central axis of a wind-turbine should result in greater air pressures being experienced on the "working sections" of the rotor sails. That is, the fairing deflects air that would normally flow through the "hub-section" of the rotor, outwards onto the sails, i.e. to where such momentum transfer should have a greater effect in producing rotation. Thus a centre-body should increase the mean mass flow rate of air through the sails of the wind-turbine, so possibly enhancing the amount of wind energy harnessed by the sails.

However, theoretical analyses of rigid (or almost rigid) propeller-type wind-turbines (6,7) suggest that there is little advantage to be gained by using centre-bodies (e.g. a spinner and a stationary after-body) to augment the power outputs. The major reasons for this assertion are alleged to be:

1. the hub-fairing increases the the local axial component of relative velocity; and

2. the frictional drag produced by the hub-fairing decreases the local tangential component of relative velocity.

Both these component changes tend to increase the local aerofoil angle of attack beyond the stall angle, so affecting adversely the wind-turbine's power harnessing capabilities. The frictional effect is believed to be less than that due to the increase in the local axial relative velocity component.

However, previous experimental investigations have not resulted in unanimous conclusions. During the nineteen-thirties, a team from New York University (8) working on the "Smith-Putnam" wind-turbine, concluded that a hub-fairing was disadvantageous. More recently Blaha (9) obtained considerable power increases by incorporating centre-bodies on a prototype wind-turbine, but there was some doubt as to whether his pocedure would "scale-up" for larger turbines.

Tip-fins and centre-bodies combined

Because air is deflected by the hub-fairing towards the "working section" of the sail, (which is near the periphery of the rotor) there will be a corresponding increase in rotor tip losses. These losses depend directly upon the sail loading (i.e. the magnitude of the pressure differential) and are expected to increase with hub-fairing diameter.

The tip losses can be reduced by employing tip-fins, so it is probable that centre-bodies will have a greater effect on wind-turbine performance if used in combination with appropriately designed tip-fins.

VERTICAL-AXIS WIND-TURBINE

Sail-type Savonius wind-turbine

In an attempt to improve the performance of the Savonius-type wind-turbine, sails rather than rigid vanes were proposed (10). The sail-type wind-rotor differs from the those previously investigated in India (11) and the Netherlands (12) in that the flexible sail rotor is designed so that the sails change their profiles accoring to their positions relative to the wind.

The employment of such sails would both:-

1. reduce the overall weight of the wind-rotor, and

2. when used with support rods, enable the sails to take up more favourable shapes, dictated by their position relative to the wind direction.

Support rods could be useful to define the sail profile as it rotates "against the wind", whereas when rotating "with the wind" the sail would "billow out" beyond the sail support rods, so allowing the wind to shape the sail pofile. Thus the sails would vary their profiles as the rotor rotates, so enabling greater rotational speeds to be attained and thereby harnessing more power from the wind than the conventional Savonius-type wind-turbine.

EXPERIMENTAL OBSERVATIONS

Tests to determine the optimal tip-fin number and position for the nine sail horizontal-axis wind-turbine, showed that nine tip-fins (i.e. one per sail) in position I (see Fig. 1) produced the greatest power augmentation. Of those systems tested, the wind-turbine augmented with tip-fin type F as in Fig. 2, produced the greatest power enhancement (see Fig. 4). Future tests will determine the optimal tip-fin design — a compromise between increased lift forces producing greater rotational speeds and increased drag forces tending to reduce the rotor's rotational speed — by varying the tilt angle and the size of the fins. Three hub-fairings and two after-bodies were tested on the horizontal-axis wind-turbine as shown in Fig. 3. Stationary hub-fairings resulted in greater power increases than rotating ones, whereas all three fairings gave virtually the same power augmentation (see Fig. 5). Preliminary tests employing both tip-fins and hub-fairings on the horizontal-axis wind-turbine show significant power increasess compared with those for a wind-turbine augmented by tip-fins only. This is currently being further investigated at Cranfield.

Data from the three-sail (0.58 m diameter by 0.58 m high) vertical-axis Savonius-type wind-turbine with a taut sail show that a relatively deep sail profile, (profile 3 in Fig. 6) gives the greatest power outputs — see Fig. 7. Initial tests on the variable sail profile concept gave relatively poor power outputs. It is suggested that the use of a deflector, see Fig. 8, will significantly increase the power harnessing capabilities. The deflector would enable smaller sails to be used and so reduce the friction between the relatively large sails and the wind-rotor end plates (see Fig. 8). This would then alow the rotor to attain higher rotational speeds and hence harness greater ammounts of the wind's energy.

Conclusions

The use of cheap, flat plate tip-fins is a more cost-effective method of augmenting the power outputs of inexpensive horizontal-axis sail-type wind-turbines, than employing a cylindrical shroud. Rotating hub-fairings are capable of augmenting the power outputs of such sail-type wind-turbines. Holding the hub-fairing stationary results in even greater power outputs. An after-body on its own (i.e. without a hub-fairing) gives approximately the same power harnessing capabilities as a conventional centre-body configuration. The combination of centre-bodies and tip-fins on small horizontal-axis sail-type wind-turbines results in even greater power outputs than employing tip-fins alone.

The variable sail profile vertical-axis Savonius-type wind-turbine deserves further investigation, with deflectors, to assess rigorously the system's power harnessing capabilities.

Acknowledgement

The authors wish to thank the Science and Engineering Reseach Council for supporting this project.

References

1. A. Wyatt, Report surveys potential for wind energy in Africa, VITA News, pp 17-18, July 1981.

2. Report of the Technical Panel on Wind Energy, Second Session, A/CONF/100/PC/24, UNCNRSE, United Nations, New York, 1981.

3. O. Igra, Research and development for shrouded wind-turbines, Energy Conservation and Management, Vol. 21, No. 1, pp 13-48, 1981.

4. P.D. Fleming and S.D. Probert, Design and performance of a small shrouded Cretan windwheel, Applied Energy, Vol. 10, pp 121-139, 1982.

5. Th. van Holten, Concentrator systems for wind energy, with emphasis on tipvanes, Wind Engineering, Vol. 5, No. 1, pp 29-45, 1981

6. R.E. Wilson, The effect of hub-fairings on wind-turbine rotor performance, Journal of Fluids Engineering, Vol. 100, No. 1, pp 120-2, 1978.

7. R.T. Griffiths, Centre-body effects on horizontal-axis wind-turbines, Applied Energy, Vol. 13, pp 183-194, 1983.

8. New York University, College of Engineering, Wind-turbine project WPB 144, War Production Board, Washington D.C. 1946.

9. R. Blaha, The effect of a centre-body on axial flow windmill performance, AMS Report No. 1266, Princeton University, March 1976.

10. P.D. Fleming and S.D. Probert, Proposed three sail Savonius type wind-rotor, Applied Energy, Vol. 12, pp 327-31, 1982.

11. S.P. Govida Raju and R. Narasimha, A low cost water pumping windmill using a sail-type Savonius rotor, Proc. Indian Accademy of Science, Vol. C2, Pt. 1, pp 67-82, 1979.

12. E.H. Lysen, H.G. Bos and E.H. Cordes, Savonius rotors for water pumping, Steering Committee for Wind Energy in Developing Countries, Amersfoort, The Netherlands, 1978.

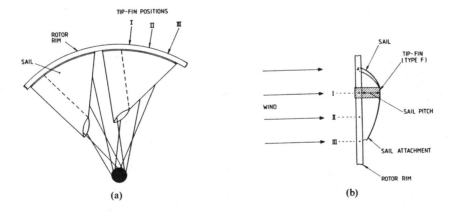

Fig. 1 (a) Rotor segment showing tip-fin position, (b) Tip-fin positions tested, and sail profile; for the horizontal-axis wind-turbine.

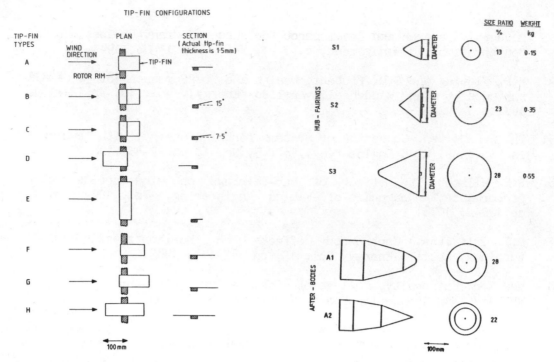

Fig. 2 Tested tip-fins for the horizontal-axis wind-turbine.

Fig. 3 Dimensions of hub-fairings and after-bodies for the horizontal-axis wind-turbine.

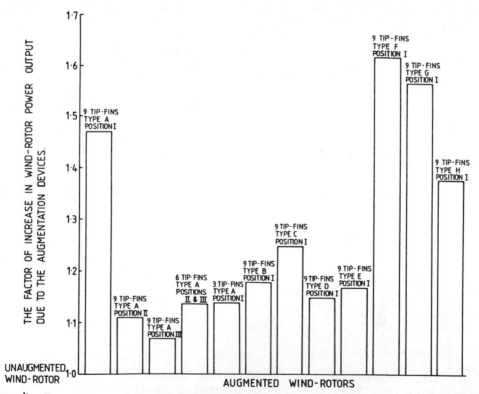

Fig. 4 Increases in the maximum power outputs for the augmented horizontal-axis wind-turbines, relative to that of an unaugmented horizontal-axis wind-turbine.

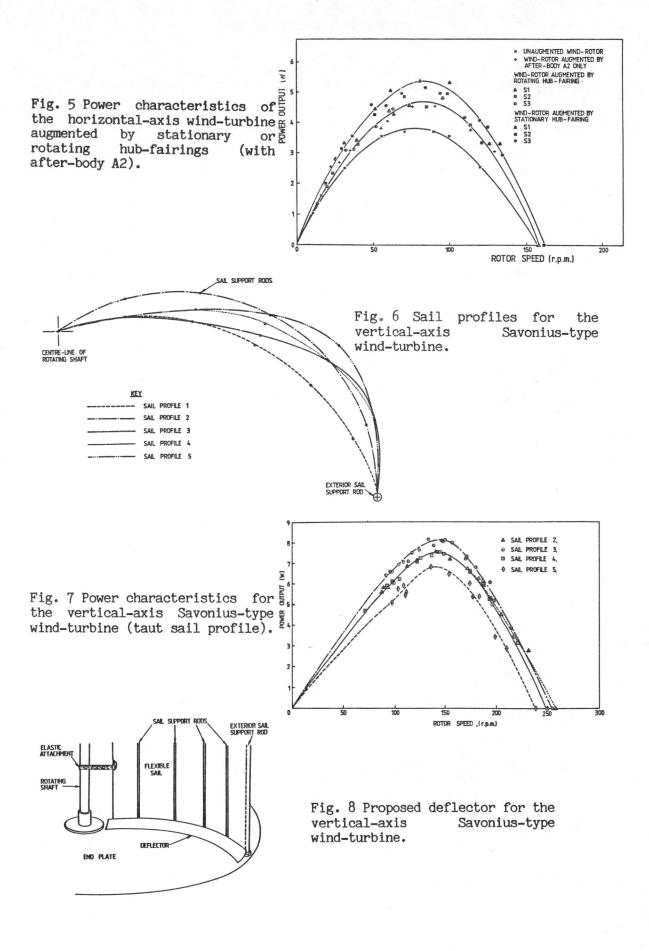

Fig. 5 Power characteristics of the horizontal-axis wind-turbine augmented by stationary or rotating hub-fairings (with after-body A2).

Fig. 6 Sail profiles for the vertical-axis Savonius-type wind-turbine.

Fig. 7 Power characteristics for the vertical-axis Savonius-type wind-turbine (taut sail profile).

Fig. 8 Proposed deflector for the vertical-axis Savonius-type wind-turbine.

WIND TURBINE RUNAWAY SPEEDS

D.J. Milborrow

CEGB - Central Electricity Research Laboratories, Leatherhead, U.K.

Abstract

There have been several incidents involving wind turbines becoming disconnected from a load source and "running away", the most recent involving the US MOD-2 machine. In order to quantify limiting tip speeds a rigorous aerodynamic analysis has been carried out for this machine, leading to the conclusion that it could never reach tip speeds higher than 276 m/s (Mach 0.82). It is shown, however, that the limit is a function of wind speed and falls to 176 m/s at 9 m/s.

The analysis makes use of transonic aerofoil data for the blade sections and shows that the rotor speed limit is reached when a negative torque contribution from the outer portion of the blade is balanced by the positive contribution from the inner portion.

The concept of a centre of pressure is used to examine "ultimate" limits, which are shown to be a function of blade aerofoil section, design lift coefficient and rotor solidity. An absolute limit of 313 m/s (M = 0.93) is derived, covering all large power-generating wind turbines of conventional design. However, most modern rotors, designed to each maximum efficiency at tip-speed ratios around 10, would not be able to reach these limiting speeds below windspeeds of at least 30 m/s.

Nomenclature

B1	Inlet relative flow angle, $\tan^{-1}(F/\Omega r)$, deg. or rad.
c	Blade chord, m
C_D	Drag coefficient
C_L	Lift coefficient
D	Drag force, N
F	Free stream wind velocity, m/s
L	Lift force, N
M	Mach Number
r	Radius, m: velocity ratio
R	Radius of rotor, m
t	Torque-producing force on rotor blade at given radius, N
z	Number of blades
α	Aerofoil angle of attack, deg.
β	Blade angle at given radius, deg.
λ	Tip speed/wind speed ratio, $\Omega R/F$
ρ	Density of air
σ	Solidity, $zc/2\pi r$
ϕ	Relative flow angle at blade, rad. or deg.
Ω	Rotational speed, rad/s

INTRODUCTION

The accident which befell the American MOD-2 wind turbine in 1981, when the rotor accelerated from 17 to 29 r.p.m., highlighted the propensity of rotors to "run-away" should they become disconnected from a load source. The MOD-2 rotor oversped by some 70%, reaching a tip speed around 140 m/s, before damage to the generator, and consequential vibration, activated safety systems which should have prevented the occurrence (ref. 1).

The implications of a rotor runaway have received little attention. Although it is known that high speeds can be attained, no attempts have been made to quantify the limits. This may be due to the complexity of the calculations, which are not warranted for small machines. Large machines, on the other hand, normally incorporate a number of safety systems which should act to control a runaway, so that the likelihood of such an occurrence is small. Nevertheless, an appreciation of likely limiting speeds is needed, so that the rotor integrity can be assessed in the event of a runaway, and also so that the likely throw distances of fragments shed from the blades can be calculated.

OBJECTIVES AND METHOD

The objective of this study was to determine the maximum speed of a wind turbine in the runaway condition and define its relationship to windspeed and other appropriate variables.

The complexity of transonic aerofoil data, with wide variations between similar sections, as shown in Fig. 1, clearly ruled out a generalised analysis and it was therefore decided to examine the performance of a rotor with known geometry and for which both subsonic and transonic aerofoil data were available. The MOD-2 rotor (ref. 2) met these criteria and its rotor power characteristic was therefore derived for a number of tip speeds in the transonic region. This enabled the maximum achievable tip speed/ wind speed ratios to be derived and from these it was possible to calculate runaway speeds, as a function of wind speed.

The rotor geometry used in the analysis is set out in Table 1. Since transonic data are not available for any sections greater than 15% thick, data for the 23015 section have been used for the thicker sections. Since drag increases with thickness this has the effect of producing pessimistic estimates of runaway speeds. Estimates of aerofoil properties for the thick sections near the root which do not reach transonic speeds have been synthesised from a number of sources and, again the values used err on the side of caution.

AEROFOIL DATA

Since the aerofoil sections used for wind turbine blades were not designed for transonic use, there is a shortage of lift and drag data, particularly at high angles of incidence. A further difficulty is that both the Reynolds Number and the Mach Number of the test data should correspond to that of the wind turbine blade. At Reynolds Numbers applicable to the normal operating speeds of modern large wind turbines (typified by MOD-2) - around 8.10^6 - no transonic data are available. The transonic aerofoil data which form the main data base of this analysis (ref. 3) apply at a

Table 1

Blade Geometry for MOD-2

	Non-dimensional radius, r/R							
	0.975	0.95	0.90	0.80	0.70	0.60	0.45	0.30
Blade chord, m	1.51	1.58	1.73	2.01	2.30	2.59	3.02	3.45
Solidity, $zc/2\pi r$	0.0108	0.0116	0.0134	0.0175	0.0229	0.0301	0.0467	0.0801
Thickness (% chord)	12	14	16	20	24	25	26	28

Reynolds Number of 2×10^6 for a Mach Number of 0.27 - which roughly corresponds to the operational Mach Number of MOD-2. Strictly speaking, therefore, the analysis applies to a 1/4 scale model of MOD-2 but in practice, in the region well below stall, aerofoil characteristics do not change markedly above 2×10^6. It may be inferred that this also applies to the transonic characteristics, from a comparison of transonic data for the 23015 section at differing Reynolds Numbers, shown in Fig. 2. The data shown for the lower Reynolds Number is quoted by Graham, Niteberg and Olson in ref. 4 and, as expected, yield lower lift and higher drag coefficients, but the discrepancy is small.

The data in ref. 3 for the 230XX sections does not cover angles of incidence above 8° and it was therefore necessary to synthesise lift and drag values at higher angles using the data of Stivers (ref. 5).

AERODYNAMIC THEORY

The maximum speed of a runaway wind turbine rotor, subject to no restraint, is reached when the driving torque falls to zero. If the rotor tip region enters the transonic region - where the performance of conventional aerofoils falls off rapidly - shock waves are set up on the upper surface of the aerofoil and the lift/drag ratio falls. As a result the torque contribution from the outer portions of the blade will become negative and, with further increases of speed, the negative and positive torque contributions will eventually balance and the rotor stops accelerating.

It should be noted that conventional subsonic performance theory predicts the existence of a limiting speed even if the runaway rotor does not enter the transonic region. This is also defined by the intersection of the performance curve (power coefficient vs tip speed/wind speed ratio or $C_p - \lambda$) with the C_p axis. Published data for the MOD-2 rotor (ref. 2) show that this occurs at a tip speed/wind speed ratio around 17. It follows that in a 10 m/s wind an unrestrained rotor would accelerate to 170 m/s. This limitation arises, again, due to a balance of forces on the rotor. At high tip speed/wind speed ratios the relative flow angle in the outer portions of the blade becomes very small; consequently aerofoil angles of attack are also small. The resolved component of the lift is therefore smaller than the drag, so that the net torque in this region is negative.

The mathematical expression for the tangential forces which produce torque is identical in both subsonic and transonic cases and is given, at any radius, by:

$$dt = dL \sin \phi - dD \cos \phi$$

Since lift and drag are functions of the same aerofoil chord, air density and velocity, we note that

$$dt < 0 \quad \text{when} \quad C_L \sin \phi < C_D \cos \phi$$

$$\text{or} \quad C_L/C_D \quad < \cot \phi$$

Evaluation of the operating conditions at any given radius necessitates an iterative procedure which takes into account tip losses, rotor geometry and aerofoil characteristics. Rotor performance is then calculated by a radial integration of the elemental torques.

Calculations of aerodynamic performance for the MOD-2 rotor have been made using standard techniques (ref. 6), with the tip loss corrections tabulated by Lock, Pankhurst and Conn in ref. 7.

The application of these techniques to rotors operating in the transonic region has been discussed by Pankhurst and Haines in ref. 8. Although they concluded that their application was valid there must be some doubt that there is no interference between radial elements when shock waves are present. Nevertheless it is unlikely that the performance would be enhanced under such conditions and so, again, pessimistic estimates are liable to be produced.

Accurate estimates of the tip speed ratio where the power falls to zero require the use of empirical relationships between elemental thrust and velocity retardation, since high thrust levels are encountered in this region and the conventional momentum theory breaks down. Most recent studies rely on the work of Glauert (ref. 9) and so the following empirical equations linking thrust coefficient (C_T) and velocity ratio (r) have been fitted to his data.

(i) $C_T < 0.96$ $\quad r = (1 + \sqrt{1 - C_T})/2$ $\quad$ (standard momemtum theory)

(ii) $0.96 < C_T < 1.39$ $\quad r = 1.49 - 0.93 \, C_T$

(iii) $C_T > 1.39$ $\quad r = 1/(30(C_T - 1.22))$

(ii) Is an empirical fit to Glauert's expression (ref. 9) and (iii) fits his data up to $C_T = 1.60$. Beyond this, high thrust coefficients are associated with low velocity ratios, with no upper limit to C_T. This enables solutions to be found for the flow conditions in a region where little is known of the behaviour of rotors in practice. If in fact the air is halted or reversed at $C_T > 2$ (as suggested but not yet demonstrated) the errors will not be large since the lift vector is almost normal to the plane of rotation at high tip speed ratios.

RESULTS

The subsonic performance curve of the rotor was calculated first, for comparison with the manufacturer's data. The close agreement between the two curves, shown in Fig. 3, supports the accuracy of the subsonic aerofoil data and the computational techniques used.

Performance curves have also been derived for tip speed Mach Numbers of 0.66, 0.76, 0.80 and 0.82. These are also shown in Fig. 3. Operation at a Mach Number of 0.82 produced no positive torque at all and consequently the calculations were not extended to higher values. The peak of the performance curve at a Mach Number of 0.82 is just below zero (at $C_p = -0.008$); since aerofoil characteristics change rapidly at these speeds this may be taken as the maximum possible runaway speed for the MOD-2 rotor, with only a small error margin.

The tip speed ratios at which the performance curves fall to zero are as follows:

Subsonic $\quad \lambda = 17.5$:

$M = 0.66, \quad \lambda = 15.2$: $M = 0.76, \quad \lambda = 13.1$

$M = 0.80, \quad \lambda = 9.4$: $M = 0.82, \quad \lambda = 7.5$

There is a slight discrepancy with the manufacturer's curve for subsonic runaway, which falls to zero at $\lambda = 16.9$. The upper limit of validity of the subsonic curve, at this tip speed ratio, was determined as follows. The angle of attack in the tip region of the blade was established, from the performance prediction program, as 4°. Reference to the transonic aerofoil data showed that, at this angle of attack, subsonic lift and drag values are maintained up to M = 0.55. This enabled the complete relationship between windspeed and runaway speed to be established, which is shown in Fig. 4. It is possible that the runaway speed could fall off from its peak value of 276 m/s at winds above 40 m/s. Estimates cannot be derived, however, due to the total lack of aerofoil data in the stalled region needed to complete the performance curves at the low tip speed ratio end. Fortunately, such high windspeeds occur very infrequently.

BEHAVIOUR OF OTHER ROTORS

A generalised analysis, as noted earlier, cannot be carried out, on account of the wide variations in rotor geometry and transonic aerofoil characteristics. However, an examination of possible variations in the limiting speed can be made by using the concept of a "centre of pressure" for the torque-wise forces on a rotor blade. This procedure has been used for propellers (ref. 10); 70% radius was suggested as an appropriate section but the data of ref. 11 implied this was too small. By plotting ($C_L \sin \phi - C_D \cos \phi$) against radius for the MOD-2 rotor at M = 0.82 and the peak of the performance curve, where $C_p \simeq 0$, shows the centre of pressure to be at 0.76R (Fig. 5). At this radius Fig. 5 shows that the lift coefficient is equal to its design value, at the peak of the subsonic performance curve. Since the design value is always close to the value at which the lift/drag ratio is a maximum, usually around $C_L = 0.80$, this provides a basis for estimates of the runaway speeds of other rotors with alternative aerofoil sections and/or design tip/speed wind speed ratios.

Alternative Aerofoil Sections

The review of transonic aerofoil data had indicated that use of the 65_2-215 section was likely to result in a higher runaway limit since the drag is lower than that of the 23015 section above M = 0.57.

Using data for the 65_2-215 aerofoil yields an estimate of M = 0.69 for zero tangential force at 76% radius, which implies a tip speed Mach Number of 0.91. This value is probably pessimistic since drag increases very sharply above M = 0.8 which, coupled with further loss of lift, would probably result in the centre of pressure moving further out.

Alternative Rotor Designs

The chief aerodynamic parameters which influence rotor design (ref. 12) are design lift coefficient and tip speed/wind speed ratio. Variations in design C_L cover a narrow band between 0.8 and 1.0; the lower value will continue to be used so as to derive upper limits for runaway speeds. A change of design tip speed/wind speed ratio alters blade solidity and blade angle. Estimates of runaway speeds have been made for design tip speed ratios of 6 and 16, for both 23015 and 65_2-215 aerofoil sections and these are shown in Table 2.

Windspeeds at Limiting Runaway Speed

In order to establish the minimum windspeed at which the maximum runaway speed can be reached a simplified analysis can also be used.

For small angles the fundamental equation relating the relative flow angles, neglecting tip losses, can be written:

$$\phi = B1 - \tan^{-1} \sigma C_L/4 \sin \phi$$

substituting $\phi = \tan \phi = C_D/C_L$ and rearranging gives:

$$B1 = C_D/C_L + \sigma C_L^2/4C_D$$

This expression contains the blade solidity. Estimates of the solidity for the 76% radius location for $\lambda = 6$ and $\lambda = 16$ may be derived by noting that $\sigma \propto 1/\lambda^2$ (ref. 10), giving $\sigma = 0.049$ for $\lambda = 6$ and $\sigma = 0.0069$ for $\lambda = 16$. This enables tip speed/wind speed ratios to be derived from which windspeeds can be calculated.

The results are summarised in Table 2, which includes all the estimates calculated in this study, assuming the velocity of sound to be 337 m/s at 10°C. The data show that "high speed" rotors would reach their limiting tip speed at much lower wind speeds, which tends to outweigh any small advantage associated with the lower limits; the wind speeds necessary to bring MOD-2 and low tip speed rotors to their limit rarely occur in most locations.

Table 2

Maximum Runaway Speeds and Corresponding Windspeeds

Design tip speed ratio, λ	Limiting Speed m/s			Minimum Windspeed for Limiting Speed (m/s)		
	6	9.4 (MOD-2)	16	6	9.4	16
Aerofoil						
23015 (MOD-2)	289	276	273	50	37	17
65_2-215	313	307	269	47	32	19

CONCLUSIONS

This analysis has shown that the limiting speed reached by a wind turbine, in the absence of any restraints, is governed by the deterioration of aerofoil characteristics in the transonic region. As there are considerable variations in these characteristics a rigorous analysis has been based on the geometry of MOD-2, yielding a maximum possible runaway speed of Mach 0.82 - in a wind of 37 m/s. The variation of runaway speed with windspeed has also been derived; at the site mean wind for which MOD-2 was designed (9 m/s) the runaway tip speed is 153 m/s.

Despite the wide variations in transonic aerofoil characteristics, most sections exhibit sharp rises of drag in the region $M = 0.8$ to 0.9 and an estimate of the limiting runaway speed for a section with higher lift/drag ratios yielded a runaway limit of $M = 0.91$. A slightly higher limit would be obtained with a rotor of higher solidity - designed for a lower tip speed/wind speed ratio. High tip speed ratio designs ($\lambda = 16$), on the other hand, have slightly lower limiting speeds but they are reached at lower wind speeds (typically 17-19 m/s).

Although transonic aerofoil data are inherently less accurate than those applicable to the subsonic regime, the very rapid increases of drag which occur in the region above $M = 0.7$ ensure that the effect of erosion the estimates of runaway speeds is small. The estimates which have been derived err on the side of caution, i.e. overestimating runaway speeds rather than the reverse.

These data provide a basis for "worst case" estimates of throw distances for particles which become detached from wind turbine blades. It may be noted that the limiting speeds quoted - typically three times the design tip speed - will not necessarily cause fracture of the rotor due to excessive centrifugal stresses. Although these will be increased by a factor of 9 they are typically only 10-20% of the stresses which dictate rotor spar geometry (ref. 12). As in the case of the MOD-2 incident, however, other components may fail at lower speeds.

Corroboration of the estimates made in this analysis is unlikely to be obtained unless an instrumented rotor, of known geometry, runs away. Wind tunnel tests would provide a corroboration of the analytical technique but lower runaway speeds would be expected due to the changes in aerofoil characteristics with Reynolds Number. However the nature of the aerofoil characteristics in the transonic region is such that large errors in the estimates quoted in this paper are unlikely.

REFERENCES

1. Little, A.D., Inc., 1982, Wind Turbine Performance Assessment, Technology Status Report No. 4, EPRI AP-2456

2. Boeing Engineering and Construction, 1979, MOD-2 Wind Turbine System Concept and Preliminary Design Report, DOE/NASA 0002-80

3. Gothert, B., 1944, High Speed Measurements on Sections of Series NACA 230 with Different Thickness Ratios, Trans. by MAP Vokenrode, No. 409C

4. Graham, D.J., Niteberg, G.F. and Olson, R.N., 1945, A Systematic Investigation of Pressure Distributions at High Speeds over Five Representative NACA Low-Drag and Conventional Airfoil Sections, NACA R832

5. Stivers, L.S., 1953, Effects of Subsonic Mach Number on the Forces and Pressure Distributions on Four NACA 64A-series Airfoil Sections at Angles of Attack as High as 28°, NACA TN 3162

6. Milborrow, D.J., 1978, Performance Prediction Methods for Horizontal Axis Wind Turbines, Wind Engng., 2, 3, 165-175

7. Lock, C.N.H., Pankhurst, R.C. and Conn, J.F.C., 1945, Strip Theory Method of Calculation for Airscrews on High Speed Aeroplanes, ARC R&M 2035

8. Pankhurst, R.C. and Haines, A.B., 1945, An Account of the Derivation of High-Speed Lift and Drag Data for Propeller Blade Sections, ARC R&M 2020

9. Glauert, H., 1926, The Analysis of Experimental Results in the Windmill Brake and Vortex Ring States of an Airscrew, ARC R&M 1026

10. Driggs, I.H., 1943, Simplified Propeller Calculations, Jnl. Aero. Sciences, 5, 9

11. Hemke, P.E., 1946, Elementary Applied Aerodynmamics, Prentice-Hall Inc.

12. Milborrow, D.J., 1982, Performance, Blade Loadings and Size Limits for Horizontal Axis Wind Turbines, Proc. 4th BWEA Workshop, Cranfield, BHRA

ACKNOWLEDGEMENT

The work was carried out at the Central Electricity Research Laboratories and is published by permission of the Central Electricity Generating Board.

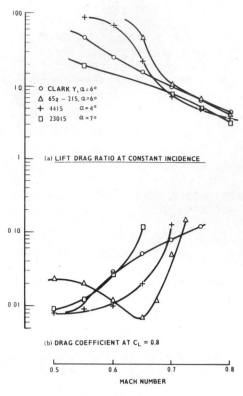

(a) LIFT DRAG RATIO AT CONSTANT INCIDENCE

○ CLARK Y, $\alpha = 6°$
△ $65_2 - 215$, $\alpha = 6°$
+ 4415 $\alpha = 4°$
□ 23015 $\alpha = 7°$

(b) DRAG COEFFICIENT AT $C_L = 0.8$

MACH NUMBER

FIG. 1 TRANSONIC AEROFOIL DATA

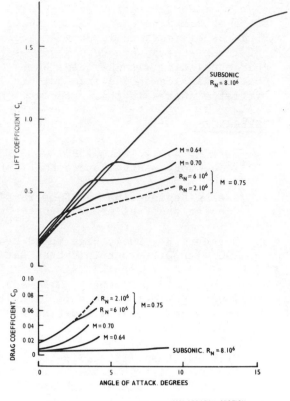

FIG. 2 CHARACTERISTICS OF THE NACA 23015 AEROFOIL SECTION

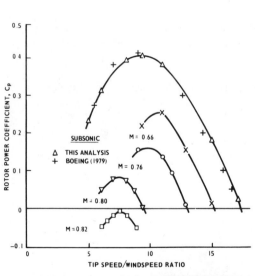

SUBSONIC
△ THIS ANALYSIS
+ BOEING (1979)

FIG. 3 INFLUENCE OF TIP MACH NUMBER ON PERFORMANCE

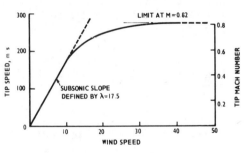

FIG. 4 RUNAWAY SPEEDS FOR THE MOD – 2 ROTOR

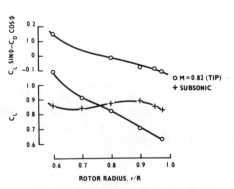

○ M = 0.82 (TIP)
+ SUBSONIC

FIG. 5 RADIAL VARIATIONS OF FORCE AND LIFT

5TH BRITISH WIND ENERGY ASSOCIATION WORKSHOP

LIST OF DELEGATES

ADAMS R. J. — Meteorological Office, London Road, Bracknell, Berks.

ALLAN J. — E.T.S.U. A.E.R.E. Harwell, Oxfordshire, OX11 ORA.

ANDERSON M.B. — Sir Robert McAlpine & Sons Ltd., 40, Bernard Street, London WC1N 1LG.

ANSELL M.P. — School of Materials Science, University of Bath, Claverton Down, Bath, BA2 7AY.

ARMSTRONG J. — Wind Energy Group, 309 Ruislip Road East, Greenford, Middlesex.

ASPLIDEN C. — Battelle, 2030 M St. N.W. Washington D.C. 20036, U.S.A.

ATTWOOD R. — Hawker Siddeley Power Plant, Thrupp, Stroud, Glos. GL5 2BW.

BAKER G. — SDRC Eng. Services Ltd., York House, Stevenage Road, Hitchin, Herts. SG4 9DY.

BANNISTER W.S. — Mech. Eng. Dept., Napier College, Colinton Road, Edinburgh EH10 5DT.

BARBOSA A. — Riso National Laboratories, Roskilde, Denmark.

BARNES J.A. — The Spinney, Church Lane, Arborfield, Reading, RG2 9JB.

BEDFORD L. — E.T.S.U. A.E.R.E. Harwell, Oxfordshire, OX11 ORA.

BENSAAD H. — Commissariat aux Energies Nouvelles, Boulevard F. Fanon, B.P. 1017 Alger-Gare, Algeria.

BOSE N. — University of Glasgow, Naval Architecture Experiment, Acre Road, Glasgow. G20 OTL.

BOSSANYI E.A. — Dept. of Eng., University of Reading, Reading, RG6 2AY.

BROWN G. — Highlands & Islands Development Board, 27 Bank Street, Inverness, IV1 1QR.

BROWN-DOUGLAS G.A. — Mansfield, Chapelton by Strathaven, Lanarkshire.

CARLI R.   c/o Tema S.p.A., Via Medici del Vascello,26
                            20138 Milano, Italy.

CARPENTIER M.   Energy Mines and Resources, Renewable
                            Energy Division, 460 0'Connor Street,
                            Ottawa, Ontario K1A OE4, Canada.

CASANOVA V.H.   Imperial College, Electrical Eng. Dept.,
                            London S.W.7 2BT.

CHADJIVASSILIADIS J.   Public Power Corporation, 10 Navarinou Str.
                            Athens 144, Greece.

CLANCY J.   Dept. of Eng. University of Reading,
                            Reading, RG6 2AY.

CLARE R.   Sir Robert McAlpine & Sons Ltd.,
                            40 Bernard Street, London WC1N 1LG.

CLARK T.   Atkins Research & Development, Woodcote
                            Grove, Ashley Road, Epsom, Surrey KT18 5BW.

CONWAY A.   BBC External Services, P.O. Box 76, Bush
                            House, Strand, London WC2B 4PH.

COURTNEY M.S.   Dept. of Mech. Eng. The City University,
                            Northampton Square, London EC1V OHB.

CRAMPTON S.J.   Institution of Mech. Eng., 1 Birdcage Walk,
                            Westminster, London SW1H 9JJ.

DAVIES G.E.   Plessey Aerospace Ltd., Abbey Works,
                            Titchfield, Fareham, Hants.

DAVIES P.   E.T.S.U., A.E.R.E. Harwell, Oxfordshire,
                            OX11 ORA.

DELNON R.J.   E.R.A. Technology Ltd., Cleeve Road,
                            Leatherhead, Surrey.

DERRY R.   Welwyn Strain Measurement Ltd., Armstrong
                            Road, Basingstoke, Hants. RG24 OQA.

DIVONE L.V.   Wind Energy Technology Div., Dept. of Energy
                            1000 Independence Avenue, Washington D.C.
                            20585.

DIXON J.C.   Technology (EM) Open University, Milton
                            Keynes, MK7 6AA.

DONE G.T.S.   Dept. of Mech. Eng., The City University,
                            Northampton Square, London EC1V OHB.

DOUBT H.   G.E.C. Energy Systems Ltd., Cambridge Road,
                            Whetstone, Leicester LE8 3LH.

EMMOTT R.B.   Hansen Transmissions Ltd., Beeston Royds
                            Industrial Estate, Gelderd Road, Leeds.
                            LS12 6EY.

ENDACOTT J.A.   Protech International Ltd., Phoenix House,
                            The Green, Southall, Middlesex UB2 4BZ.

ESCARDA A.	Construcciones Aeronauticas S.A. Division Espacial, Getafe/Madrid, Spain.
EVANS J.H.	Beech House, 33 Trull Road, Taunton, Somerset TA1 4QQ.
EWENS M.	Rural Industries, Innovation Centre, Private Bag 11, Kanye, Botswana.
FAWKES J.F.	Marlec Eng. Co. Ltd., Unit 5, Pillings Road Industrial Estate, Oakham, Rutland, Leics. LE15 6QF.
FICENEC I.P.	Beckermonds, 5 Creskeld Crescent, Bramhope, Leeds, W. Yorks.
FLEMING P.	School of Mech. Eng., Cranfield Institute of Technology, Cranfield, Beds. MK43 0AL.
FORDHAM E.J.	Cavendish Lab., Madingley Road, Cambridge.
FRAENKEL P.L.	I.T. Power Ltd., Mortimer Hill, Reading, Berks. RG 3PG.
FRERIS L.L.	Elect. Eng. Dept., Imperial College, Exhibition Road, London S.W.7.
FULLER E.B.	A.B. Fuller Ltd., 196 Morland Road, Croydon CRO 6NF.
GAIR S.	Mech. Eng. Dept., Napier College, Edinburgh.
GAMMIDGE A.	Northumbrian Energy Workshop Ltd., Tanner's Yard, Gilesgate, Hexham, Northumberland.
GARRAD A.D.	23 Denbigh Terrace, London W.11.
GARSIDE A.J.	Cranfield Inst. of Technology, Cranfield, Bedford. MK43 0AL.
GASKELL T.	Marlec Eng. Co. Ltd., Pillings Road Industrial Estate, Oakham, Rutland, Leics.
GAUDIOSI G.	E.N.E.A. CSN Casaccia, 00100 Roma, P.O. 2400 Italy.
GILBERT J.	Cilwg, Llandyfaelog, Kidwelly, Dyfed, Wales.
GORDON D.	
GRANT W.T.	Hawker Siddeley Power Plant, Thrupp, Stroud, Glos. GL5 2BW.
GRIFFITHS P.E.	S.E.R.C. Ruherford Appleton Labs. Chilton Didcot, Oxon. OX11 0QX.
GUSTAFSSON A.	Aero Research Institute of Sweden, Box 11021 S-16111 Bromma, Sweden.
HAERS F.	State University at Gent, Dept. of Machinery St. Pietersnieuwstraat, 41, B-9000 GENT.
HALES R.L.	School of Mech. Eng. Cranfield Institute of Technology, Cranfield, Bedford, MK43 0AL.

HALLIDAY J.	ERSU, Rutherford & Appleton Labs, Didcot, Oxon. OX11 0QX.
HANCOCK M.	Gifford Technology Ltd., Carlton House, Ringwood Road, Woodlands, Southampton SO4 2HT.
HAU E.	Man Neue Technologie, Dachauer Str 667 8 Munchen 50, Germany.
HAUGE-MADSEN P.	Riso National Laboratories, Roskilde, Denmark.
HURLEY B.	College of Technology, Bolton Street, Dublin 1, Ireland.
INFIELD D.G.	Rutherford Appleton Laboratories, Chilton, Didcot, Oxon. OX11 0QX.
JEWELL P.	17 Church Road, Southbourne, Bournemouth, BH8 4AS.
JONES K.C.	Hamilton Standard (UK Rep) c/o AEL, Station Approach, Bicester, Oxon OX6 7BZ.
JOUGHIN M.	North of Scotland Hydro Electric Board, 16 Rothesay Terrace, Edinburgh, EH3 7SE.
KILVINGTON T.	European Power News, Queensway House, Redhill, Surrey. RH1 1QS.
KROPACSY C.C.J.	Sir William Halcrow & Partners, Burderop Park, Swindon, Wiltshire SN4 0QD.
LACK L.W.	7 Hamilton Road, Reading, Berks. RG1 5RA.
LAWSON-TANCRED H.	Aldborough Manor, Boroughbridge, N. Yorks.
LEICESTER R.J.	Crown Agents, 4 Millbank, London SW1P 3JD.
LEIDNER J.R.	c/o Electrowatt Eng. Services Ltd., P.O. Box CH 8022 Zurich, Switzerland.
LELY V.D.	DP Enterprises (North Sea) Ltd., 6 Kingshill Avenue, Aberdeen AB2 4HD.
LENEL U.	Fulmer Research Inst., Stoke Poges, Slough, SL2 4QD.
LINDERS J.	Dept. of Elect. Machinery, Chalmers University of Technology, S-412 96, Goteborg, Sweden.
LINDLEY D.	Wind Energy Group, Taylor Woodrow, 309 Ruislip Road East, Greenford, Middlesex.
LIPMAN N.H.	Rutherford Appleton Laboratories, Chilton, Didcot, Oxon. OX11 0QX.
LLOYD R.	Dept. of Applied Physics, University of Strathclyde, Glasgow G4 0NG.

LOVIJESTEIJN C.   Netherlands Energy Research Foundation Westerduinweg 3, P.O. Box 1, 1755 ZG Petten, Netherlands.

LYONS R.A.   W.S. Atkins Ltd., Consulting Engineers, Woodcote Grove, Ashley Road, Epsom, Surrey.

McANULTY K.   E.T.S.U. A.E.R.E. Harwell, Oxon. OX11 0RA.

McLEAN N.   James Howden & Co. Ltd., 195 Scotland Street, Glasgow G5 8PJ.

McLEISH D.   James Howden & Co. Ltd., 195 Scotland Street, Glasgow G5 8PJ.

MAGRAW J.E.   The Bungalow, Townend, Ardington, Oxon.

MASKELL C.   E.T.S.U. A.E.R.E. Harwell, Oxon. OX11 0RA.

MASSINI G.   E.N.E.A. CSN Casaccia, 00100 Roma, P.O. 2400 Italy.

MATTINGLEY L.A.   c/o J & S Marine Ltd., Pottington Industrial Estate, Barnstaple, N. Devon.

MAYS I.D.   Sir Robert McAlpine & Sons Ltd., 40 Bernard Street, London WC1N 1LG.

MENDONCA J.M. de A.B. Imperial College, Electrical Eng. Dept.

MENSFORTH T.

MILBORROW D.J.   Central Electricity Research Labs. Kelvin Avenue, Leatherhead, Surrey.

MILLER G.G.   Stork Fans Limited, 4 Hercies Road, Hillingdon, Middlesex UB10 9NA.

MOGHADDAM T.   21 Bartok House, Lansdowne Walk, London, W11 3LT.

MORRIS C.J.E.   W.S. Atkins & Partners, Woodcote Grove, Ashley Road, Epsom, Surrey KT18 5BW.

MOWFORTH E.   Dept. of Mech. Eng. University of Surrey, Guildford, Surrey GU 5XH.

MUSGROVE P.   Dept. of Eng. Reading University, Reading, RG6 2AY.

NANCE C.T.   Medina Yacht Co. Ltd., Cowes, Isle of Wight. PO31 8BL.

NEWMAN R.A.A.   25 Winchester Gardens, Canterbury, Kent. CT1 3NA.

NICHOLSON G.   Northumbrian Energy Workshop Ltd., Tanner's Yard, Gilesgate, Hexham, Northumberland.

NILSEN F.P.   Institute of Energy Technology, Postboks 40, 2007 Kjeller, Norway.

O'DONNELL I.H.D.	North of Scotland Hydro Electric Board, 16 Rothesay Terrace, Edinburgh.
O'FLAHERTY T.	Kinsealy Research Centre, Malahide Road, Dublin 5. Ireland.
PAGE I.	E.T.S.U. A.E.R.E. Harwell, Oxon, OX11 ORA.
PAGE R.	Halfway House, Scoulton, Norfolk NR9 4NZ.
PEDERSEN M.	Technical University, Building 404, 2800 Lyngby, Denmark.
PEDERSEN P.D.	Institute of Energy Technology, Box 40, N-2007, Kjeller, Norway.
PERCIVAL M.D.	Rutherford Appleton Laboratories, Chilton, Didcot, Oxon. OX11 OQX.
PETERS M.C.	21 Burntwood Road, Sevenoaks, Kent.
PETERS W.A.	Modern Power Systems
PETERSEN H.	Riso National Laboratories, Roskilde, Denmark.
PIEPERS G.	Netherlands Energy Research Foundation, BEOP Westerduinweg 3, P.O. Box 1, 1755 ZG, Petten, Netherlands.
PLATTS M.J.	Gifford Technology, Carlton House, Ringwood Road, Woodlands, Southampton, SO4 2HT.
PONTIN G. W-W.	Wind Energy Supply Co. Ltd., Bolney Avenue, Peacehaven, Sussex BN9 8HQ.
POOLEY D.	U.K.A.E.A. Harwell, Didcot, Oxon OX11 ORA.
POWLES S.J.R.	The Cavendish Laboratory, Madingley Road, Cambridge.
PRETLOVE A.J.	Dept. of Eng. Reading University, Reading, RG6 2AY.
PRIETO J.	Construcciones Aeronauticas S.A., Division Espacial, Getafe/Madrid, Spain.
PUDDY J.	Lundy, Bristol Channel, Via Ilfracombe, N. Devon. EX34 8LA.
PYBUS D.	The Torrington Co. Ltd., Yarm Road, Darlington, Co. Durham. DL1 4PP.
QUARTON D.	Wind Energy Group, Taylor Woodrow, 309 Ruislip Road East, Greenford, Middlesex.
RANGI R.	Wind Energy R & D Dept., NRC, Montreal Road, Ottawa, Ontario, Canada K1A OR6.
REES D.M.	4 Woodbine Cottages, Melincourt, Resolven, Nr. Neath, West Glamorgan.

RENDALL P.L.   British Aerospace Dynamics Group, Manor
                       Road, Hatfield, AL10 9LL.

RHODES H.   10 Dale Park Close, Cookridge, Leeds 16,
                       LS16 7PR.

RICHARDSON J.   Balfour Beatty Power Construction Ltd.,
                       7 Mayday Road, Thornton Heath, Surrey.

RIDDELL J.C.   J.C. Riddell & Associates, Ardyll House,
                       Hollybush Green, Collingham, Wetherby, Yorks.

RILEY P.R.H.   24A Highlands Road, Fareham, Hants. PO15 6AA.

RODWELL B.   Trinity House, Tower Hill, London EC3 4DH.

SAIA A.   NEI Cranes Ltd., Smith House, Town Street,
                       Rodley, Leeds LS13 1HN.

SEXON B.A.   Dept. of Eng. Reading University, Reading,
                       RG6 2AY.

SHARPE D.J.   Dept. of Aeronautical Science, Queen Mary
                       College, Mile End Road, London E1 4NS.

SHI P.   Dept. of Eng. Reading University, Reading,
                       RG6 2AY.

SIX D.   Vrije Universiteit Brussel, Dienst
                       Stromingsmechanica, Pleinlaan 2, 1050
                       Brussel, Belgium.

SKIPPER R.G.S.   Dept. of Energy, Thames House South,
                       Millbank, London SW1P 4QJ.

SLACK G.   Research Student, Dept. of Eng. Reading
                       University, Reading, RG6 2AY.

SLATER B.   Yard Limited, Charing Cross Tower, Glasgow,
                       Scotland.

SOMERVILLE M.W.   International Research & Development Co. Ltd.
                       Fossway, Newcastle upon Tyne, NE6 2YD.

STACEY G.   Dept. of Eng. Reading University, Reading,
                       RG6 2AY.

STEVENSON J.   E.P.D. Consultants Ltd., Marlowe House,
                       Station Road, Sidcup, Kent.

STEVENSON W.G.   North of Scotland Hydro Electric Board,
                       16 Rothesay Terrace, Edinburgh 3.

SWANSBOROUGH   E.R.A. Technology Ltd., Cleeve Road,
                       Leatherhead, Surrey.

SWIFT R.   Engg Mechs. Venables Building, Open
                       University, Milton Keynes MK7 6AA.

SWIFT-HOOK D.T.   CERL, Kelvin Avenue, Leatherhead, Surrey.

SZELESS A.   Verbund Gesellschaft, A-1011 Wien, Am Hof
                       6a, Fach 67, Austria.

TALBOT J.R.W.	C.E.G.B. 15 Newgate Street, London EC1.
TAYLOR R.	C.E.G.B., 15 Newgate Street, London EC1.
THEYSE F.H.	Theyse Energieberatung, Alte Wipperfurter Strabe 138, 5060 Bergisch Gladbach 2. Fed. Rep. of Germany.
TODD R.W.	Centre for Alternative Technology, Machynlleth, Powys, Wales.
TORIELLI	Tema, Via Marconi 29/1, 40122 Bologna, Italy.
TSITSOVITS A.J.	Imperial College, Dept. of Electrical Engineering, Exhibition Road, London SW7.
TUNSTALL S.R.H.	33 Winter Road, Norwich, NR2 3RR Norfolk.
TWIDELL J.W.	Dept. of Applied Physics, University of Strathclyde, Glasgow G4 ONG.
WALTHAM M.R.	Dept. of Eng. University of Reading, Reading, RG6 2AY.
WARNER P.C.	Northern Eng. Industries plc, Nei House, Regent Centre, Newcastle upon Tyne, NE3 3SB.
WATERTON J.C.	South of Scotland Electricity Board, Cathcart House, 1 Spean Street, Glasgow G44 4BE.
WATSON G.	Northumbrian Energy Workshop Ltd., Tanner's Yard, Gilesgate, Hexham, Northumberland.
WEBB J.	Cranfield Institute of Technology, Cranfield, Bedford.
WERSBY J.	Aircraft Designs (Bembridge) Ltd., Embassy Way, Isle of Wight Airport, Sandown, Isle of Wight.
WEST M.	
WHALLEY L.	Dept. of Industry, Research & Technology Policy Division, 29 Bressenden Place, London SW1E 5DT.
WHITE R.	International Power Generation
WILLIAMS P.B.	John Laing Construction Ltd., Alternative Energy Group, Page Street, Mill Hill, London NW7 2ER.
WILSON D.M.A.	Cavendish Laboratory, Madingley Road, Cambridge, CB3 OHE.
WILSON R.	c/o James Howden & Co. Ltd., 195 Scotland Street, Glasgow G5 8PJ.
WRIGHT B.	271 Camden High Street, London NW1.

YOUNG T.

c/o James Howden, 195 Scotland Street, Glasgow G5 8PJ. Scotland.

ZWEYGBERGK VON S.

Chalmers Technical University, Dept. of Electrical Machinery, 41296 Gotenburg, Sweden.